STRANGERS DEVOUR THE LAND

BOYCE RICHARDSON

Chelsea Green Publishing Co.
Post Mills, Vermont

Library of Congress Cataloging-in-Publication Data
Richardson, Boyce. Strangers devour the land : a chronicle of the assault upon the last coherent hunting culture in North America, the Cree Indians of northern Quebec, and their vast primeval homelands / Boyce Richardson.
p. cm.
Reprint. Originally published: New York : Knopf, 1976.
Includes bibliographical references and index.
ISBN 0-930031-40-7 (pbk. : alk. paper) : $14.95
1. Cree Indians—Government relations. 2. Cree Indians—Politics and government.
3. Cree Indians—Land tenure. 4. Hydroelectric power plants—James Bay Region
(Ont. and Québec) 5. James Bay Hydroelectric Project. I. Title.
E99.C88R52 1991   90-26332
CIP
Printed on recycled paper.
Printed and bound in the United States of America.

For
Willie Awashish,
who should have inherited
a great tradition

When the dams are built where will the animals go? The caribou won't know which way to go.

—*Samson Nahacappo, Cree hunter, Fort George*

We are not thinking only of ourselves but of all those young kids who are just starting to hunt, and those that have yet to be born.

—*Mary Bearskin, wife of a Cree hunter, Fort George*

If you set fire to the land, the land remains, and life returns to it. If you set fire to a piece of paper, like a dollar bill, it burns away to the end, and nothing is left.

—*Charlie Gunner, Cree hunter, Mistassini*

We are told that we own the land. But really nobody can own it, the land. For eventually everyone dies.

—*Sam Blacksmith, Cree hunter, Mistassini*

It is the white man who has the money, and on the other hand the Indian has the land. The white man will always have the money and will always want to have the land.

—*William Rat, Cree hunter, Fort George*

There will never be enough money to pay for the damage that has been done. I'd rather think about the land and think about the children. What will they have when that land is destroyed? The money means nothing.

—*Job Bearskin, Cree hunter, Fort George*

Objectively, unemotionally, one can state that the development of James Bay will set Quebec on a new road to progress. It is an undertaking which once again will furnish tangible proof of Quebec's vitality and spirit of enterprise, for the development of James Bay is the most daring project in Quebec's history. James Bay is the land of tomorrow.

—*James Bay Development Corporation initial-phase plan*

CONTENTS

1 ISAIAH AND HIS SONS   *3*

2 COURTROOM   *18*

3 THE INDIAN RHYTHM   *33*

4 1969: WASWANIPI   *47*

5 1969: MISTASSINI   *61*

6 1971: MISTASSINI   *80*

7 1972: MISTASSINI   *85*

8 1972: RUPERT HOUSE   *101*

9 JOB'S GARDEN   *118*

10 1972: LAC TREFART   *198*

11 COURTROOM   *242*

12 1973: LAC TREFART   *260*

13 ROSIE   *288*

14 JUDGMENT   *296*

15 NEGOTIATING TABLE   *303*

Contents

16 BACK TO COURT   *310*

17 SETTLEMENT   *318*

Epilogue   *328*

Notes   *363*

Index   *367*

ACKNOWLEDGMENTS

For a white man to penetrate into the world of a Cree hunter is not easy. Ideally, one would wish to be inside the mind of a hunter as he walks through the bush, able to see his universe as he sees it—an impossible feat. But I would not have been able to achieve even the incomplete understanding that I have without the help of very many friends to whom I am deeply indebted. First among these I must mention Philip Awashish, who through five or more years has borne with and encouraged my curiosity, and has been most generous in opening my way into Cree communities, translating for me in more ways than the obvious. In these communities I met many hundreds of people, young and old, too numerous to mention, who have responded sometimes with enthusiasm, sometimes with patience, but always with great courtesy, to my persistence. I am indebted to Chief Smally Petawabano, of Mistassini, for the patience with which he bore my many visits to his people's village; to Buckley Petawabano, of Mistassini and Montreal, for making available transcripts of interviews with his father and other hunters; to Glen Speers, manager of the Hudson's Bay Company in Mistassini, for helping smooth my way many times on journeys into the Cree lands. I am also grateful to James O'Reilly, of O'Reilly, Alain and Hudon, Montreal, for having made available to me the voluminous transcripts of the evidence in the case of Robert Kanatewat *et al.* versus the James Bay Development Corporation and others; and to the anthropologists Harvey Feit, Adrian Tanner and Ignatius LaRusic, who have often shared with me their profound knowledge of the Cree world. These men have done much to restore the good name of anthropologists, for they have repaid the Cree people in devoted labor on their behalf. Dr. John Spence, of Montreal, has helped me with valuable environmental advice, and it was Noel Mostert, of Tangier, who, when I had given up the idea of writing a book, persuaded me that I must make the time to commit to paper the material I had gathered. Without his urging I would never have got going again.

Naturally, the views expressed in the book are my own, and should not be attributed to any of the above-mentioned people. Finally, I will never be able to repay adequately Job Bearskin, of Fort George, and Sam Blacksmith, of Mistassini, who welcomed me on their land, and looked after me, and taught me that life has other dimensions than those I had previously comprehended.

FOREWORD

As I sit at my desk I can almost see the shore of James Bay—ten miles downstream from the island of Moose Factory in the Moose River. Almost Arctic Ocean beachfront, some would say, yet I'm coming to think of it as a Hydro viewpoint. As a matter of fact, if all the plans go ahead I may no longer live on an island but on a peninsula of the mainland overlooking a huge freshwater reservoir, where presently I've become accustomed to the saltwater of James Bay, which at the base of Hudson's Bay is the largest bay and estuary system on the continent. Dams upstream from this small village promise to lower water levels and change the ecosystem permanently. And amazing as it may sound, a powerful group of engineers has a proposal to place a dike over the mouth of James Bay to create a vast reservoir of fresh water to sell to cities in the southwestern U.S.

There are many things Cree people have taken for granted over countless generations: that the rivers will always flow, the sun and moon will alternate, and there will be six seasons of the year—yes, six. That is how time is counted here in the North, in seasons based on the migrations of caribou, geese, sturgeons, and other relations, and on the ebb and flow of ice and water. The Cree also have assumed that there will always be food from the land, so long as the Eeu, the Cree, do not abuse their part of the relationship to the animals and the land. Now, the rivers do not always flow, the animals are not always there, and strange as it may seem, there are no longer six seasons in some parts of this land. Hydro-Quebec has made sure of that.

In *Strangers Devour the Land* Boyce Richardson provides an intimate look into the people and communities of James Bay, particularly the Cree of the east coast of James Bay, those most heavily affected by the first stage of the Hydro-Quebec project. This book, originally published at the onset of James Bay I, is the testimony of people and the land. Richardson brings the reader to the hunting territories, the traplines, and the powerful rivers that are the lives of these people. The book is a moving chronicle of the resistance of people to the dams, the story of James Bay I, and how Hydro-Quebec came to begin the largest single hydroelectric project in North America.

This is also a story about all of us, about how industrial society is consuming the lifeblood of this continent. Unfortunately, we must now tell not only the story of James Bay I, the project which devoured the traplines and hunting territories of many of the people written about in this book, but we must tell of the proposal of James Bay II.

As people contemplate the destruction of the Amazon rainforest, global warming, "the greenhouse effect," and other climatic changes, this $60 billion mega-project brings it all home to U.S. and Canadian consumers. The new dams, water diversions, and hydroelectric projects at issue will, according to the National Audubon Society, "make James Bay and some of Hudson's Bay uninhabitable for much of the wildlife now dependent upon it." Audubon senior staff scientist Jan Beyea reports that the Society is "convinced that in fifty years [this entire] . . . ecosystem will be lost. . . ." The ecosystem at stake is as large as California, and includes the central flyway of most of the migratory birds in North America, the drainage of most northern-running river systems in the central part of the continent, a number of endangered species of animals, and Inuit, Cree, and Naskapi/Innu people, who have lived here for at least nine thousand years.

There are no longer "strangers" who devour the land. They are entrenched in the North, in the form of Hydro-Quebec, which put 4,400 square miles of land under water and wreaked ecological havoc in an additional 67,954 square miles. Hydro-Quebec and its counterparts in Ontario and Manitoba are taking a vast territory notable for running water, and essentially proposing to turn it into a vast territory of stagnant reservoirs — virtual toxic sinks.

Already there is spreading mercury contamination from James Bay I. Methane from decomposing plants and trees, which have been drowned in the flooding, converts the inorganic mercury already present in the soil into organic methyl mercury, a lethal poison. Because the process is enhanced in acidic conditions, the mercury levels in the reservoir system are up to six times the level considered safe for humans.

In the village of Chisasibi, downstream from one set of reservoirs (LG 1–4), scientists tested for mercury poisoning several years ago. Two of every three people were found to have excessive levels of methyl mercury already present in their bodies—30 milligrams per kilogram of body weight. Some elders registered twenty times the level deemed acceptable, and had developed symptoms of mercury poisoning such as shaking, numbness of the limbs, loss of peripheral vision, and neurological damage. Hydro-Quebec advised the Cree to stop eating river fish and instead to harvest fish from James and Hudson's bays. These fish, which are still relatively free from methyl mercury, are frequently contaminated with PCBs, a result of other "development" projects in the region and contamination now moving into the arctic food chain from industries to the south.

The Cree call it *nimass aksiwin*, "fish disease," and no other two words could have such a devastating effect on people. "*Nimass aksiwin* strikes at the very heart of our society. It's like being told that armageddon has started, and people are scared as hell," says George Lameboy, a Cree fisherman and trapper. "The scientists come in here and tell us we're getting better [by eating less fish], but hey, you can't measure the effects of *nimass aksiwin* by taking hair samples. How can you measure a man's fear? How can you measure your way of life coming to an end?"

As if the methyl mercury were not enough, the change in water levels in the rivers had devastating results in at least one case. Normally rivers run highest in the spring melt and lowest in winter. Since the flow of rivers is now determined not by nature but by the electrical demands of southern consumers, the order has been reversed; many times, it is increased or decreased dramatically to respond to the "power grid" of the south. In 1984 a sudden release proved deadly, as water was released out of the Caniapiscau Reservoir (now the largest lake in Quebec at 1,865 square miles) precisely during the seasonal migration of the George's River caribou herd. Ten thousand caribou drowned. Hydro-Quebec officials called the disaster "mainly an act of God."

The ongoing environmental problems have solidified Cree opposition to any more development in their territory, and strengthened their calls for a comprehensive environmental review of the first phase of the project prior to construction of any new dams. The Cree call to halt the project is now supported by a growing number of local, national, and international environmental and consumer groups who are deeply concerned about possible long-term consequences of the development. The Cree and other groups have joined in extensive and seemingly endless legal challenges to the project, which at this point have resulted in a court decision calling for an environmental inquiry. Unfortunately, neither the scope of the review nor its weight (that is, whether findings will be binding to the utility) have been determined, leaving many Cree and environmentalists frustrated and skeptical.

The Cree and other groups have consistently called for federal intervention. Cree Chief Mathew Coon Come points out the irony, saying, "When you have the largest project of the century in your backyard, and no environmental assessment . . . not one person monitoring the impact, there is an obvious failure of federal responsibility. . . ." Bill Namagoose, of the Cree Regional Authority, echoes his words, calling the federal sidestepping of the issues "environmental racism." Can you imagine a man who has lived his whole life in Paris — and one day awakens, looks out his window, and Paris is underwater? It just wouldn't happen. The Cree, Inuit, and Innu are far away, dark and different. That is one reason this project, like the exploitation of the Amazon and other rainforests, is planned to go ahead. If Hydro-Quebec proposed to flood the villages, farms, homes, and gravesites of thousands of French-speaking white people, well, it just wouldn't happen.

If the second phase goes ahead, the new dams would greatly accentuate the present environmental damage. At Great Whale, four smaller rivers will be diverted into a single large one. On the Nottaway, Broadback, and Rupert river systems, eleven dams would be built — with the Nottaway being diverted into the Broadback, then the Broadback into the Rupert. In total, the reservoirs will cover more than ten thousand square miles, an area the size of Lake Erie. The project, according to the National Audubon Society, "is the northern equivalent to the destruction of the tropical rainforest."

What is worse is that the Quebec dams are only one set of proposals for James Bay. Another huge hydroelectric project has been put in place on the Nelson River in northern Manitoba, draining into Hudson's Bay, and an undetermined number of dams are planned for the rivers in northern Ontario. Virtually every single river flowing into James and Hudson's bays is now proposed for some hydroelectric or diversion scheme.

This worries not only the native people, but also environmentalists and other people to the south. For although an environmental impact assessment is pending for the Quebec projects, there is no proposal as yet for a cumulative impact assessment for all projects in what is essentially one, unified ecosystem: James and Hudson's bays. As Alan Penn, an environmental advisor to the Grand Council of the Crees of Quebec points out, "There is no precedent for the manipulation of a subarctic watershed elsewhere in the world on the scale proposed here. The project represents a natural experiment, both ecological and sociological, on a massive scale."

Perhaps most horrendous is that this massive experiment is all about making money. Hydro-Quebec is the provincial government's chief economic tool for capitalizing its economy. Although the 125,000 jobs promised never materialized from James Bay I, Hydro-Quebec has all in all done well from its huge investments. In 1970 Hydro-Quebec had 12,000 employees, assets of $3.5 billion, and debts of $2.6 billion. Today the provincial utility has 23,000 employees, assets of $34 billion, and a debt of $23 billion. This corporation accounts for 20 percent of all new investments in Quebec.

A great portion of the scheme is designed to service electrical markets in the U.S. A number of U.S. utilities have accepted Hydro-Quebec's promotion of its power as a cheap, clean alternative to coal and nuclear generating. New York, for instance, has purchased Hydro-Quebec power, and the purchase has accounted for about 9 percent of the state's electricity supply since 1970. This figure is expected to rise to 30 percent by the year 2000. Seven U.S. utilities—the New England Power Pool, the New York Power Authority, Vermont Joint Owners, Massachusetts Power Authority, Citizens Utilities, Consumers Power, and Detroit Edison— have entered into long-term contracts with Hydro-Quebec and Ontario-Hydro to secure power for the next twenty years or more. These contracts, of course, enhance the utility's ability to raise the huge investments required for the new phase of development. In other words, U.S. consumers are clearly implicated in the destruction of this ecosystem.

Canadians, however, are far from innocent. According to Tom Adams of the Toronto-based Energy Probe, "We are the single most inefficient consumers of electricity in the world. We are twice as inefficient as even the next in line—the U.S." And that inefficiency is buttressed by low rates: industries in Ontario, for instance, pay six times less for electricity than would their counterparts in Japan. Not only do provincial electric corporations subsidize the "hidden costs and dis-economies" of power production, but these very "cheap" rates discourage conservation and undermine any incentive to plan realistically. Energy analysts like Amory Lovins have frequently pointed out that conservation of electricity would make the dams not only unnecessary for projected demand, but cost a great deal less in hard cash. It is outrageous that "cheap electrical rates" are a justification to destroy an entire ecosystem and way of life.

As politicians, environmentalists, and economists speak of the future, "sustainable development" is the phrase most in vogue. While the meaning of that phrase varies with the person using it, the concept has validity for me. Some days I listen to my father-in-law talk when he has come in from his trapline—which is, incidently, just west of the proposed Nottaway–Broadback–Rupert (NBR) project. He explains that he walked five miles one way to check his rabbit snares and his traps. And he tells me of reaching his hand into a beaver house, to count the number of beavers in the house. There is even a word for this counting in Cree. The point of the counting is so that no person will take more beavers than should be taken from a certain area. There is no word for this in English, only a long description. And it makes no sense whatsoever to explain to a Cree the concept of "sustainable development." My father-in-law and his ancestors have been harvesting and hunting this same area for thousands of years. It appears to me that "sustainable development" and a "sustainable economy" are scheduled for destruction only so that twenty years from now some southern expert can "reinvent" a sustainable economy for this region.

The problem is not Hydro-Quebec, Ontario-Hydro, and the U.S. energy contracts. The problem is "development," and the structure of Canada's (and for that matter, the U.S.) industrial economy. The Canadian economy has always been based on the exploitation of raw materials and resources from the "frontier." The North has always been the frontier, and continues in that role today. The Canadian economy requires this exploitation to prosper. The James Bay dams and diversions are only a small set of many such mega-projects presently underway or proposed for the North. All share a common denominator—a development policy based on capital-intensive, resource-extractive industries. The promise is jobs and prosperity but, as evidenced in James Bay I, the reality is stark and destructive.

At some point, there will be no more "frontiers" to conquer. There will be no more resources to mine, rivers to dam, trees to fell, or capital to invest. As we approach the year 2000, those who have an interest in surviving to the next century would say that point in time is now. And as I sit on my Arctic Ocean beachfront I think about that. I think about the testimony that is in this powerful book, and I hope that by a collective act of conscience, sanity, political and economic change, James Bay will remain saltwater and free of methyl mercury.

Winona LaDuke Kapashesit
Moose Factory, Ontario
January, 1991

STRANGERS
DEVOUR
THE LAND

The James Bay Development Project

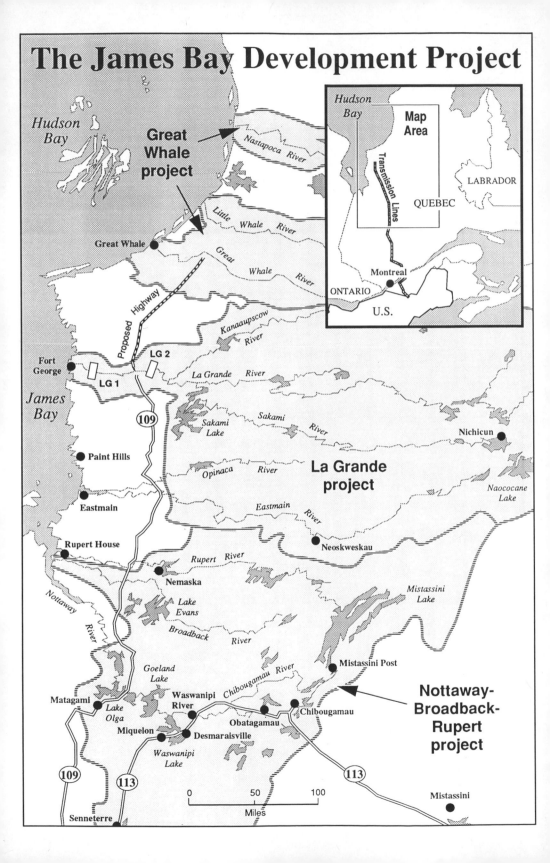

Hudson Bay

Great Whale project

Little Whale River

Nastapoca River

Great Whale

Great Whale River

Map Area

Hudson Bay

Transmission Lines

LABRADOR

QUEBEC

Montreal

ONTARIO

U.S.

Proposed Highway

Kanaaupscow River

Fort George

LG 2

LG 1

La Grande River

James Bay

109

Sakami Lake

Sakami River

Nichicun

Paint Hills

Opinaca River

La Grande project

Naococane Lake

Eastmain

Eastmain River

Neoskweskau

Rupert House

Rupert River

Nottaway River

Nemaska

Lake Evans

Mistassini Lake

Broadback River

Goeland Lake

Matagami

Lake Olga

Waswanipi River

Chibougamau River

Mistassini Post

Nottaway-Broadback-Rupert project

Miquelon

Obatagamau

Chibougamau

Desmaraisville

Waswanipi Lake

109

113

113

Mistassini

0   50   100

Miles

Senneterre

ISAIAH
AND HIS SONS

The rock is north, as far as the mind can stretch. Until 8,000 years ago this rock was covered by ice that for millennia had ground and creaked painfully and slowly over the rock, scratching, gouging, advancing, retreating, advancing again, creating with its irresistible force the depressions which today form millions of lakes. When the ice retreated for the last time, plants and trees migrated north, drawing sustenance from the meager earth that lay on the rock. They were followed by animals and fish; and finally men, hunting men who fed themselves from the other animals. Together, men and animals clustered around the lakes and the great rivers that drained the land.

It is the same today. In the spring when the sun melts snow and ice, the water runs everywhere, flowing across the rocks in every direction, filling every depression, scouring every slope, finally gathering into mighty rivers which roar across the rocks, lifting the remaining ice from the shorelines as they beat their way across the land toward the sea.

This is the awesome wilderness of northern Quebec. Rocks, trees and lakes, stretching north for 1,000 miles from the flood plain of the St. Lawrence valley into a wilderness at first marked here and there by roads and towns, the so-far feeble scratchings on the surface of the land by the white man and his omnivorous, devouring civilization, but soon running off into the mysterious silence, far north until the trees diminish in size and number as the earth becomes thinner, finally giving way to the springy, soft muskeg, the rootless lichen plant which, abandoning the miserly rewards to be drawn from the earth, takes all its sustenance from the air.

Animal and human life is not plentiful in this part of the world, but it does exist, both below and above the treeline. However sparse the vegetation, in tundra or in forest, it nourishes animals of many different kinds living in a hierarchy from the ant up to the black bear—and always at the top of the hierarchy, but part of it, the men. In the millions of lakes and countless streams, the hardy fish—pike, whitefish, sturgeon, trout among them—have adjusted to the intense cold, the long winter, the great spring flood. Undisturbed, they grow for many years, and to a great size.

Around the wetlands, the rodents—mink and ermine, mice, muskrat, groundhogs—the otters, marten, rabbits, porcupine, beaver, the lynx and fox and wolf, the moose, caribou and bear, herbivores and carnivores, live together in that exquisite balance arranged by nature, each depending on other animals and plants to sustain it, each in its turn part of the chain of survival for some other animal.

And then, of course, at the end of the food chain come the men— Cree hunters who moved into the forest 1,000 or 2,000 years after the retreat of the ice, and have been there ever since.

A HUNTER ON HIS LAND

Along one of the mighty rivers a man and his son make their way downstream by canoe to check the beaver on a small lake on their hunting territory. It is only a few days before freeze-up. The lakes will freeze first and after them the rivers. The ice has already formed along the shores and in the rock pools along the river. For six weeks the pair have been traveling across their territory, locating the beaver lodges before the ice locks them into the banks and the snow hides them to all but the most practiced eye. For this trip down the big river they are using an outboard motor. They have started early in the morning because on all of these rivers they have to negotiate many ferocious rapids.

The father is Isaiah Awashish, tall and slightly stooped, with a long, melancholy face that suggests a man with little concern for the external world. Born in these forests nearly sixty years ago, he has lived in them ever since, every day earning his survival from his skill in hunting the animals that live around him. His son is Willie, seventeen, the only one of his four sons who has followed the ways of his father. For six years the two have moved everywhere together around the 1,000 square miles of his territory, traversing its streams and lakes. Every fall they set up a camp in a different part of their land. Today they are far from their camp on this last voyage of reconnaissance before the arrival of the winter.

The two do not speak together much, for they have a perfect understanding of their roles, and anyway, the father is not a man of many words. As the father expertly brings the canoe up on a small island in the middle of the river at the safe point closest to the rapids, Willie jumps out and holds it. They have performed this many times, but they are always careful, for two tiny figures in the huge wilderness cannot afford to make mistakes. The water churns and boils over the rapids parallel with the foot of the island, and then moves swiftly downstream to another set of rapids not far away. They ease the canoe into the rapids, the father at the front, Willie holding it firm with a rope from behind.

"I was about halfway through the rapids, and I was holding the rope," said Willie. "I couldn't see my father very well where he was standing, and I thought he had a hold of it, because the current wasn't pulling it at all. The rope went slack, and I let it go, and the canoe just went floating right away across the river.

"We had had to get up early in the morning so that we could shoot those rapids and then get back over them and return to the camp by nightfall, so it was still quite early in the morning when this happened."

Left standing on the island in the middle of the torrential Rupert River, they were stripped of all their equipment. The unthinkable had happened: an accident, one of the very few that had ever occurred to the old man in his lifetime as a hunter. They could not swim off the island because the waters in this part of the world are always too cold for swimming. No prudent man would plunge into a northern river in October. Fortunately, the cross-wind was strong, and it carried the canoe over to the shore before it reached the second set of rapids.

"We had nothing there, no paddles, no axes, everything was in the canoe. All we had was our knives, and I was getting a bit panicky." But for a man like Isaiah Awashish, who sees himself as a participant in the awesome processes of nature and is in touch with the spirits which control these processes, the situation was not altogether without hope. Nature had not been entirely unkind. For they were stranded on an island that had been completely burned over. "There had been a big fire several years ago," said Willie. "The trees were all dry, and most of them were solid. My father said we were going to make a raft."

The father led his son deep into the bush on the island, looking for trees that they could break by hand, trees still standing but slightly rotten. It is something that Indians are used to doing, for they gather their firewood, which keeps them heated throughout the winter, by choosing the dead standing trees, trees that are not too knotted, not too gnarled, not so alive as still to contain sap, but not so dead as to be completely rotten. They pushed over the trees that were just right for the job, and carried them laboriously down to the river. Nature also came to the rescue in another way. A forest fire is not an unmitigated disaster for an ecosystem, and the Cree know it well. Growth returns in the wake of a fire—young, fresh growth that is excellent for the moose, and nourishes the animals better than the old growth did. "When a place burns, the jackpine and black spruce start to grow again," said Willie, "little trees whose roots really grow long, as much as twenty-five feet. So we dug them up and used them to tie the logs together." It took them all morning to do that. When the job was finished, the father told his son that he would go alone. They stood at the head of the rapids, the water racing past them. The father walked up to his son, shook his hand and began to sing.

"I knew from this water that even if there were two of us in the canoe, paddling for all we were worth, we would be carried down almost to the next rapids. But the logs in our raft were really long: the current would have ten times as strong a grip on them as it would have on a canoe. And yet here was my father intending to go into the rapids alone, on this raft. He came up to me and he sang this song about the otter. He said—well, I don't want to tell you what he said because my father doesn't want to tell people about that. It is too personal for him."

The hunter who, through his lifetime of experience, is in touch with the hunting spirit can call on that spirit to aid him: he has, as it were, through his deep respect for the animals with which he lives, other powers than his own to depend upon. He gave a promise to the otter, in return for its help in extricating him from this difficult situation. He promised the otter that he would not kill so many this year if the otter would help him now.

"He had just a small piece of stick, that's all he used for a paddle," said Willie, "and he went straight across the river on the raft. The current didn't bother him at all, he went straight across the rapids.

"I should have had the sense to know," said Willie, "that my father would know what to do to get out of that situation."

HARD TIMES FOR A HUNTER

The spiritual and mystical qualities of a man like Isaiah Awashish are developed through a lifetime of suffering, of work, of hunting, a lifetime during which the hunter has learned to keep his proper place among the natural forces developed by the Great Spirit for controlling all of life.

Unlike any other kind of human being, the man who earns his subsistence from hunting, who survives, as the Indians say, from the land, depends on knowing where he must stand in the strangely efficient and mysterious balance that is arranged for the propagation of all life. He is one of the animals which roam the land. He is the predator on all others, but just as a wolf pack depends for its continued existence on the survival of the caribou herd, so this predatory man will not survive unless the animals continue to flourish. The balance between the two is the most important thing in his life. And right down to our own day, right down to the generation of men who were born in the 1930s, these Cree hunters have followed a spiritual system designed to maintain that balance. In this scheme of things the man is not dominant; he is a mere survivor, like every other form of life. His personality is not more marked than those of other forms of life. All animals and plants, all natural forces are personalized in the Cree mind and are spoken of in the Cree language in the personal form. These natural forces make decisions, just as people do; and if their personal qualities are not respected, they can make life impossible for the hunter. For example, *chuetenshu*, the north wind, gives the animals to the Indians in January by making the snow accumulate to such a depth that the moose's belly drags and the hunters can easily catch him. *Chuetenshu* also blows the snow thin on high ground so that the moose will cluster there, digging through the thin snow for winter food, and the hunter can easily locate them at that time of year. But *chuetenshu* can be capricious, like any person, and when he brings weather that is too cold, the snow hardens and the Indian's snowshoes make such a clatter on the crust that the moose is warned of his coming and is hard to catch. *Chuetenshu* can bring warmer weather as the winter nears its end, the snow becoming slushy and heavy and difficult for the Indian to penetrate. If colder weather returns, the surface of the snow then freezes; the moose cannot run over this thin crust, and again it becomes simple for the Indian to catch him.

Everything in nature behaves in this complex way, and after a lifetime of learning his proper place in this personalized universe, a Cree hunter like Isaiah Awashish develops *metew* (pronounced "medeo")—powers derived from the exceptional relationship that he has established with the animals and with the hunting spirit which controls all of these forces. Then the man is powerful indeed. Isaiah Awashish suffered for the land when he was a lad. As a teenager he lived through the hardest times the Cree people have ever known. It was long before the days of the airplane or even the outboard motor. In those days hunters would take off from Mistassini for their hunting grounds in the north and travel for as long

as a month, making great portages from lake to lake, carrying huge
quantities of equipment back and forth laboriously each time, with the
stoic patience that has been the mark of their civilization. As a young man
just married who had learned everything his father could teach, Isaiah
had to take his father into the bush with him when the old man was no
longer capable of moving around. It was before the days of government
assistance, baby bonuses, welfare payments. The Indians had little if any
money, and Isaiah can remember his whole family going into the bush
with only five pounds of flour, a bit of lard and sugar, nothing to buttress
them against the possibility of failing to catch animals.

And there were no animals. The animals go in cycles, and occasionally in the millennial history of the Cree all of the cycles have coincided
and all of the animals have disappeared. For two years everywhere he
moved on his snowshoes he pulled behind him on his sled his crippled
father as he searched for animals which never seemed to appear. Then
for four years the old man stayed around the lodge, unable to move out
even on his son's sled.

There were seven in the family, and Isaiah, the eldest boy, was
responsible for raising his brothers and sisters, and had to look after his
father, too. Most of them died in the hard times; only three remained.
Fortunately, the fish remained stable, not affected by the cycles, and they
were able to survive mostly on fish. He remembers one occasion on which
they were sure they had just about got to the end. Everyone in camp was
thin and undernourished. But just then Isaiah was able to catch five
rabbits, enough to give sufficient sustenance to the men to enable them
to carry on. Once the brothers came upon a camp of Nemaska people on
the verge of starvation: so poor were they that they did not even have
tents or clothing. They were living under boughs and were wearing robes
they had made from rabbit skins. The brothers stayed with them awhile,
helped them to catch enough rabbits to keep them alive. Isaiah did not
actually see people starve to death, but he knew that some people had.

The hunter and his drum

Through such experiences the hunter proves himself, demonstrates his
affection for the land. And after a lifetime on the land, Isaiah Awashish
had established a contact with the hunting spirit, a relationship with the
animals, through dreams and trances and most of all through the medium
of his drum. "The hunter talks to his drum," says Isaiah, "and the drum
decides what is possible and what is right. The drum is like a person, you
can talk to it and it will reply. But you cannot force the drum, the drum

will not do just what you want it to do. If you try to force something on the drum, then the drum will make life difficult for anybody else hunting on that land.

"The hunter secures his relationship with the land through his drum. The land, the trees have to be respected. The animals live off the trees, and if there are no trees, there are no animals and the Indians suffer.

"A hunter cannot just go and demand of a tree that it give him something, help him, aid him, cure him from sickness. You have to give something back for what it gives you.

"If you are in a dangerous situation, you cannot just demand of the otter or some other animal that he get you out of it. You have to give something back. So you cannot ask that the otter will get you out of a dangerous situation, as when Willie and I lost our canoe. Or that the otter will cure someone as I once asked it to do when Willie was sick. You cannot demand of a tree that it will cure someone who is ill without giving something back. Some people know they are going to be sick. You can dream about that, since you can also be contacted through dreams as well as through the drum. If you dream that you will be sick, you can prepare for that, you can prepare by offering something to the sickness. You might give a small piece of meat, so that when your sickness comes someone will cook it for you; that's your offering. But it's not just a piece of meat you are offering: it is a piece of your hunting spirit.

"If you do not dream about it, and you get sick, then that is the end of you. People who ask the trees or animals to cure them, who do not offer anything in exchange, they allow all of this to pile up, and sometimes the hunting spirit will have to leave that guy. That's when he gets sick and nobody can help him.

"Since I have been a trapper I have been able to see the future in a way that I really could not understand what it meant. Only after, when things had happened, did I understand that I had seen it and had known it would happen. During the time I was in contact spiritually with the hunting spirit, what I could see was all about this, what is happening now. A hunter must always watch his dreams, for from them he can tell where the animals are. He may see in his dreams a map of the land and on that land he can see where he will find the animals. Now that I am coming to the end of my hunting life, now that my moccasins wear out every week or two because it takes me twice as much work to do as much as I used to do, the animals in my dreams are becoming smaller, so I know I am coming to the end."

A HUNTER'S DREAM

Years ago, before his sons were born, Isaiah Awashish had dreamed that
he would have three sons, of whom the eldest would dress in suits and
carry briefcases, the second would engage in some sort of business and
the third would be a hunter, like himself.

To a remarkable extent, life followed his dream, though he had four
sons, not three. His oldest son, Philip, was the first boy from Mistassini
to enter a university, which is to say that he underwent a process aimed
at assimilating him into the dominant Euro-Canadian culture. For a while
it worked, for when Philip entered McGill University in Montreal he
hoped to emerge as an engineer. He could not, however, stand the white
society and the hustling atmosphere of the cities for long. He dropped
out of the university and returned to the reserve to try to reintegrate
himself with the Cree culture. In the nearby town of Chibougamau he
founded an Indian Friendship Center and literally became a briefcase
Indian, sought out for translation and interpreting work by anthropologists, government officials and journalists, engaging in the ceaseless
round of meetings that were the primary point of contact between the
government and his people.

Isaiah's second son, Solomon, as the dream foresaw, though never
exactly a businessman, turned away from the Cree life and worked for a
group of hydrologists who were studying the waters of the Cree homelands for Hydro-Quebec, the provincial electricity-generating agency. He
flew with them all over the wilderness in which his forefathers had hunted
for many generations. He helped them take soundings and make readings
about the extent and quality of the waters his ancestors had always known
intimately. Eventually he decided he would like to fly the little bush
planes himself, and he undertook a course as a pilot.

And, as the dream foresaw, it was Willie, the third son, who chose
to follow the ways of his father: he left school at the age of eleven and
never returned, but learned from his father everything there was to know
about how to survive in the northern wilderness in winter.

REALLY PROUD OF MY FATHER

"When I first started trapping with my father, I was really amazed by what
he did, and what he knew about the bush, and what he could tell you,"
said Willie. "He could tell you, In two days from now, it is going to be
snowing. It would be a clear, fine day, not even a cloud in the sky, but

he'd tell you it's gonna snow, you know. And you could see it happen. When I first went into the bush, I never thought there was such a man as that. I was really proud of my father. At first he used to always send me home early, tell me, It's gonna take you about an hour to go home, half an hour to cut the wood, build the fire, by the time you'll ever be inside it'll be seven o'clock. Well, when I would get home and all finished, it'd be seven o'clock. It was like he could time everything, he knew how long everything would take you. . . .

"Lots of times he would tell me where to go to kill animals. One time I remember, it was before I had even shot a moose, it was going to be the first time I shot a moose. We had a certain area there, I went there and there were no moose around. There had been moose there at the first fall of snow. You could see where they'd eaten, but you couldn't see their tracks, you couldn't follow them where they'd disappeared to. And I'd been back there and there was no moose. Sometimes when you see signs of a moose, you can't even see the track, you can only see the willows where they ate, and you have to start judging how old that willow is, how long ago the moose ate it, and you have to try to follow where he started from and where was the last time he ate it, that's how you know which way he has disappeared to.

"So I told him, I'm gonna go and do that on the mountains. And he says, Aye, do that tomorrow, he says. So the next day is a very nice day, and my father wakes me up. He says, If you're gonna go out there you better go now, he says. It was early in the morning. He tells me, There's a lake there. There was an old beaver lodge there, he says, and right from the beaver lodge, straight north, there's a little creek going up the mountain, he says, and sometimes when you never find a moose anywhere he'll be in that creek there going up the mountain, just in that small part of the area, that particular place. Make sure you go out there, he says, if you don't see any signs of him, you go to this mountain. So I went over there, and right in that little creek where he told me, I was going along and I saw these two moose. He just told me exactly where they were. He knew they were there.

"Once he was going around on one of his weekend drunks when we were in Mistassini and he came up to each one of us and he wanted something from us, like a shirt, or anything that he could wear, a coat or anything; but nobody gave him anything because everybody thought he'd just forget it when he sobers up. So we went in the bush and the last airplane trip that came up, Anne-Marie, Philip's wife, had bought him a pair of trousers. He told my mother, Finally somebody gave me something to wear: keep my trousers, I will ask for them when I need them.

In the bush he always wears these really scrabby pants, full of patches. My mother used to tell him, Why don't you wear those new trousers Anne-Marie gave you? There's gonna come a time when I need them, he says. And all fall, all freeze-up, he never killed no bear, no moose. So after freeze-up, when it was really safe to walk on the ice, he said, Where's those trousers that Anne-Marie sent me? And he took them and put them on. He brought home two beavers that evening. He told my mother to cook one of them. So we had one of them. He invited the other family to come and eat with us. Everybody was sittin' there. And my father said, There was a time in Mistassini when I was drinking, and everybody thought I didn't mean what I told them, that I wanted something from them, something to wear. All fall I always wanted somebody to give me something, like a shirt at least or a T-shirt, something I can wear. Now I have got my trousers. Now somebody has given me something.

"He told me to go out and get his drum. It is not too often that my father will ask me to get his drum. You cannot bring the drum into the lodge unless you want to use it. So I went out to get it. He played his drum, and the next morning he was gone when I woke up, he was gone two hours before I woke up. He didn't come back until about nine o'clock in the evening. He said he had killed a bear, he had followed the bear's tracks all the way to its den. He told me, Lots of people when I was in Mistassini, they think I'm just a big drunk when I walk around and I tell them to buy clothes for me. But it's not that. All year, he says, we wouldn't have killed nothing, not a bear nor a moose, unless somebody had given me something."

The beaver that won't be caught

Isaiah tells a strange tale: "There are three lakes on my territory where I have never been able to catch the adult beaver. They are in the corners of my territory, and there are many creeks running into these lakes where I have often enough caught young beaver. But I have tried all my life, in every way I know how, to catch adult beaver in these lakes and creeks and I have never succeeded. My father tried before me, and his father, and no one has ever caught the beaver in these three lakes. During all of that time the adult beaver have maintained the population around these three lakes. But they have not wished to be caught, and so we have never been able to catch them."

How the hunter finds his way

"When you're trapping," says Willie, "if you watch the land, where you're going, two years later you can go through the same place and remember that you went past this tree and between these two trees. You don't carry a map or nothin' like that. My father can tell me, There's a bear's den here, and he can describe the area to me. There's a little hill there, and right on top of it there's a rock there, but it's not that rock, from that rock you can see another rock. It's been ten years since the last time he's seen it, he can remember back years and years, you can't even identify the spot where the bear's den was, because it's all grown in, there's nothing left of it. But he can still say, It was here, right here. All there will be is as if something has fallen in, like a ditch or something, that'll be it.

"I guess you could say my father knows the bush like the back of his hand. One time I had a Skidoo [snowmobile] in the bush and I decided to go to Mistassini with my Skidoo and leave it there and come back by plane so that we don't have to pay for an airplane to come and get my Skidoo. I told my father, Let's go to Mistassini. It was during the spring So I went over to see my Uncle Sam, and he said, I'm gonna take my Skidoo down too, I'll go with you people. It's been several years, about fifteen, since the last time my father walked through to Mistassini, since before the planes came in. When you walk and you know you're going a long ways, you can't walk over all these mountains, you have to go on the lakes, where it's flat. So one morning Sam came over and we took off. Along the way to Mistassini there were blazes on the trees. We would get onto a lake and I'd stop the Skidoo. My father was riding behind me. I'd ask him, Which way now? He'd say, See that point there, you go around it, and you go into a little bay, and at the bay there's gonna be a rock. You go right along the rock, and the first tree you see there's gonna be a blaze on it. So we would go there and go along the rock, and the first tree we'd see would be dead, rotten, but you can still see the old blaze on it. And all the way to Mistassini, going through all the lakes, he made only one mistake, when I didn't even know where the fuck I was going, I could have been going north or west or south, and that one place, we had to go through a whole bunch of bush, and he said, This is not the way, there's another way around, but we may as well just go straight through here. He would just say, Go this way.

"The reason my father knows the country so well is that when he was young they used to have this mail service, the snowshoe express, I guess you could call it. Someone would start with it from Rupert House to Nemaska; and then the second guy would bring it up from Nemaska to my dad's hunting grounds, and then my dad or one of his late brothers would take it on all the way to Mistassini. It would be just a piece of paper containing the price of fur. And from Mistassini someone would take it up to Neoskweskau, and from there right up to Nichicun.

"I don't think it would be too easy for me to have that kind of knowledge. Even on our trapping ground I had trouble trying to get around the area, I missed out on a hell of a lot of lakes. I could remember most of the lakes where we caught the beaver, you can remember that, when you go back there it all comes up to you once you see it.

"It was only the last year I was trapping that my father allowed me to go out alone. When you're out alone like that, you've got no time to think about whether you're happy or not. You've got nobody to talk to, but you do feel proud that you can do it, and that you're doin' it for your mother and father. All you think about is, Am I man enough to do it? and when you're doin' it, it makes you feel good. When you're walking you're not by yourself, you are watching your shadow, and the dogs, too, you can tell they like it. When you're out like that you talk to the dogs, take them as human beings, they eat when you eat, what you eat. You don't feel scared, children are taught not to feel scared, and you're not by yourself, there's always. . . . Well, when you're trapping, you know, you're totally in a different world, not like the world outside where you have to depend on machinery and all that. When they're trapping, the Indian people have their own world. It's a totally spiritual world you live in when you're trapping."

Willie goes away

"What I can say is that my father has really taught me good. He has taught me that there is more to life than trapping, or working for wages. What I can see now is that I will go back trapping with my father until he doesn't want to go trapping any more. There will come a time when he tells me to stop, because there will be no more animals. But my father has given me a choice. When we were trapping once he told me, You can go if you want to go and work, he said. I will be doing okay this year. So I told him, I'm gonna go and try it out for a while. He wanted me to go, too. He wanted me to go out and do what I can do without him. He told me, I'm getting old, and I cannot always be there to tell you what to do, you've got to know what you have to do, and you've got to know different sorts of work you can do yourself.

"My mother didn't want me to go. So finally I killed this moose, you know, I handed everything over to my mother. I told her, You can have the moose hide and everything. We were eating, and my father said, Amongst all my sons, he said, you've been the only one that's been with us all the time, and there's something I always wanted you to do, and that is go out and experience something else different from what you experienced with me. My mother started talking about not letting me go. But my father said, Look at us now, we are getting on, me and your mother, we can't always be there to tell you what to do. You've got to start making your own decisions."

A HUNTER'S PRESENTIMENT

In the first half of 1974, when Philip Awashish was twenty-five, he was one of three young men to whom the Cree Indian chiefs of James Bay turned when they had to appoint someone to negotiate on their behalf with the Quebec government as they began a struggle to defend their culture and way of life against a huge hydro-electric project which represented a ferocious onslaught on their traditions. Philip had, it is true, been alienated from his culture by his many years of white man's schooling. But he always remained deeply respectful of the values and capacities of his father, and it was that respect which was to carry him through more than twelve months of countless meetings with countless bureaucrats. To him fell the task of trying to negotiate with outsiders (who really did not begin to understand the values of his people) conditions for hunting, trapping and fishing in the future which would permit the ancient culture to survive. It was a grave responsibility for so young a man.

At about the same time, the second son, Solomon, overturned his plane as he was bringing it into Chibougamau airport from a training flight. He was severely injured about the face and head and was rushed to a hospital in Montreal.

"I will have to fly into the bush to tell Isaiah," said Sam, Isaiah's brother. "Of course, he will already know about the accident, but he will be worrying and I must relieve him of his worry."

Solomon was still in the hospital when Sam flew into Isaiah's camp. The old man's first question was, "How is Solomon?" He had sensed that an accident had occurred to his second son.

BANNED FOR TWO MONTHS

Two years after being nearly stranded with his father on the river in the wilderness, Willie Awashish was cutting lines for the Chibougamau Mining and Smelting Company, doing seismic work for them. Four young Indians would go into the bush together to do the work, work that was by now pretty well given over to Indians because of their superior skills in the bush. They would stay in the bush for as long as thirty days at a time, and when they returned to the little mining town of Chibougamau, they would do quite a lot of drinking, moving from the Waconichi Hotel over to the Obalski (occasional venue of Baby Honda, a 300-pound stripper who always packed the joint in this masculine mining town), to the Monaco and ending up, usually, early in the morning, like eight or ten o'clock, at Alice the bootlegger's.

"One night there was this gang of French guys, the Sex-Fox, who beat up a bunch of Indian kids at Alice's," said Willie. "So the next night ten of us went into the Monaco and took three tables. Then these Sex-Fox started to pick on us.

"Well, we beat the shit out of them, and then we went up to the bar and the whole place emptied. The bouncer came along and suggested we leave. We asked him if he would like to try to make us. We bundled him out to the head of the stairs and rolled him down. Then someone called the police. They suggested we leave, too. We invited them to make us, so they pulled their guns on us.

"Christ, I was banned from that place for two months after that."

NOBODY ELSE LIKE ME

"Ever since I have been trapping," said Isaiah, "younger people than myself have been changing and changing every year. There used to be more people in touch with the hunting spirit. Now there are not so many. Some still are, because they have followed their father's footsteps, they have taken up everything their father has taught them. But now the young men are starting to change. I am not against what the teenagers now think, that is their business.

"But when I was Willie's age there was no beer, or anything that could corrupt my thinking. I know that Willie cannot have the same attitude to everything around here that I had when every person was here in the hunting grounds and there was nobody else and we all had the same attitude. As I grew up, I never drank anything until I was thirty.

There will be nobody else like me. But I have tried to teach Willie everything I know. I appreciate his being with me, helping me, catching all these animals for his mother and the others in the camp to eat. As I grow older I do not do certain things that I used to do. When I was young a blizzard would not stop me. But now, if it is snowing hard, I will just go back to the lodge.

"Now that I am at the end, it is through the drum that I have handed on my relationship to the animals, and to the land, and to the hunting spirit. I have never allowed my drum to be taken off my territory, for the drum would not want that. But I have now given my drum to my younger brother Sam. He is the man who is worthy of accepting it. And with my drum I have given my land. I no longer own the land that was handed down to me by my father."

2

COURTROOM

On a damp, cold December morning in 1972 a group of fifty or so rough-clad men marched along narrow, traffic-jammed Notre Dame Street in old Montreal, pushed open the swinging glass doors of the huge and impersonal Palais de Justice and stood uncertainly in the cavernous spaces of the lobby, an open space surrounded by balconies soaring four stories above their heads. Even in this building, which every morning sees a procession of some of the stranger people in this large city—pimps and whores, petty thieves and con men, unhappy husbands and wives, debtors and drunks, shyster lawyers and pillars of society—even here their presence created a stir of interest.

They were a strange group, in their great, plain working boots, the laces tied with no concern for appearance, their wind jackets weather-beaten and torn, their shirts hanging below their jackets, their faces curiously open beneath their rude, home-fashioned haircuts. But they looked no stranger than they felt, for they were Cree and Inuit (Eskimo) hunters from the huge Ungava Peninsula of northern Quebec, come to the city for the first time to undertake the audacious and apparently hopeless task of asserting their rights as occupants of their hunting lands from time immemorial, and bringing to a halt a hydro-electric project on which the provincial government had staked the entire economic future of Quebec. The older men regarded the building with dignified incomprehension, for they had never seen such huge enclosed spaces before, and they waited patiently to be told what to do, whether to go into the strange room which moves up and down and which was causing them much trouble at their hotel, or whether to go to the other side of the lobby and try to mount that moving staircase as many other people were doing, and so be carried upstairs in the strange manner of the white man, without having to bother to walk a step. Not all of them were elderly men: the group included some younger men with fashionably long hair falling blackly to their shoulders, wearing purple jeans and the thin, high-heel boots fashionable in the city, and one or two girls with classic high-cheekboned faces, their eyes large and slanting, their noses straight, their foreheads high, their long black hair falling gloriously down their backs. The group was being ushered upstairs by some southern Indians, barely distinguishable from white men, members of the new Indian establishment that had arisen in the last decade, officials of the Indians of Quebec Association, cautious bureaucrats who after eighteen months of delay had agreed to help the Cree hunters go to court in their effort to stop the white man from damming the rivers around which they had lived for longer than anyone in North America could remember.

Just a month before, the lawyers for these hunters had filed a petition for an interlocutory injunction designed to bring all work being done on the James Bay hydro-electric project to a halt until the question of their rights in their hunting territories could be decided by the courts, and it was the hearing of this petition that was now about to start in Quebec Superior Court.

"I have told their lawyers that it is hardly necessary for their clients to appear in court at this stage," said Mr. Justice Albert-H. Malouf, a day or two before the opening of the proceedings, "but he has said they are so eager that he cannot stop them."

In two or three trips the men went upstairs in the elevators, plodded across the thick-piled carpets in the upper corridors and filed into a small courtroom, occupying almost every seat and standing around the walls.

Many of them were not clear as to what was expected of them in these strange surroundings: they knew only that they had been brought south to answer questions that would be put to them, and to tell people in the south about how they lived so that they might defend their way of life. Certainly today they could not have realized that the whole day—in fact, the whole week—would be occupied by legal argument so complicated and dry that even most people who understood the French and English languages in which the argument was conducted would find it difficult to follow what the lawyers were talking about. But the hunters were not men for whom time had great importance: they had patience, and they stayed in the overcrowded and overheated room throughout the day,

looking for all the world as if they were following the argument closely
as the lawyers tossed obscure legal references back and forth before the
judge.

It was the first day—December 5—of a case which was expected to
be over by the end of January. But the case become so complex, involved
so many areas of evidence, demanded so much expertise, that it occupied
78 days to hear 167 witnesses who produced 312 exhibits and spoke
10,000 pages of transcribed evidence. The hearing did not end until June
21, and it was almost a year before the judgment was handed down.

The case was one of the most remarkable ever heard in a Canadian
court. It had profound implications for almost every aspect of Canadian
society. At the heart of it was the confrontation between the non-acquisi-
tive, sharing ethic of the hunting culture and the individualist ethic of the
technological culture of North America. That confrontation called into
question not only the survival of the hunters but also the human qualities
(or lack of them) of the powerful technology embraced unquestioningly
by the white man.

THE PROJECT

The scheme which the hunters felt would destroy their way of life was
enormous even by the standards of North America. It would create dams
or diversions on seven of the great wild rivers running through the Cree
and Inuit hunting grounds, establish huge reservoirs across the low-lying
wilderness and generate 30 percent of the power that the whole of
Canada was now producing each year. An area of 133,000 square miles
(three times the size of New York State) east of James Bay had been
created a single municipality and placed under the control of a Crown
corporation, the James Bay Development Corporation, set up by the
Quebec government. The corporation had ignored the protests of the
Indians and had already built roads, airports, construction sites, was
preparing the way for the first dam and was actively encouraging mineral
exploration all over the territory.

Phase One of the hydro-electric project was already under way. Basically, it was to create along the La Grande River, running into the north
of James Bay, a series of seven vast reservoirs, into which would be
diverted waters from the Great Whale River from the north, the Eastmain
and Opinaca rivers from the south and, from far inland, the Caniapiscau
River, which runs directly north from the center of the Ungava Peninsula
across the tundra to the sea at Ungava Bay. With only a few miles between
them, these reservoirs would stretch in an almost unbroken line 500 miles

inland—equivalent to the distance from Washington, D.C., to Portland, Maine; or from New York clear to Detroit.

This first phase of the project, to be completed by 1983, was supposed to create 125,000 jobs, stimulate the normally somewhat retarded economy of the province of Quebec and provide the cornerstone for a great economic leap forward that would save Quebec from the political threat of separatism. Its cost was at first estimated at $5.8 billion (a figure which, before two years were up, would be almost trebled). Thus, politically and economically, the stakes were high. They could scarcely have been higher. But the Indians knew only that already their hunting life had been interfered with by the preliminary work, and that when Phase One was completed, one of their major rivers, the La Grande, would have disappeared.

Phase Two, which would follow presumably in the eighties and nineties, would complete the job in the southern part of the Indian hunting lands. Under this scheme, the Nottaway, Broadback and Rupert rivers would be dammed and diverted, the waters to be transferred from one river to another through a series of interlocking man-made channels. Seven power stations would be built on the Rupert River; Mistassini Lake, far inland, would be turned into the basic reservoir for the system; and dozens of major lakes in the Cree homelands—Waswanipi, Matagami, Soscumica, Nemaska, Sakami, Opinaca lakes among them—would disappear in an unimaginable flood. The size of the territory affected would approach that of the entire state of California.

In this court case the Indians would not be allowed to talk about their fears for the distant future. They would have to prove that the work already done had damaged them, and that the work contemplated in the next twelve months or so would disrupt their lives to such a degree as to justify the suspension of work while the question of the Indian rights was decided by the courts in a long-term action. Never before in Canadian history had so politically powerless a group of people tried to stop so huge a scheme.

A COLONY WITHIN A COLONY?

The courtroom was the physical embodiment of a sharp and bizarre cultural dilemma present in this case. The persecutors of the hunters in James Bay were themselves the country's largest and most vocal cultural minority, the French-Canadians, or Quebecois, as they now like to be called, whose battle for cultural and linguistic survival had in the last decade become so intense that apparently about a third of them now favored the separation of their province from Canada and the setting up of their own nation-state. This minority had survived in North America by the application through 200 years of an absolutely fierce determination to propagate and maintain themselves as an entity. From 70,000 at the conquest of Quebec in 1759, they had become 5 million in Quebec, with as many again who had emigrated to the United States. Though always culturally firm and sure of itself, politically and economically Quebec was until 1960 one of the most backward areas on the continent.

After a decade of awakening, the Quebecois had shed much of the xenophobia they developed during their fierce struggle for survival. But still every aspect of public policy among them was seen through a nationalist perspective—that is, through the perspective of the only French-speaking nation in North America. Every major economic advance or project had come to be regarded as the triumphant assertion of the national will. But, oddly enough, while this cultural revival had been going on among them, it had become increasingly difficult to tell the difference between the Quebecois and the rest of the North Americans, apart from the fact that they spoke French. Like all other North Americans, they wished to embrace the technological dream, and shrewd politicians among them had found the means to pursue policies of resource sellout on a gigantic scale while making it sound as if they were all necessary in defense of French-language cultural and national survival. It was this nationalist ethic, harnessed to a technological dream, which the Indians and Inuit now had to fight. The construction of big dams and the generation behind them of electricity from the waters of the teeming rivers of Quebec were the major economic achievements of the Quebecois in the 1960s, and the James Bay hydro-electric project, which the Indians were now opposing, was merely more of the same thing: from its inception it had been clothed by provincial political leaders in the rhetoric of national development. Though the scheme was ill-prepared economically, it would create jobs and was therefore acceptable to a public already suffering from Canada's most persistent unemployment. Though it was an offense to environmentalists in its proposed destruction of a great wilderness, most of the environmentalists were English-speaking and therefore didn't really count. Though it was imposed on the Cree hunters as if they were of no account, their cries of minority outrage could be depended upon not to move Quebec public opinion because these, too, when they were not uttered in Cree, were uttered in English. Also, the Cree were looking to the federal Canadian government—another English-speaking agency—to defend their rights.

Given this kind of reasoning, it is perhaps not strange that the

Quebecois should have shown themselves insensitive to the needs of another minority also struggling for cultural and linguistic survival. With every issue decided on grounds of the Quebec national interest, it followed that any other interest must be inferior, or at least of lesser importance. As it happens, this doctrine was eventually spelled out in so many words by one level of Quebec court.

But meantime the courtroom physically provided a fascinating view of these lines of battle. The case was fought, interchangeably, in both English and French, the lawyers, witnesses and even the judge sometimes switching from one language to the other in mid-sentence. But there was no doubt about the battle array. Lawyers for the native people (with one exception) were English-speaking, though capable of pleading in French; while those for the Quebec government, the James Bay Development Corporation and the James Bay Energy Corporation (the Crown corporations entrusted with the work on the project) and the various contractors being proceeded against were French-speaking, though capable of pleading in English. The Indians and Inuit gave evidence in their own languages, translated by one or other of several young men who had been educated in the southern schools and who knew English. On the native side almost all of the expert witnesses spoke English. Geographers, hydrologists, engineers, anthropologists, brought to Montreal from all over the continent, almost all gave evidence as to their expert knowledge in English; while when the corporations put people in evidence, nearly all were Quebecers, their evidence given in French.

From the beginning it was a dialogue of the deaf as between the native witnesses and the French-speaking lawyers for the corporations, who began the case without really thinking it was necessary and only with difficulty were brought to the point of taking it seriously when the native people were about halfway through their evidence. Legally speaking, the battle was focused on the dogged and persistent effort by the tenacious little lawyer leading the native case, James O'Reilly, to persuade the judge that Indian rights did exist, had been the subject of legislation in Canada through at least 200 years and were, though perhaps anachronistic, real. The corporation side began by treating this argument with contempt. Its leading lawyer, who disappeared after the first few days, said it was "absurd." This man, C.-Antoine Geoffrion, perfectly bilingual, one of the wealthiest lawyers and an influential political figure in Quebec, was the only French lawyer who elected to argue in English. A perfect *assimilado* (as the Portuguese call those Africans who have made it into colonial society), Geoffrion was fat, amiable, with a round, cheerful face under a head of curly white hair, a bumbling, stuttering figure who mean-

dered his way through a collection of Indian cases with which he sought
to prove that Indians in northern Quebec did not have, and never had
had, any rights of any kind.

No need for a hearing?

The Cree and Inuit people sat in court listening impassively to the government argument that since the Indians had no rights in James Bay,
there was no need for a hearing. The government lawyers had argued this
in preliminary skirmishes in November, after which both sides handed to
the judge a huge documentation of past case law which he had since been
studying.

Unlike the United States, Canada does not have a vast body of Indian
law. Until the recent formation of political organizations financed by the
federal government, Indians in Canada had never had money to hire
lawyers. So it was only in the last five or ten years that they had been able
to bring any great number of cases into Canadian courts in defense of
their traditional rights. During these recent years James O'Reilly had
become one of the foremost experts in such Indian law as does exist, first
as lawyer for a community of Indians who live just across the St. Lawrence
River from Montreal, and later as lawyer for the Indians of Quebec
Association, the government-financed political organization which had
now undertaken to bring the fight of the Cree people to court (using for
the purpose money granted to it by the Canadian government). In some
ways O'Reilly cut an odd figure: short, balding, with a long wisp of hair
usually in disarray, a small, round, fleshy face and large lips, full of
slightly irritating mannerisms, disheveled, always short of time, he was
the sort of man who would hurry late into meetings with a big briefcase
and an air of never having quite finished whatever he was last doing.
Formerly an employee in one of the province's two most prestigious law
firms, O'Reilly had wrestled with his conscience for months when his firm
was engaged as legal representative for the James Bay Development
Corporation, which was given control over the land occupied by Indian
people for whom he was the leading lawyer in the province. Eventually
he had resolved his dilemma in the only possible way by leaving the firm
and setting up his own firm with two colleagues, he himself continuing
to specialize in Indian law.

Until now Indian cases had always been small stuff. But O'Reilly
knew that any attempt to stop the James Bay project would be on a
different scale from anything he or the Indians of Canada had undertaken
before. For months he had advised delay and caution as the Cree people

and their representatives had tried to persuade the Indian Association to take action to stop the project. But now that the decision had been taken and he had the bit between his teeth, O'Reilly showed himself to be a tireless worker to the point of being almost boring in his determination to leave nothing to chance in proving that the Indians, for whom he had by this time developed a considerable feeling, had real rights which should not, in a society based on law, simply be trodden over.

In this preliminary argument he had only to persuade the judge that the Indians apparently did have some rights and were therefore entitled to bring evidence to court to support their petition for an injunction. He told the court that the petitioners represented virtually all of the people living in all of the Cree settlements of Mistassini and Waswanipi (inland from James Bay) and Rupert House, Eastmain, Paint Hills, Fort George and Great Whale River along the James Bay coast, and the Inuit populations of Great Whale River and Fort Chimo in Ungava Bay, who were supporting the Cree because their lives as hunters would be affected by the damming of two of their rivers.

"Petitioners are invoking not only their collective territorial rights, also known at times as aboriginal rights, in the land; they are also invoking rights presently recognized by the province of Quebec to hunt and trap exclusively in all of the territory which will be affected, and the rights to use the navigable waters, the highways of the north in summer as in winter," said O'Reilly.

He spoke for nearly five hours, running through the long catalogue of legal references to Indian rights, from the first mention of them in the Royal Proclamation issued in 1763 following the British takeover of Quebec after Wolfe's victory, right up through a bewildering array of instructions to governors, orders-in-council, legal *obiter dicta,* minority and majority judgments of the English Privy Council and the American and Canadian supreme courts, through several laws in which Indian rights were safeguarded. This was all familiar ground to O'Reilly, and even to me, a journalist who had been writing about Indian affairs in Canada for three or four years, but it was clearly unfamiliar to the judge, and to the opposition lawyers, who had difficulty in taking it seriously.

INSTRUCTIONS TO GOVERNOR MURRAY

The main lines of his argument were that recognition of Indian rights flowed from a concept of aboriginal rights enjoyed by the occupants of North America at the time of the white conquest. From the earliest times it was clear, argued O'Reilly, that pains were taken to maintain good relations with the Indians for practical reasons as much as anything. He quoted instructions sent to Governor Murray by King George III on December 7, 1763, to the effect that "whereas our Province of Quebec is Inhabited and Possessed by several Nations and Tribes of Indians with whom it is both necessary and expedient to Cultivate and Maintain a strict friendship and good Correspondence. . . . You are upon no Account to molest or disturb them in the Possession of such Parts of the said province as they at present occupy or possess; but to use the best means you can for conciliating their Affections, and uniting them to our Government. . . ."

In a later session the King forebade anyone to take possession of any of the lands reserved to Indians "on Pain of Our Displeasure without our Especial Leave for that Purpose first obtained." What did it all mean? asked O'Reilly. "At that time the Indians occupied vast tracts of land for hunting and fishing and it's absolutely clear that they were meant to continue their same way of life. Why? Because in those same instructions they set out specific regulations for the fur trade. Who did they depend upon for the fur trade, but the Indians? No hunting and fishing by the Indians, no fur trade. Therefore the English merchants at the time could not make an economic go."

From these basic concepts, argued O'Reilly, the recognition of Indian rights in Canada led to the signing of treaties in many parts of the country, though not in Quebec. The lands owned by the Hudson's Bay Company (most of northern Canada, including the northern parts of all provinces from Quebec west) came to the Canadian government at confederation, and were transferred to the provinces bit by bit. Quebec was given some of them in 1898, when no reference was made to the Indian title. By 1910, however, the federal government was insisting in an order-in-council, that "the province of Quebec, to perfect its title to this domain, should adopt the established practice and agree with the Dominion upon the terms and conditions of a treaty with the Indians for the formal cession of their title in the lands." By the Quebec Boundaries Extension Act of 1912, in taking possession of all of northern Quebec, the province undertook "to recognize the rights of the Indian inhabitants of the territory above described to the same extent, and will obtain surrenders of such rights in the same manner, as the government of Canada has heretofore recognized such rights and has obtained surrender thereof. . . ."

A PERFUNCTORY APPROACH

These documents, O'Reilly argued, proved that the Indians apparently had rights. That was all they needed to prove at this stage. All of this was obviously hard for the French-Canadian lawyers on the other side to swallow: unfamiliar with Indians, unfamiliar with vague and strange concepts of Indian law, certain that ownership was something that could be proven only by the possession of a deed, they had to grapple with a massive documentation produced by their untidy and fervent little adversary. Classically educated, middle-aging, they showed every evidence of having minds so closed on this subject that they could not figure out how to deal with the arguments before them. Roger Thibodeau—a spare and austere figure representing the James Bay Development Corporation, a perfect specimen of the image of the civilized bourgeois, familiar to us from French films—had come down from Quebec city for the day and was in a hurry to return there. He paused just long enough to say that O'Reilly's eloquence was completely wasted, since the case was an open-and-shut one, and simple to boot: the law under which the work in James Bay was being done was constitutional until proven otherwise, and was therefore legal. That was all there was to be said about it. He raced off before the day's hearing was over, needed, no doubt, for more important business in Quebec city.

It was left to Maître Geoffrion to argue in a rambling dissertation that the Indian title could not be said to exist because it was unascertainable, that if any recognition had been given to it in the past by the government of Canada, this recognition was so vague that it did not amount to any sort of recognition under English public law, common law or civil law; and, lastly, since this so-called aboriginal or Indian title differed in different parts of Canada, whatever else might be said, it was clear that it had never been recognized in northern Quebec. Rights, to be meaningful, had to be specified. And all these acts and orders-in-council did not specify the nature of the rights. They merely recognized, perhaps, a moral obligation, but certainly not rights. He argued that when the government of Canada had accepted from the Indians on the Ontario side of James Bay the ceding, releasing, surrendering and yielding up of all rights, titles and privileges to the Crown, this did not imply that the Crown recognized that the Indians had any rights whatsoever to cede, release and yield up.

Maître Geoffrion was difficult to follow, even for those who knew the legal jargon. But what he was saying was straightforward: to the claim of

the Cree people that the government was in the process of destroying
their culture and their way of life, he replied with great good humor that
every solemn reference to Indian rights written by the white man into his
own laws in the last 200 years was not intended to have any meaning at
all. In the 1970s it was an astonishing official response from a society
apparently preoccupied with its own cultural survival, a frank and hearty
shrug of indifference about the Cree and Inuit dilemma. So the Cree
would be destroyed? So what?

PALLID, BLOODLESS FOOD

The judge expressed to O'Reilly his concern that he would have to keep
the hunters in Montreal while he made up his mind whether to grant an
inquiry or not, but he promised to decide as soon as possible and, in the
event that he decided to allow the case to proceed, to go ahead immediately. It was the first of many occasions during the case on which he
showed a specially solicitous understanding of the people who had come
from the north to testify, and it was an early indication that, whatever fate
the white man's institutions might have in store for them, the hunters
were at least appearing before a man with decent human instincts.

The main problem the hunters faced in the city was the food: the
meat they could buy was pallid and lacking in blood compared with the
rich moose, caribou and beaver that they were used to eating. To an
extraordinary extent, these people, in their own setting, lived on a diet
of meat—wild, gamey meat, of which they could consume enormous
quantities. When hunting in the bush throughout the winter, they would
eat almost nothing else but meat (apparently obtaining a sufficient supply
of Vitamin C by eating every part of the animal, including the intestines
and their contents). They would eat more of the white man's food when
living in their villages, close to shops, but even there they were never
separated from the country food, as they called it, for which they felt far
more than just a physical craving. For such hunters there is a spiritual
dimension to this country food, and it is always the lack of this food which
first weighs upon them when they have to stay for any time in the city (as
some sometimes do for medical treatment).

They were astounded, of course, by the high buildings, but it was not
the open-mouthed astonishment of a rubbernecking tourist: they felt
vaguely threatened by these buildings forty stories high that you could
see through and that looked as though they might fall any minute. They
could not venture out of their hotels without someone who could speak

one of the European languages to take them around, for all the signs were strange, and machinery was everywhere. All their lives they had used their legs for transportation on the ground and now they found that people seldom walked, but moved from one machine to another—from a bus onto a moving sidewalk, from that onto an underground train, and then onto the moving staircase, and then into the mobile rooms. They were living high up, fourteen, fifteen floors, sharing their rooms with other men, lacking women to sleep with, and when they went out they would look up and fail to see the topmost window in their hotel, so that they wondered if anybody would ever be able to reach the top floor of the building, no matter how hard he tried.

Their lawyers were operating on the assumption that they would be granted a hearing, and the work of preparing the witnesses went on for hours over the weekend. It was difficult work. Even people sympathetic to the hunters, people trying to help them, like Monique Caron, the pretty young French-Canadian lawyer in O'Reilly's firm who did most of the preparatory work, found it hard to think on the same wavelength as the hunters. The interrogators wanted precise information, facts about distances and times and places, things which the hunters never thought about and took completely for granted.

CARON: Can you show me on this map here?

SIDNEY LOON [Mistassini hunter]: It's not so good, but it's got the numbers.

CARON: Number five? That's the number of your trapline? How much time do you spend in the bush during one year?

LOON: The whole year, twelve months.

CARON: You live twelve months in the bush?

LOON: Yes, the whole year.

CARON: Are you there with your family? What kind of trapping or hunting, what do you trap or hunt?

LOON: We hunt caribou, moose, there's moose, we trap beaver, otter.

CARON: Beaver and . . . ?

LOON: Otter and lynx and marten.

CARON: Do you fish also? What's the lake?

LOON: Michicous.

CARON: Where is it on this map? [Laughter]You spend twelve months here or . . .

LOON: In the fall and then come back in June.

CARON: How many months altogether in there?

LOON: Well, we start in September and we come back at the break-up.
CARON: Which is? April? March, no?
LOON: June.

To get the precise detail they needed, the lawyers had to go over and over the same questions, and then they had to decide the relevance of a particular hunter's knowledge to what they had to prove in court.

DARK-COMPLEXIONED JUDGE

This laborious preparation was rewarded. Mr. Justice Malouf decided to grant the Indians and Inuit the hearing they had asked for. And once again they filed into the courtroom, where they sat in the body of the court facing the judge, who sat on a well-upholstered, long-backed chair slightly higher than the rest of the room, looking over the heads of the court officials and the advocates who faced each other on either side of the room. Mr. Justice Malouf wore a black robe with two red panels down his chest: a man of Middle Eastern origins, he was, ironically, as dark-complexioned as many of the Indian witnesses. He had a long, sensitive face which could move bewilderingly from warm laughter to an expression of the utmost severity. That Malouf himself came from a minority, from a family which had emigrated from the Middle East several generations before and now comprised ten families in the Montreal telephone book, was another piquant element added to the cultural clash in the courtroom. His first language appeared to be English, though he spoke French well.

He read off his reasons for granting a hearing in the slow and deliberate tones which became the mark of his behavior during the subsequent months. His judgment was an almost complete vindication of O'Reilly's five-hour exposition of Indian rights. The judge ran through the many orders-in-council and acts of Parliament in which reference was made to Indian rights and concluded: "Petitioners have therefore established that they have some apparent rights in the territory and having succeeded in so doing are entitled to present before this court such evidence and proof as may be required in the circumstances."

The attorneys for the government had taken the Indian challenge so lightly that they did not expect to have to answer a case. For one thing, they were convinced that the Crown or any agent or mandatory of the Crown cannot be sued, and they were not prepared that the judge would accept the argument of O'Reilly's supporting lawyer, Max Bernard, appearing for the Inuit, that jurisprudence existed to justify an individual suing the government if the government was acting beyond its powers. O'Reilly now said—and the words must have sounded ominous to the corporation's lawyers—that he was taking it for granted that the Indians and Inuit must prove every element and aspect of their case, and "we are so prepared to do."

O'Reilly said: "There are many things which we will have the burden of proving. We will attempt to show that the areas in the north covered by Bill 50 are occupied fully by Indian bands and to some extent by Inuit settlements. They have been occupied traditionally for hundreds of years in exactly the same way. We will show the way of life of the people, anchored on a hunting, fishing and trapping economy, and depending on all those resources." They would show the effects of the work already done, and produce expert opinion as to the likely future effects. They would produce the plans of the corporation, and prove the damage already done to the environment through the cutting of trees, the building of roads, the interference with rivers and streams.

Then Maître Geoffrion, the leading lawyer for the Crown, lumbered to his feet to make almost his last appearance in the case. With all these witnesses to be heard for the petitioners, he said glumly, they were going to be there for forty days. They would do everything they could to shorten the matter as much as possible. "My confrere wants to get rid of as many of his witnesses from the north as possible," he said. "Well, since they are here, they might as well be heard." Personally, he could not be around to hear them. "I don't think we will really get down to cases," he said, as if the proposed evidence of the Indians and Inuit were some kind of irrelevancy in the case, "until late January or early February." He proposed that they should sit for a few days in the following week and then adjourn until February 15. Geoffrion then sat down. Somewhere along the way, as the Crown's case drifted aimlessly out of control, he was replaced.

"We must proceed as rapidly as possible," said O'Reilly. "The work is going on on the project. Every day something happens which might affect this litigation. It is imperative in the interest of justice to proceed as quickly as possible."

The judge agreed: an adjournment of more than a few days for Christmas would not be appropriate, he said. "What is of the utmost importance is that adjournments should not be for an excessive period of time."

With that, O'Reilly called his first witness, Chief Billy Diamond, the twenty-four-year-old chief of the Rupert House band which lives in the little settlement at the foot of Rupert Bay, the settlement that was established in 1670, when the first ship sent into James Bay by the newly formed Hudson's Bay Company arrived to do business in furs with the Indians of the area.

3

THE INDIAN
RHYTHM

The story told by the Indian and Inuit witnesses, and their manner of telling it, made a considerable impression on at least a section of the Canadian public in the next two months. Their experience of the sophisticated world of the white man was limited; their experience of court procedure was nil. But their experience of the forests of the north was vast; and their knowledge of the animals, fish and plants with which they lived in a symbiotic relationship was immense. They spoke of what they knew and of nothing else. For many thousands of years their ancestors had been organized in groups of families, no more. So there was no great scope in their daily lives for deceit, which would have been too easily found out. It seemed to come naturally to them to speak simply and truthfully, and this perhaps more than anything else was what impressed both the judge and those people who sat in the court watching the drama unfold.

O'Reilly used Billy Diamond to outline the basic facts about the yearly cycle of life among the people of James Bay. These facts varied only in detail from village to village. In September the people of Rupert House prepared themselves to go goose hunting, either along the banks of the Rupert River, at the mouth of which lay their settlement, or along the shore of James Bay as far north as Loon Point, fifteen miles south of the next village along the bay, Eastmain. These were the feeding grounds for the blue geese, Canada geese and various species of duck which would stop there on their fall migration from the Arctic to the Gulf of Mexico.

They would hunt these birds in September and October, and return to the settlements late in October to prepare to go inland to their traplines. They would set up camps on their traplines with two or three families in each camp, and remain there hunting big game and trapping beaver until late in March. Then they would return to Rupert House and prepare for the spring goose hunt along the same shores of James Bay as the geese returned to the Arctic. This hunt would last until the ice broke up toward the end of May, when they would return to Rupert House to put their catch in the community refrigerator. They would spend July and August fishing along the banks of the Nottaway, Pontax and Broadback rivers and by the end of August would be again preparing for the fall goose hunt.

When he was young, said Billy Diamond, his family would leave to paddle up the Rupert River in a canoe, perhaps four canoes traveling together, each with one family. At the rapids the men would pole the canoes through while the women would walk along the shore carrying their children on their backs. They would return to Rupert House in February, walking on snowshoes.

Before going into the bush the hunters would buy from the store a supply of basic foodstuffs, such as sugar, tea, flour, lard, to supplement the diet they would catch from the bush, of rabbit, beaver, marten, caribou, moose, fox and fish—sturgeon, whitefish, trout and northern pike. Still to this day his people depended on the bush for 90 percent of their food, since even those families who stayed in the village set snares for small game, and trapped ptarmigan in their nets.

BUT THINGS HAVE CHANGED?

The leading lawyer for the Crown, Maître Jacques LeBel, was a plump, fastidious and cheerful, man whose appointment to represent the James Bay Development Corporation may not have been unconnected with the fact that he was the brother-in-law of the premier of the province, Robert Bourassa. From the first moments of his cross-examination of Billy Diamond and throughout the case, LeBel's main concern was to establish that Indians in James Bay led a life pretty much the same as that of other Canadians.

LE BEL: Isn't it a fact, Chief, that things have changed a lot in the last 15 or 20 years? Instead of paddling up the rivers in their canoes, they use outboard motors now. I'm speaking of hunters, real hunters going to a hunting expedition, leaving for two, three, four months.

Isn't it a fact they use outboard motors now, it has changed over the years?

DIAMOND: There are people that use outboard motors; there are people that paddle up the river, still this very year.

LE BEL: Yes, but most of the hunters, isn't it a fact that they use outboard motors now and that they have been using outboard motors for the last 10, 15 years?

DIAMOND: Most of them, yes.

LE BEL: Instead of using dogsleds for the last 10, 15 years, they have been using Skidoos?

DIAMOND: No, I have to answer you no, because when the people leave for their trapline, they still take their dogs with them, and they come down by dogsleds and they walk with their snowshoes.

LE BEL: Okay, so I repeat my question. For the last five years, Chief, isn't it true that they are using skidoos now and that they don't use their dogsleds any more?

DIAMOND: For the last five years, I don't even see three trappers using Skidoos.

LE BEL: Is it correct to say, Chief, that there were more hunters in your childhood than there are now?

DIAMOND: If you are speaking of hunting, no, I think there has been an increase of hunters. People like me have returned to the north and continue the hunting.

LE BEL: But there has been a decrease of game?

DIAMOND: I can speak about goose and duck and ptarmigan hunting, and I feel that there has not been a decrease in the game, there's just an increase of hunters.

LE BEL: And would you say there has been a decline in trapping?

DIAMOND: No, I would say there has been an increase in it. In June 1969 when I was band manager there were 55 trappers. I would say since then the number has almost doubled.

Trying to shake Diamond's evidence that his people depend for 90 percent of their food on the bush, Maître LeBel merely gave the young chief another chance to run through the details: the people ate geese, including the intestines ("nothing is wasted"), in September and October; from November to March they caught and ate beaver and ptarmigan and finished off the geese they had preserved from the fall hunt; in March, because the deep snow slows down the moose and caribou, they lived on big game; they caught more geese in the spring, and during the summer lived on fish and geese until September rolled round again.

AN OLD CHIEF

The next witness was Chief Matthew Shanush, of Eastmain, sixty-one years of age, a round, jolly figure, speaking only Cree. Apparently somewhat suspicious of all these questions,. he deeply pondered each one before vouchsafing it a reply in a tone that rose little above a whisper. His reply would then be translated into English by Ted Moses, a tall, saturnine young Cree from Eastmain, whose long, immaculate hair, lean, thoughtful face and mod clothes made a startling contrast with his chief.

LeBel and Chief Shanush were quickly entangled in misunderstandings, starting with a disagreement about whether his village had an Indian reserve or not. In Canada a reserve is usually the land, sometimes a fairly large piece several hundred square miles, which was "granted" to the Indians in return for their surrender of the rest of the country under the treaties entered into with the Crown. But in some parts of the country where treaties were never made, such as British Columbia and Quebec, reserves were nevertheless granted to various groups of Indians by the provincial governments and administered, as is demanded in the British North America Act (Canada's nearest thing to a written constitution), by the federal government. These reserves are usually quite small. Of the people in court, only one group, those from Mistassini, lived on such a reserve, an area of land and water five miles by eight miles on the banks of Lake Mistassini, which was created as the site of the Mistassini Indians' village in the 1960s. The other Indian communities of James Bay were so far from white populations that they had never felt the need of such protection, and merely lived at random on the land they had always occupied.

But Maître LeBel had discovered, apparently, that Eastmain, like Mistassini, had a reserve, just a few hundred yards from the settlement. If such a reserve existed, the Eastmain chief didn't know about it. "It's pretty hard to translate reserve into Cree," said the interpreter. "Such a thing does not exist in our language."

The lawyer spoke and acted throughout as if it had never occurred to him that the James Bay project could do other than bring the greatest of benefits to everyone who lived in the area. Was it not obvious that the more work available, the better off everyone would be? The more roads and machines, the easier life would be? The more fire protection, the safer the hunters and their animals? The more government, the greater the state of order and peace?

None of these things was obvious to the hunters. They seemed amazed at the ridiculous questions they were being asked (sometimes even by their own lawyers), and began by being rather affronted by the implication of the cross-examination that they had been telling lies. That was definitely unfriendly.

For most of the time, the exchanges between hunters and lawyers were a dialogue of the deaf. When LeBel asked if the chief knew that one Eastmain hunter had been paid compensation for some equipment that was damaged by white men working in the area, the old man answered: "I am aware of the compensation, in the way of money. But not in the way of what was destroyed. And the money given the trapper is nothing compared with what he would have obtained from these traps, what he would have got for the future if he still had these traps."

I DID NOT SEE EVERY FOREST FIRE

The James Bay Development Corporation had started a forest-fire protection service and this became a dominant theme in Maître LeBel's line of questioning of successive witnesses, who regarded his enthusiasm for forest-fire protection as extremely droll.

LE BEL: In the last 15 years are you aware of large forest fires in this region?

SHANUSH: I am aware of forest fires. This area has already regrown.

LE BEL: How many forest fires have there been in your region in the last 10, 15 years?

SHANUSH: In the last 15 or 20 years, there were forest fires, but now the trees have already regrown. I do not know exactly how many forest fires there were.

LE BEL: What is the portion of your territory affected by forest fires ?

SHANUSH: [sitting with his cheek in his hand for a long time pondering this knotty problem]: I cannot answer this question. (But [says the interpreter] he indicates small forest fires.)

LE BEL: Only small fires?

SHANUSH: I cannot indicate this, because a forest fire may start anywhere. And I cannot give you the exact spot of all the forest fires because I have not seen all of them. I might be able to show you if I had gone up in a plane.

LE BEL: Are you aware that your band council has asked for forest protection?

SHANUSH: No.

Planes, outboard motors, canned foods, even store-bought clothes

—all of these things if used by the Indians were proof that they had become more or less like white people. The lawyers hammered at this theme for the next few months, as the case unfolded, apparently unable to accept that these people could continue to live as hunters and trappers even though they did use certain of the accouterments of white civilization which helped them in their subsistence life.

LE BEL: Is not it a fact that most of the trappers for the last few years have used a plane to go to their traplines?

SHANUSH: Not all of them use the plane. Those people who do not take the plane, they paddle up the river, and for those people that have a trapping territory close to the settlement, well, they wait until the river is frozen and then they will walk to their traplines.

LE BEL: And instead of paddling up the river for the last few years have they been using outboard motors to go to their traplines?

SHANUSH: I do not know any of my people that use outboard motors to go to the river, mainly because if they did they would have to take a lot of gas with them, and since you cannot eat gas, these people prefer to paddle up the river so that they could get more of their equipment into the canoe and a bit of supplies to take up with them.

LE BEL: So would you say that there is no outboard motor in your settlement?

SHANUSH: I am not saying there's no outboard motors in the settlement, but I do not know any of my people that use Skidoos to go to their trapline mainly because if they use the Skidoo it would only frighten the game away, and also it would be difficult if they used the Skidoo to get around their trapline because of the dense forests. They prefer to walk.

HOW MANY? HOW MUCH?

As the days rolled by and one native witness followed another, the cultural gap in the courtroom became more and more obvious. The lawyers on both sides were, perhaps understandably, obsessed with facts and figures. How many miles did they travel? How many fish did they catch? How many Skidoos in their village? How much money did they make? How big were the trees growing back after the forest fires?

I cannot tell how many miles I traveled because I did not measure it.

I cannot say how long it takes because I do not keep track of the time.

I do not know how many children there are, because I have not counted them.

Still the demand for statistics poured over the amazed hunters.

LE BEL: Is the population increasing or decreasing?

SHANUSH: It is increasing. Babies are born every year.

LE BEL: Are there any seals in the region?

SHANUSH: Yes, there are seals in the bay.

LE BEL: Do you hunt them?

SHANUSH: Yes, I hunt them, for myself and also to feed my dogs.

LE BEL: How many seals have been caught last year?

SHANUSH: I cannot tell you, because I did not go round and count them.

To William Gull, fifty-six, of Waswanipi River, who occasionally worked for mining companies cutting lines through the bush, or for the government clearing the brush along the main road from Senneterre to Chibougamau:

LE BEL: How much money did you earn last year?

GULL: I do not know. About $300.

LE BEL: Out of this, how much from trapping?

GULL: I never kept track of it.

Though the Indians of James Bay have used white men's clothing for at least 100 years, it still seemed important to Mr. LeBel to prove that it came from the store.

LE BEL: You mentioned in your testimony, Mr. Gull, that when you are at the settlement, sometimes you buy food at the store. What kind of food do you buy?

GULL [through interpreter, his son Jacob]: He'll buy just the basic needs, the flour, lard, sugar, coffee, tea and then sometimes if he has money and he doesn't kill anything during his hunt, he'll buy a little bit of cheap meat from the store.

LE BEL: What do you buy from the store when you're in the city?

GULL: He'll buy what he likes to buy.

LE BEL: But what kind of meat? Does he buy meat, does he?

GULL: He said hamburger.

LE BEL: He likes hamburger?

GULL: Yes.

LE BEL: Do you buy canned food?

GULL: No, he doesn't like anything that is in cans.

LE BEL: What about clothing? Do you buy clothing at the store?

GULL: Yes, he'll buy clothing from the store.

LE BEL: For himself and for his children?

GULL: Yes, for everyone in his family.

LE BEL: Other members of your tribe, they also buy clothing?

MAX BERNARD [lawyer for the Inuit people]: No!

LE BEL [persisting]: To your personal knowledge, Mr. Gull, the other
    members of your band, are they also buying clothing at the store?

GULL: Yes.

Every concept of the meaning of life held by these people arose from
their relationship with the animals, from their life as subsistence hunters
in the bush, and most of the concepts being put to them by the lawyers
were either difficult for them to comprehend or quite meaningless. Matthew Neeposh, for example, sixty-one, a hunter from Mistassini, could
not answer when asked how far he had traveled by canoe on the Rupert
River, for the question was mathematical and therefore unanswerable.
But when asked by O'Reilly if he had seen the Rupert River (a question
which, posed to a man who had lived along the river all his life, in our
language seems rather redundant) he was able to reply: "Yes, leaving
from Mistassini I have seen the Rupert all the way to Rupert House"—
in other words, from its source to its mouth. It was Mr. Neeposh, too, who
set Maître LeBel back with a neat description of the cultural differences
between them when under pressure to admit that his way of life was so
corroded that his children would never go with him on the trapline again:

LE BEL: Do you go alone on your trapline?

NEEPOSH: Other people go with me.

LE BEL: How many?

NEEPOSH: Six people, two families who go with me.

LE BEL: Including women and children?

NEEPOSH: Yes.

LE BEL: Kids of what age?

NEEPOSH: I do not know.

LE BEL: Do they not go to school, these children?

NEEPOSH: Yes, kids are in school now, as of this year.

LE BEL: Does it mean they will not go to your trapline in the very near
    future?

NEEPOSH: Eventually they will go back to the trapline.

LE BEL: Are you going to take them with you to your trapline next
    year?

NEEPOSH: The children who want to go with me will go with me.

Behind that dignified answer lies the tremendous drama of an education system—the white man's—which does not educate; but more than
that, the reply indicated a concept of education which is not confined to
the classroom but takes place in the daily acts of life itself.

Understanding judge

The Indians showed themselves sharp-witted in their answers: one man, asked how many snowshoe hares he caught in a particular year, replied (though the translation never got completed to the court) that the white man writes down what he thinks is important, the Indians remember what is important. He could not remember how many snowshoe hares he had caught because it was not important. And George Pachano, forty-nine-year-old hunter of Fort George, asked by LeBel if he kept track of the fish he caught, sternly replied that "it was not intended by the Creator who created the fish that the Indian should keep track of all the fish that he kills." That ended the cross-examination, smartly.

In the first days O'Reilly was on his feet time after time, objecting to the nature of the questions being put to the native witnesses by LeBel, on the grounds that they were beyond the knowledge and competence of the witnesses, were confusing to them because of their lack of knowledge of court procedure, impossible for them to answer and generally unfair. The question first came up when LeBel succeeded in getting Billy Diamond to answer "None" when asked if there would be effects on his Rupert House people of work done by the corporation outside the Rupert House hunting territory. Since it was a major part of the Indian case that the work would have spread effects throughout the territory, O'Reilly on re-examination tried to get this answer changed, which, strictly speaking, is against the rules. He complained that the cross-examiners were wanting the Indian witnesses to be like computers, to produce all sorts of figures and documents that should more properly be sought elsewhere, and that the lawyers were going on fishing expeditions and trying to confuse the native witnesses. He claimed that the Code of Civil Procedure authorized special latitude for witnesses who were unfamiliar with court procedure. The judge agreed with him: "The court will take into consideration the type of person who is testifying . . . whether he has had previous experience . . . is familiar with court procedure . . . and so on." He allowed Billy Diamond to "explain" his previous answer.

Throughout the case the judge lived up to this early decision to be respectful of the native people: if he felt that their view of the proceedings was being blocked by the many charts and maps that were put up from time to time, he would ask that they be moved. Once he pleaded with lawyers to be more patient with Billy Diamond as a witness, and asked him: "Have you finished your answer?"

"It's okay," said Diamond.

"It's not okay," said the judge. "If you have not finished it, you will be given the opportunity to finish it. That's why we are here."

O'Reilly was anxious to stretch the laws of evidence to embrace states of feeling as well as the hard, dry facts usually elicited by legal procedure. Sometimes he would begin his examination of witnesses on a chatty, personal level, as when he asked little old John Kawapit, a sixty-nine-year-old hunter from Great Whale River: "Mr. Kawapit, how are you enjoying your visit to Montreal?"

KAWAPIT: Uh, not so well.
O'REILLY: So you want to stay in Montreal?
KAWAPIT: Not very well.

And later: "Mr. Kawapit, what are you eating in Montreal?"

KAWAPIT: I eat the white man's food. The food the white man eats.
LE BEL [facetiously]: Du filet mignon?
KAWAPIT: I have come up to the stage where I can hardly eat this food.
O'REILLY: What do you mean, you can hardly eat this food?
KAWAPIT: When I go back home to Great Whale River I'll be able to eat
    better, because I will be eating the food that I have been eating in
    the past, and that I prefer to eat, not like the food that I have been
    eating here in Montreal.

YOU'RE TALKING ABOUT MY WIFE

Some of the hunters, of course, came to the court determined not to give an inch to these men who were trying, as they felt, to destroy their land, and some, such as John Weapiniccapo, forty-nine, of Eastmain, seemed almost determined not to give clear answers, and were quite belligerent, though in an amused and humorous fashion:

LE BEL: How much did you earn in the year 1972?
WEAPINICCAPO: I can't tell you the amount I made because I don't know.
    Even if I were to tell you a number you probably wouldn't believe
    it.
LE BEL: Well, let's try. Why don't you take the chance? A lot of money,
    sir?
WEAPINICCAPO: No, I didn't make a lot of money.
LE BEL: How much, approximately? Give me an approximate figure.
WEAPINICCAPO: I don't keep records of how much I earned or write it
    down on a piece of paper like the white people do.

LE BEL: Without keeping records you don't have a small idea of how much you made?

WEAPINICCAPO: My memory is not that good to remember how much money I made.

When LeBel suggested he made $30 a day working for a company that was doing blasting on his territory, the hunter commented dryly: "You would probably be right, because you probably have it on record. As for myself, I don't have a record of it." But the lawyer fared little better when trying to find out what this particular man eats.

LE BEL: I am interested in knowing what you eat for breakfast. Could you tell me that?

WEAPINICCAPO: Normally for breakfast I eat what I used to eat in the past, the food that I was raised from.

LE BEL: Do you eat toast in the morning, like we do? Bread and butter?

WEAPINICCAPO: Whenever I have bread . . . well, if you were staying in my house you would probably be forced to eat what I eat.

LE BEL: But do you eat bread sometimes, and butter?

WEAPINICCAPO: Sometimes I eat bread and butter, and if I were to give you some bannock if you were staying in my house you would probably have to take it.

LE BEL: Do you sometimes eat chicken, beer, pork chops?

WEAPINICCAPO: Don't, don't mention the white man's food to me, because I'm still living on the food that I used to eat in the past, which is country food.

And so it went on, the hunter parrying the lawyer's questions with an expertise that not many of the hunters showed in these unfamiliar surroundings.

LE BEL: What about your wife and children, do they eat vegetables, fruits, beef, chicken, pork chops?

WEAPINICCAPO: Sometimes they eat food from the store, but they eat more country food, rather than food from the store. Well, if your wife was up on the settlement I imagine she would want to eat the country food also.

LE BEL: Never mind about my wife. Let's talk about yours.

WEAPINICCAPO: How come you're talking about my wife also?

LE BEL [to interpreter]: Can you tell the witness that I have, being a lawyer, the privilege of questioning the witnesses.

WEAPINICCAPO: I know that, but whatever question you were asking me, I am trying to answer.

How big is paint hills?

The *reductio ad absurdum* of this dialogue came in Maître LeBel's cross-examination of Clifford Shashaweskum, a thirty-eight-year-old trapper from Paint Hills, who had been called by O'Reilly to give evidence that he had seen trees being cut down to make way for a road. In his cross-examination LeBel never did get around to the question of the trees:

LE BEL: Your trapline, sir, is number 20, right?

SHASHAWESKUM: It's number 20.

LE BEL: Is it not a fact, sir, that the superficies of your trapline is very big?

TED MOSES [interpreter]: I can't translate into Cree.

LE BEL: The area of your trapline is very . . . big?

SHASHAWESKUM: It's fairly big, it goes past the lake and on the east side of the lake there is another person's trapline.

LE BEL: Isn't it a fact, sir, that the area of your trapline covers 758 square miles?

INTERPRETER: Square miles, we don't have.

LE BEL: That does not exist?

INTERPRETER: I can't translate square miles. . . .

LE BEL: Isn't it a fact that the area covered by your trapline is about a thousand times the size of Paint Hills settlement? That you know, I suppose?

SHASHAWESKUM: Of course, my trapline is bigger than the settlement of Paint Hills because the settlement is not very big.

LE BEL: I said a thousand times bigger.

SHASHAWESKUM: But I don't know how many times bigger it is, because I didn't measure it.

LE BEL: With snowshoes how long will it take you to go from one end to the other end of your trapline? Of course, with your snowshoes walking from one end of your trapline to the other end?

SHASHAWESKUM: It will take me at least three days to get from one end, if I took my time. But a white man would probably take about 10 days.

LE BEL: Well, okay, you walk fast, eh?

SHASHAWESKUM: . . .

LE BEL: What does he say?

INTERPRETER: He says: why I said the white man it would take him longer was because I went around with a white man on my trapline and he was not very . . . he was very weak.

LE BEL: Okay, that will be all. Thank you.

AN EPIC QUALITY OF LIFE

But even within the limits of court procedure the native witnesses succeeded in giving a remarkable account of their lives as they roamed across the snow-blanketed frozen wilderness, covering vast distances on foot, unerringly finding their way through a landscape that seems featureless to an outsider, living a life stripped to the essential element of survival, dependent entirely on their own skills and energy and their profound knowledge of the land and the animals. It amounted to perhaps the most exotic body of evidence ever presented in a Canadian court.

No one hearing them could remain indifferent to the undramatic way in which they tossed off accounts of enormous journeys. For instance, the old Eskimo hunter Thomas Kudluk had walked south from Fort Chimo and had "seen" Great Whale River—a mere 500 miles away, as the crow flies, over some of the world's most unfriendly terrain. And a laconic description was given by the tall, gaunt Fort George hunter Stephen Tapiatic of how, twice, he had "seen" the Labrador coast, 1,000 miles from Fort George. Twenty years ago, the hunter told LeBel, four families had left Fort George by canoe, paddled up the La Grande until freeze-up and then walked inland to Caniapiscau Lake, beyond the headwaters of the 500-mile-long river. They walked on snowshoes, and found three families (presumably of Naskapi Indians) from Labrador there in almost the geographical center of the Ungava Peninsula (an area as large as Western Europe). All then walked together northward along the Caniapiscau and Koksoak rivers to the now-abandoned inland post of Fort McKenzie. It took them three months to paddle and walk inland to Caniapiscau, another month north to Fort McKenzie and another month to walk to the Labrador coast.

He was talking about an epic journey, the equivalent of walking from one side of Europe to the other in the middle of winter, but a totally uninhabited Europe in which the voyageurs depended on their hunting skills for their survival. A similar journey made by a white man would have made headlines around the world: the survival of a white man lost in northern Canada is always a sensation. But these people survive every year and cover the entire country, all in the day's work, as it were.

LE BEL: Did you go back to the Labrador coast since that long trip 20 years ago?

TAPIATIC: I would say more than 10 years ago.

LE BEL: How often have you been to Caniapiscau lake?

TAPIATIC: I don't know the exact number of times I've been to Caniapiscau lake, but I would say I've been there more than 10 times. The
first time I was there I was just little, that high. I was still with my
parents when I first saw this.

LE BEL: In the last five years how often did you go to that lake?

TAPIATIC: I would say approximately four times in the last five years.

A SCENE TO WONDER AT

But the elements brought to the courtroom by the hunters that imposed
themselves most strongly on their listeners were their simple directness
and concern for truth-telling. There was no way of denying the authen-
ticity of the account they were giving. They went into the courtroom
experience with the wide-eyed innocence of the unsophisticated.

"It was a scene to wonder at in Montreal in the morning when we
would go to eat," said François Mianscum, a Mistassini hunter, when he
told me how he felt about it. "It would be very unhealthy for children to
be walking around at that particular time, as the cars do not make room
for the people to walk. And the cars are a wonder to me, too, because
there are simply too many of them there, in the morning and also in the
evening when people that have been working are going home. Though
some streets have four lanes, each lane is crowded because they literally
touch each other, these cars. And the people are scattered all over the
sidewalks. There are certainly a lot of people there."

The forty-two-year-old hunter had been out on his trapline when he
was called by his chief to go to Montreal to give evidence about the effects
of a road that had been built by the James Bay Development Corporation
through his trapline. He had left his bush camp only a few days before
he appeared in court. "When I was first told to touch the book, my first
reaction was to wonder what this book is for," he said. "Until I was told
to touch it, the book, so that I could speak the truth."

In the courtroom the round-faced, cheerful François, deeply serious
as he was asked if he would tell the truth, the whole truth and nothing
but the truth, placed his hand on the Bible. Before answering he began
to talk to the translator. After some conversation the translator looked up
at the judge and said: "He does not know whether he can tell the truth.
He can tell only what he knows."

1969: WASWANIPI

In 1968, the year before I went to the Mistassini reserve for the first time, I had visited Indians in various parts of Canada. I had found them living in many beautiful places, on the edges of silent lakes, often just around the corner from the last place to which the white man could penetrate in his automobile. I had found them living in small shacks, rude lean-tos, without water, toilets, electricity, jammed six in a bed, eleven in a room. I had found them suffering, as the saying goes, from poverty. Yet I'd had a lot of fun going around with them.

I had begun by making a trip for a newspaper across northern Ontario, through the rocks and trees and lakes with a little fat Ojibway called Willie John, who looked like Charlie Chan. The Indians here were said to be in a parlous state, and Willie was working with them trying to help them. To follow him around was to take a trip through the Siberia of Dostoevsky. "Knock-kneed and pigeon-toed," he'd say, "all Indian women. I can tell them a mile off." When it was over each day we would sink back on our cots in whatever joint we were holed up in for the night with a bottle of whiskey. "Want a shot?"

"Man," he'd say as we rode the train across the Canadian shield, endlessly silent and unchanging, "man, I'd like to come out here sometime by myself on a snowmobile." He would find his way, he said, by just pointing his nose in the right direction. In Armstrong, a stop along the Canadian National line that crosses northern Ontario, the Indians were jammed into filthy shacks, but up on the hill the armed services (like the

Indians, the responsibility of the Canadian government) had everything
money could buy.

"You remember Sagan Kushkan," Willie asked an old man who sat
bowed and silent in the bar. "Sagan Kushkan, the man who appears from
the clouds. My father."

The old man nodded. He had never met Sagan Kushkan, but he had
heard of him. The community was in a state of shock, because the old
man's son Johnny had just been murdered in a drunken fight, murdered
by a quiet boy, a good boy who just couldn't handle his liquor. The quiet
boy was in jail, and Johnny's body was in Thunder Bay, 250 miles south
by rail and road, where police had taken it, without permission, and then
just released it there, where no one was around to pick it up. The old man
had to pay $78 to get the body back. He didn't have $78. He needed
Willie's help. "You want a shot?" asked Willie wearily, leaning back on
the pillows. Then Cecilia Kwandibens came into the room, tucked her
feet under the chair, a young woman already looking old, and tried to
explain how confused she was about the family allowance check and the
food chits, and how sick her husband was, and how they needed Willie
to straighten it all out.

I had learned that you need somebody like Willie to take you around
Indian communities or no one will talk to you at all. You must enter an
Indian house with a friend, and for a while say nothing, giving the people
time to feel you out. In time they will open up, but they have to feel you
out first. I had got kind of used to the smells, and the overcrowding, and
the too-much drinking, and the men hanging around doing nothing, and
the cardboard partitions in the houses, and the beds piled high with rugs
and parcels and clothes and children. But I could see that behind this
confusion and squalor lay something else. Willie John had that something
else: he was deeply experienced, though he'd never had much schooling.
He could handle any job or any situation. He knew his way around the
country like nobody I'd ever met. And the people he was trying to help
—now that he'd decided to give up his heavy equipment operating, long-
distance truck driving, tugboat captaincy, construction work and all those
other jobs he'd proven master of, including bellhop in a tourist hotel—
they, too, had a lot of experience and knowledge, patience and perhaps
too much resignation.

How about right here at home?

I had written about Indians and Eskimos in several provinces of Canada, the Northwest Territories and Alaska when I received a letter from an Anglican priest in Chibougamau in northern Quebec, the Rev. Hugo Muller, telling me that this concern for far-off Indians was all very well, but I should go north in our own province to see what was happening among the Cree, and if I wanted to go he would be only too happy to show me around at any time. That seemed reasonable enough and I decided to go: I was a little worried, for I had seen elsewhere that missionaries were none too popular among Indians, and a missionary might be by no means the ideal guide.

But the missionary who awaited me when I flew to Chapais, a landing strip cut out of the wilderness about 300 miles north of Montreal, was a plump and amiable fellow, dressed in the traditional sober, dark clothes of the priest. Surprisingly, he was a Dutchman who had worked for the Hudson's Bay Company when he first emigrated to Canada but eventually had decided to abandon the commercial arm of the northern colonial enterprise and join its spiritual arm, the Anglican Church. Chapais and Chibougamau, twenty-six miles away along a curving, paved bush road, turned out to be towns cut from the same coin, though Chibougamau was much the bigger: a muscular mining town built by men on the make who did not have time to stop and think before throwing together their straight streets lined by plain and functional buildings—concrete blocks thrown up hurriedly to make a town, so that people could get on with the all-important business of making money. When you drive to the end of the main street, you are back in the wilderness, and you can then drive for hundreds of miles before you get back into what might be called inhabited territory, land that bears signs of having been shaped by human hands. All this was Indian territory until the last twenty years or so. Now, very definitely, the white man is in command, except out there in the wilderness among the trees and lakes, where the Indians disappeared to in the winters. Of course, as far as the white man could see, the Indians were a bit of a mess. They would come into town, move silently through the Hudson's Bay store and hang around by the big window at the end of the check-out counters. Then they'd move along the road to the Waconichi Hotel, the green one that usually had at least one broken window and a big bit of cardboard in the door to keep out the draft. And there, often enough, they would drink and fall asleep at their tables or even on the floor.

o'reilly: What did the influx of white people mean to you?

jimmy mianscum [fifty-three, Cree hunter]: The town of Chibougamau is located on my trapping territory. Nearly 20 miles away there is another town called Chapais and all these towns are on my trapping area.

o'reilly: Are there any animals left in the center of your trapline?

mianscum: There are some on the fringes of my territory, away from the roads.

o'reilly: Before the establishment of Chibougamau were there a lot of animals on your territory?

mianscum: Before the town was established there was much game.

Doré lake

Some Indians were usually hanging around Chibougamau. Most of these lived in a collection of dirty tents on the banks of Doré Lake, fifteen miles from town, down the road to the south, a little reserve of which Jimmy Mianscum was at that time regarded as the chief. After I had checked into my hotel, the Rev. Muller drove me down there. The Indians had lived there a long time, he said, and did not want to move, but the government was insisting that if they wanted a house, they would have to move for it to the main reserve about seventy miles north, on Lake Mistassini, since the government wanted all the Indians to live in one place.

We got out of our car on the edge of the clearing, which was ringed by the tents. The Reverend adopted his most cheerful bedside-visiting manner and took me into several of the tents, jammed full of cardboard boxes, bedclothes and children. These Indians had been living at Doré Lake in the summers long before the white men came north. They would return from their winter's hunting in the spring and summer to rest up. But the building of the road, the sinking of the mines, the establishment of Chibougamau had interrupted their hunting cycle. More than that: it had brought to their doorstep the delights and terrors of the white man's civilization. Doré Lake was no longer one of those charming settlements just around the corner from the white man's influence, but a collection of mean huts and tents into which a bunch of destroyed men and women crawled most nights after the bars closed.

Fifteen houses, if houses is the word: 150 people living in them. Each one big enough for maybe ten people to stand up in. Two beautiful sisters of nineteen and seventeen, with thin little bodies and soft, round faces framed in long black hair, talked to us outside one of the tents. They had just married brothers, Muller said. The seventeen-year-old already had a baby. They were moving into an apartment in Chibougamau, where they would inevitably cause trouble. "When they set up house," said Muller, "the family moves in. Before we know it, fifteen people are living in the apartment and the landlord is on the phone to me asking me what to do about it."

I looked over at the nearest tent, whose overcrowding had been somewhat eased by the fact that the grandparents in Mistassini had taken one of the women's six children. Muller smiled rather heavily, spoke portentously. "They are sharing their worldly goods with their brothers, as our faith tells us we all should do. Am I to tell the landlord to throw them out?"

In the community hall a plumpish, cheerful woman, Lucy, chatted about family problems. She had recently lived through the death of her husband, who had fallen into the rapids, caught pneumonia and died, leaving her with eight children. With the welfare payments, widow's allowance and the child allowance for eight, she was bringing in $250 a month, a princely sum, especially since four of the children were living elsewhere. She no longer had a husband to drink it up, so she was drinking up quite a bit of it herself. She had been able to shake off the squalor of Doré Lake, had moved into an apartment in Chibougamau, leaving the children behind in the tent for someone else to look after. And there were behind the Reverend's explanations some dark, unspoken assumptions that the lady was meeting a fate worse than death in nightly performances. "I don't know what she's going to do," said the Reverend, as if she were borne down by the weight of some temporary disorientation brought on by her excessive grief. "Perhaps she will go into the bush this season with her mother." Going into the bush, it seemed, was some sort of solution. It was a year or two before I realized how true this was.

We drove farther along the main road for a few miles, occasionally bumping down a dirt side road, but without finding the Indians that Muller was looking for. "They were here," he said, "but I'm not sure of the road." Finally, he gave it up. They were a group of Mistassini men who were on a Manpower Department training course designed to fit them for operating heavy equipment. "I think the training program has dispersed," he said apologetically. "They all went to a wedding last weekend, and they haven't come back." I expressed some surprise at this cavalier behavior. "They're not really interested, you know," said the Reverend. "These programs are our idea, not theirs. But you can see what happens when we come among them interfering. If only we would learn to leave them alone, they'd be fine. They regard themselves as participants in nature. I went out with some of them one day and something happened that I will never forget. Some ducks flew over our group, and long before I knew the ducks were there, every man in our group had frozen. It came to them quite automatically, like a reflex action, an instinct. They are so close to nature that they didn't even have to think how to react. They have this wonderful capacity, until we come along with our training programs and plans for them."

As politely as possible, I suggested that missionaries themselves had imposed a heavy alien load on Indian people. But all that, he said, had been done in the past. Nowadays the Indians were all Anglicans and he wasn't out to convert anyone. The danger of that sort came from the Pentecostals, a fanatical sect of Americans who had set up in Chapais, where they were making heavy inroads among the Anglican Indian faithful, imposing on these children of nature a rigid fundamentalist faith, joyless and barren, the most disruptive emotional baggage that could have been designed for them.

"I have to admit that the Pentecosts have stopped them from drinking," said the Reverend sadly. "That is one thing they are able to achieve. But at what a cost!"

THE WACONICHI: INDIAN BAR

That night I went for the first time into the Waconichi, the Indians' favorite bar. A big room painted dull green, not too clean, with simple kitchen chairs and tables, a big rectangular bar in the center, plenty of room for people to wander from table to table, talking. I met the Mistassini chief, Smally Petawabano, a large young man with a dark, round face and under his ready smile a curious reserve that I never succeeded in penetrating in the next five years. The Reverend had a polite drink, talked to Smally for a while and left early. I stayed for a few more drinks. Jimmy Mianscum was there, a large, strong man with a sensitive, open face from which his hair was pulled back into a queue. He had a high-pitched voice which seldom rose above a whisper and a soft, almost feminine smile. A part-time hunter, but a full-time drinker, he had his reasons.

JIMMY'S TESTAMENT

Six months before, Jimmy Mianscum had written out in halting English, translated for him from Cree by one of the young men of Doré Lake, his complaints about the way he and his people were being treated. Jimmy Mianscum is not one of the prophets of the Cree people, but his message did prove prophetic. In the simplest language, it took up most of the major themes of the Cree resistance to the James Bay project three years later.

> This is the thing we ask for [he wrote]. It is the ground we want to own to live on. We would be very pleased if we have the ground we ask for to be given to us. So we could be comfortable to live on it. We think we got the right to own a ground to be given to us. For we have been living here always. It is right for us to live in the same ground for all of us. For we have never left this ground at Lake Doré since we lived on. So we asked the government to give us a ground to live together on and on. When the old Hudson's Bay Company was closed down at Chibougamau in the year 1942 we have never left this ground we stayed on this ground just the same. And where we lived on then, the company hired us and gave us jobs to earn money, so we never want to leave then. For as far as we can remember, we have lived here since the year 1926, on and on.
>
> Now we will talk about our hunting grounds. The Indians who are here at Doré Lake we know that the white men working all around our hunting grounds. And of course the game is not like it used to be any more. Same like if there was hunting, even when the hunter was not hunting at his hunting ground. [He means: it is as if the hunting went on even in the years in which we allowed our land to rest.] Some game, like it's been poison, fish and game, some of the game animals get poisoned by smelling something different.
>
> So we are waiting for a ground to be given to us, for there has been so much of our game and our hunting ground spoiled away. We know the government gets a lot of money by our hunting grounds. We think that we will be the first one who cannot use our grounds. Another thing we want to know if we have the houses we ask for to be given to us. We like to know how much it would cost us a month, by electricity and water, or if the government can help us to pay a little bit. The government does not help us much at Doré Lake, just we get help by food a little.
>
> Yours truly Jimmy Mianscum saying that this is my hunting ground: all around water and land. This is the reason we ask for all this. And I know that I cannot use my hunting ground, and I never get anything for spoiling my hunting ground, and of course, not even money to be given to me.

He then outlined an amazing situation, one which created indignation when I wrote about it in the newspaper on my return to the city, and which demonstrated the total lack of interest of the Quebec government in enabling the Cree hunters to continue in their traditional ways. Far from aiding them, as they could have done, the government was systematically harassing them in their hunting camps. Under the white man's laws, the moose-hunting season, designed for the convenience of tourists, was in September. But a moose is useless to an Indian hunter in September. He needs it during the winter when he is deep in the wilderness trying to feed his family from what he can catch on the land. And at this time the game wardens of the Quebec government were occupying their time by flying around the forest, landing at Indian hunting camps, seizing any meat killed outside the limits laid down by the white man's laws and taking it away. Jimmy Mianscum's testament gives a terrible picture of this cruel practice:

> Now the game wardens. I did not know why the game wardens came and landed at my hunting ground by aircraft while I was hunting in the bush. I did not know who send them to come at my camp. They ask for my meat. They took all my moose meat and when I come out from my trapping ground, when I come by aircraft they have also waited for me at Cache Lake, Fecteau Transport Ltd [at Chibougamau]. And of course they took all my moose meat again, what I had left.
>
> I was 40 miles out in the bush, that is where I brought my meat along for me to keep going while I am at Doré Lake. I also told them that I was a sick man, not in good health. This is the reason why I brought my food along with me, because I had to go in hospital and my family needed that food to eat. They told me their boss or manager want that meat.
>
> And so I am short of food to eat. The food they took from me, it would keep me going at least one month. I did not understand any of this, why they took my meat. They did not even tell me what's the reason they took my food. And it is right for a hunter to be given a permit, what he's got the right to do, same like the game wardens, the permits they have. And I don't know who told them to watch the Indian hunting there. Is it the government, or someone else? We would all of us like to know, and we'd like to know why this happen.
>
> From Jimmy Mianscum, chief, Doré Lake, this meeting was helped and written by David Bosum.

On his way home from the bar that night, Jimmy smashed up his car, was arrested and was thrown in jail.

ALONG THE NORTHERN ROAD

The Indians in this part of the world are divided into two main groups: those centered on the Indian reserve of Mistassini, around a Hudson's Bay trading post on the lake, fifty-five miles north of Chibougamau; and those who had been living around the Hudson's Bay post on Waswanipi Lake. When the company closed this post in 1965 these Indians scattered into nine or ten different locations, most of them strung along the northern roads. These were Muller's flock, and we set off the next day along one of the loneliest roads in Canada the 270 miles from Chibougamau to Senneterre. Here were Indians whose lives, it seemed, had been completely disoriented as a result of the arrival of the white man. I was to discover later that they had already been the subject of a study by a Montreal anthropologist, Ignatius La Rusic, a nervously intense red-haired Nova Scotian, who had written a paper chronicling the search of the Waswanipi people for what he called *The New Auchimau*—"a study," as he subtitled it with scholarly detachment, "of Patron-Client Relations Among the Waswanipi Cree." For two centuries, said La Rusic, the Waswanipi people, who had lived by hunting the animals in the 14,000 square miles of the headwaters of the Broadback River, had been the clients of one major patron, the Hudson's Bay Company, which they called *au-chimau*—roughly translated as "boss." When the post closed, the Indians had been forced "to seek new intermediaries with the white world." Most of them had tried "to find a single individual to handle all intermediary functions," La Rusic wrote, but by 1967 they had already been transformed from a coherent hunting people to "a landless proletariat."

Muller didn't like the anthropologists, a swarm of whom had had his flock under the microscope for several years. (A few years later in a little book of poetry published by his bishop, he wrote:

> The anthropologists arrive
> when spring comes to the land
> they pry into our ways and lives
> with notebooks in their hand
> and for a study on the Cree
> some even give us beer
> to make us talk for their degree
> and then they disappear.)

You have to drive far to get anywhere in the Canadian north, and the

Reverend drove with the dogged determination of a veteran, sparing neither himself nor his car. We bumped for 150 miles over the dusty road before pulling in beside a bridge over the broad and handsome Waswanipi River, along the banks of which lived some Indian families in log cabins that they had themselves hewn out of the wilderness. The center of this settlement was the four Gull brothers, who had set up their homes on a spot of open land right at the middle of one of those extraordinary northern water systems. The water flowing past their door in the Waswanipi River originated in two major sources farther east, the Obataga-mau and Chibougamau rivers, both flowing west through a network of lakes and joining into one stream a mile or two east of where the Gulls were living. The river then flows farther east into Waswanipi Lake, where the reserve once existed, and then north through Goeland Lake, Olga Lake, Matagami Lake, Soscumica Lake and so into the broad Nottaway, which goes into Rupert Bay. This was the very heartland of the James Bay hydro-electric project as first conceived. The early maps indicated that the spot on which the Gulls' houses stood would be flooded as the waters backed up from the huge series of dams that the government was propos-ing to build in the 100-mile-wide triangle of country to the north of Waswanipi Lake.

This had been the trapping and hunting ground of the Gull brothers' father, Jacob, who had died at the age of 101 and been buried on a hill on the other side of the river, looking down over the great wilderness toward the rising sun. We went into the house of the Rev. John Gull, a Cree preacher who had lost his captive flock when the reserve was dis-banded and was having some difficulty in getting them back into church. He had one big room whose walls were lined with beds, on one of which I sat as he showed me the register of births and deaths that he had brought with him from Waswanipi Lake. The laconic entries, which cer-tainly left plenty for the imagination, perhaps indicated why Waswanipi Lake had been abandoned. "November 27, 1961, May Blacksmith, thirty-five years, baby birth; Maggie Blacksmith, one month, starvation; April 1962, Jane Gull, eighty-six years, starves; Charlotte Otter, one year, cry-ing of death [cried herself to death]; 1963, Lizzie Happyjack, two months, chickenpox; Abel Icebound, two months, chickenpox; Charlie Neeposh, two months, chickenpox." And so on: "eighty-nine years old, dead of tuberculosis; forty-four years, dead of baby born; twenty years, dead of accident. . . . " The list told of a subculture to which the dominant society, for all its obsession with proteins, hygiene, calories, consumption and production, had never bothered, even in the last ten years, to export even its minimal benefits. Jacob Gull, John's brother, a living example of the hard times, hobbled around on his peg leg: he had got gangrene as a boy and his father had chopped his leg off with an ax to save his life.

We took William Gull, fifty-three, on with us (the next time I saw him was in court three years later). A cheerful, talkative man, he had just returned from a month or two linecutting in Ontario. A complete master of the bush, he took a job when he needed it and the rest of the time lived by fishing and hunting. William had brought up four sons and was looking after the son of his daughter, who had recently been killed in an accident. Two of his children were still in residential school, where they were learning the ways of the white man, ways very different from William's life. The anthropologists had been investigating Cree children, too, and had discovered that the contrast between the values of their parents and the values they were taught in school was so disturbing that two thirds of the teenagers were suffering from depression, anxiety and other mental and emotional disorders as a result of the identity conflict set up for them. These "discontinuities in the enculturation of the children" (to use the jargon of anthropologists Peter Sindell and Ronald Wintrob) were deliberately created by government policy. And few youngsters were coming out of school equipped either to command the bush as their parents could do, or to move smoothly into Euro-Canadian society.

ONE OF THE NEW AUCHIMAUS

We drove through a place called Desmaraisville, twenty miles past Waswanipi River, little more than a mine and a small store by the roadside. This, according to La Rusic, is the only place along the road where the Indians did not settle when their reserve broke up in 1965. The store owner refused to give them credit and the mine manager did not offer them work. La Rusic relates a story that the mineral traces on which the mine at Desmaraisville was based were discovered by an Indian who handed them to the present owner of the mine in expectation of reward. He got $1,000. In view of the great subsequent value of the minerals, the Indians considered he had been cheated.

Thirty-five miles farther along the road we stopped at Miquelon, where a voluble little French-Canadian, Roger Ratté, who had married a Waswanipi woman, had set up a small tourist camp. Indians had, even before the abandonment of the post, clustered around his camp, using him as a source of work and an intermediary between themselves and other whites, even the Indian Affairs Department. Ratté, according to La Rusic, was one of the five most important new *auchimaus,* but his popularity had fallen when he tried to influence Indian Affairs to ensure that if the Waswanipi were granted a new reserve, it should be on land on which he held mineral rights. For some time the Indians at Miquelon had been asking for land, just enough to build some houses on, so that they could cease being squatters at Ratté's place, but the request was always denied by the Quebec government, though anyone who wanted to build a mine in the area could get a land grant immediately on application.

"The disadvantages of the Indian are manifold as he moves into the Euro-Canadian society," reported La Rusic. "He is a member of a minority group who is discriminated against in the region. He speaks a language which whites cannot understand and the language he has been taught in order to operate in the white world, English, is not functional in the French-speaking region in which he lives. His cultural background and his preference for interpersonal relations combine to make interaction with the white industrial society almost impossible." Functionally illiterate . . . landless . . . lacking resources and job skills . . . it is no wonder, concludes La Rusic, that the Indians sought intermediaries to help them along, make their contacts, provide them with credit, help to keep them afloat. But so long as they remained in the shelter of the new *auchimau,* they would remain a marginal people.

Before we turned back toward Chibougamau we were told stories of prostitution among the Indian women and of excessive drinking among all the Indians along the road. The Reverend wearily accepted it all as the inevitable consequence of our interference in what used to be a perfectly viable way of life. Surrounded by the French, the developers, the explorers, the anthropologists, all the interfering busybodies on the make, it seemed there was little hope for the Reverend's tiny flock. On the way back we stopped at a tent an Indian family had set up beside the road. A few years later I was able to see these tents with different eyes, but at this time I was carrying around with me the not untypical mental baggage of a metropolitan newspaperman. These tents seemed to me the worst housing I had run across since I left Bombay in 1951, when I had seen millions of people living on sidewalks under cardboard lean-tos. It is a deeply ingrained idea among us that the purpose of life has been largely fulfilled when a family can move into a comfortable home with lots of kitchen gadgetry, a pushbutton system for whirling one's excreta into the common collector sewer, and a nice, neat lawn all around. It takes an effort of comprehension to get inside the mind of the man who glories in a patch of wild grass, and who sees the neat lawn as a depressing effort to impose clinical patterns on nature. Years later when I came across these tents beside the northern roads I felt envious of the freedom of their occupants, people able and ready to pick up their homes and move whenever they felt like it, capable of feeding themselves from their command over the nature that lay around them. Quite possibly the Gull brothers in their poor little houses beside the broad river were among the richest people on the continent. It took me quite a time to learn this.

A TENACIOUS CULTURE

Curiously enough, it was among these same apparently disoriented and degenerated people, who, according to La Rusic, were showing a definite preference for white man's employment, that Harvey Feit, a McGill University anthropologist, gathered in the next two or three years powerful evidence of the tenacity with which the Cree clung to the bush life. Though these people had been affected more than any other group in James Bay by the incursions of the white men, Feit found that 52.6 percent of the family heads were still following the traditional hunting pattern by going into the bush to set up a camp in October and staying until May. Another 17 percent would stay for part of this time, and nearly 9 percent who lived the entire winter in the settlements along the road would trap on a part-time basis, day by day. So 78.6 percent of the Waswanipi family heads, according to Feit, were still engaged in their traditional pursuits, and only about 4 percent of them took jobs in the winter. In the summer the percentages were reversed: some 74 percent took jobs, but nearly every family "in some way gathered fish or game."

O'REILLY: What kind of food do they eat?
FEIT: Big game, moose, caribou, beaver, bear. Some eat otter, marten, muskrat, weasel, mink, lynx, hare. Partridge, geese, loons, ducks. Pike, doré, sturgeon, northern pike, whitefish, red and white sucker, walleye.

So evidently the people in these rude roadside tents were at least eating like kings. In all, Feit concluded, the Waswanipi families who were hunting and trapping were getting 82.4 percent of their food from the bush in the winter. Though they were apparently dependent on the white man, only 20 percent of their income came from wages earned, said Feit. The greatest proportion of their income by far—52 percent—came from the cash equivalent of food caught in the bush. Only 10 percent came from fur sales, 9.6 percent from transfer payments (old-age pensions and family allowances) and 8.1 percent from welfare.

He had studied the young people, too, to try to verify at first hand the common assumption that they no longer hunted and trapped. He found that of those born between 1940 and 1949—persons between twenty-two and thirty-two—45 percent were full-time hunters and trappers during the winter and 20.4 percent part-time. Only about 5 percent had had no experience whatever of bush life, and "my conclusion is that there has been no major swing away from hunting and trapping among the young people, though there is a slight decline." Of this same group of young people, only 10 percent had full-time jobs, and though all of them had some experience of wage employment, more than 22 percent were without work.

5

1969: MISTASSINI

It is a common experience in many parts of the world that the mental and physical health of an indigenous people that has come into contact with a powerful technology declines in direct proportion to the degree of that contact. The first Cree Indians I had met were those living on the edges of towns built by the white man. It was among them that the disturbance created by the invading civilization was at its maximum. Mistassini, the main Indian reserve, was one step farther away from the white influences; and a long step yet farther north the traditional life was being carried on in the isolation of the wilderness in conditions of emotional and physical well-being rare among North American Indians. It was to be three years before I would see this life for myself.

But before I went to Mistassini the next day, I adopted the Rev. Muller's suggestion and went to see Edna Neeposh (daughter of the hunter Matthew), who was working in Chibougamau with a Quebec government social agency dealing with the social problems of Indians in the area. Edna was one of only three young Indians from Mistassini and Waswanipi who at this time could be said to have "made it," as the saying goes, in white society. (The others were Philip Awashish, then at McGill University, and Peter Gull, son of William, who was working as a draftsman for a Chapais mine.) Edna was looking after twenty-four Indian children, of whom eight had been put in government care because their mothers had neglected them, a couple were illegitimate and some had been abandoned. It was something new for the Cree to produce such children, because in a traditional Indian community, unwanted children were unknown. Even Mistassini, however, was feeling the pressures of change. "When I was young I never noticed so many people staggering about the reserve with beer bottles in their hands, fighting," she said. "The drinking is getting worse. It seems everyone is drinking, men, women and the young people, too. They are drinking to forget."

It was, perhaps, not surprising. Many of Edna's school friends had been brought up to be ashamed of the way Indian people dressed and behaved, and did not want to be identified as Indians. When such children returned home, the parents felt helpless. Traditionally, the parents had always been thinking about their children, had taught them everything. But now the children would laugh at their parents, for they had already been taught different things. Edna herself, under the strong influence of an eighty-year-old grandmother, had gone into the bush for a year after finishing school. "It is really a good life," she told me. "I think every Indian child should do it."

When we went to Mistassini the next day, we bumped for nearly fifty miles over a rough dirt road north of Chibougamau along the east side of Mistassini Lake. When the road gave out, we went down to the lakeshore where someone was waiting with a canoe to take us the two miles or so across to the point where the reserve stands. (Within a year this canoe trip became no longer necessary, when the last ten miles of road linking the reserve to Chibougamau were finished.)

The Hudson's Bay store, a nondescript yellow-and-green prefabricated building identical with other Hudson's Bay stores I had seen throughout the Canadian north, stood just up the hill from the dock at which we tied up. There I met Glen Speers, the manager, a man with whom I had quite a bit to do in the next few years. The Hudson's Bay Company was founded in 1670 in England especially to trade for furs with the Indians of James Bay, and had since grown rich and become one of the biggest commercial enterprises in Canada. It was in 1668 that the little ship *Nonsuch* arrived at Rupert House to test out the market for fine furs, a test that proved so successful that within a few years the company was able to import £950 worth of tools and trinkets in the summer and exchange them within a few weeks for £19,000 worth of fine James Bay furs.

The Royal Charter granted to the Governor and Company of Adventurers of England Trading into Hudson's Bay was, to say the least, generous in the amplitude of lands given to the company by a king who had no authority over any of them.

GENEROUS CHARTER FOR THE GENTLEMEN

Charles the Second, by the grace of God, King of England, Scotland, France and Ireland, Defender of the Faith, etc., to all to whom these presents shall come, greeting: Whereas Our dear and entirely beloved Cousin, Prince Rupert, Count Palatine of the Rhine, Duke of Bavaria and Cumberland, etc., Christopher Duke of Albemarle, William Earl of Craven, Henry Lord Arlington, Anthony Lord Ashley, Sir John Robinson and Sir Robert Vyner, Knights and Baronets, Sir Peter Colleton, Baronet, Sir Edward Hungerford, Knight of the Bath, Sir Paul Neele, Knight, Sir John Griffith and Sir Philip Carteret, Knights, James Hayes, John Kirke, Francis Millington, William Prettyman, John Fenn, Esquires, and John Portman, Citizen and Goldsmith of London, have, at their own great Cost and Charges, undertaken an Expedition for Hudson's Bay in the Northwest Part of America, for the Discovery of a new Passage to the South Sea, and for the finding some Trade for Furs, Minerals and other considerable Commodities, and by such their Undertaking, have already made such Discoveries as do encourage them to proceed further in Pursuance of their said Design, by means whereof there may probably arise very great Advantage to Us and Our Kingdom.

And whereas the said Undertakers, for their further Encouragement in the said Design, have humbly besought Us to incorporate them, and grant unto them, and their Successors, the sole Trade and Commerce of all these Seas, Streights, Bays, Rivers, Lakes, Creeks and Sounds, in whatever Latitude they shall be, that lie within the entrance of the Streights commonly called Hudson's Streights, together with all the Lands, Countries and Territories, upon the Coasts and Confines of the Seas, Streights, Bays, Lakes, Rivers, Creeks and Sounds aforesaid, which are not now actually possessed by any of our Subjects, or by the Subjects of any other Christian Prince or State.

Now know ye, that We being desirous to promote all Endeavours tending to the publick Good of our People, and to encourage the said Undertaking, have of Our especial Grace, certain Knowledge and mere Motion, given, granted, ratified, and confirmed, and by these Presents for Us, Our Heirs and Successors, do give, grant, ratify and confirm unto Our said Cousin Prince Rupert, etc., that they, and such others as shall be admitted into the said Society as is hereinafter expressed, shall be one Body Corporate and Politique, in Deed and in Name, by the Name of The Governor and Company of Adventurers of England, trading into Hudson's Bay. . . .

The company was designed on the great monopolistic model of the Dutch and British East India Companies, which not only took over all trade, local and international, in the East Indies, but governed great empires as well. The Gentlemen Proprietors—or shareholders—of the company were given by the charter a monopoly over all the trade they could drum up and ownership of all the lands they could find. The King deemed the land in question to be "reputed as one of our plantations or colonies in America, the Company deemed to be true and absolute Lords and Proprietors of the same territories . . . yielding and paying yearly to us, our heirs, etc., two elks and two black beavers, whensoever and so often as We shall happen to enter into the said Countryes." (In 1970 the Hudson's Bay Company celebrated its tercentenary, and newspapers all over Canada, all of them growing rich from the company's huge advertising for its great retail stores, waxed eloquent about "the honorable company," "the admirable company," "the great company." The Queen in that year happened to enter the "said Countryes" of the Company's former territories, and, sure enough, the Governor of the company, Lord Amory—formerly Mr. Derek Heathcote-Amory, once British Chancellor of the Exchequer under Harold Macmillan—flew to Winnipeg to make a presentation in a richly comic ceremony to Her Majesty of two elks and two black beavers, live.)

For its first 200 years, until the company's territories were bought by the Canadian government in 1870, the monopolistic charter of the Hudson's Bay Company was challenged many times in the British Parliament, but never shaken. The Adventurers and Gentlemen were given rights to "all mines royal of gold, silver, gems and precious stones" (there is still a Hudson's Bay Mining and Smelting Company, now owned by Harry Oppenheimer of South Africa); and the company was also given the right to "send shippes of war, men or amunicion unto their Plantacions, Fortes, Factories or Places of Trade," or to "build Castles, Fortifications, Fortes, Garrisons" at which trespassers who "would incurr our Indignacion" should be seized and brought "to this Realme of England."

The company set up trading posts around Hudson Bay and for 100 years waited there for the Indians to come down the rivers to the coast each summer to trade their winter's catch of fur. French traders later went inland by canoe from Montreal through the river-and-lake system and tried to intercept the Indians on their way to the coast, to pull off a deal first. For a century the company had to deal with the rivalry of the French; they were freed from it only in 1763, but the rivalry was renewed when their trade was challenged by the daring Northwest Company from Montreal in the early years of the nineteenth century. Since then through normally aggressive business methods the company had continued to grow richer and richer, and, as I had found in the last year or so by going into Indian communities in different parts of the country, had stored up for itself a great deal of animosity among Indians. In 1970 one former company store manager in western Canada confessed in an interview that he had quit the company because an Indian band council had passed a resolution to lynch him and from then on he had been forced to carry a pistol at all times—perhaps an extreme example, but not surprising to anyone who has spoken to Indians about the company.

To some the company is "a dear old institution which has done well by the Indians" (as Duncan Pryde, a Northwest Territories councillor and former company manager among the Eskimos has argued in his book *Nunaga*). To others it is a rapacious commercial enterprise which has mercilessly exploited the Indians under the guise of looking after them. Almost everyone agrees, however, that there have been Hudson's Bay men who were sincerely devoted to the welfare of the Indians. And in Glen Speers, I found over the years, they had a manager who combined a taciturn and businesslike quality with a real concern for the welfare of Indian life in James Bay. Three years later, toward the end of a filmed interview, I managed to get Speers to admit that "you won't find finer men anywhere" than the Indians of James Bay. It was a remarkable moment, for he seemed to screw the admission out of himself under intense pressure, and tears sprang into his eyes as he said it. More typical of him, though, was his remark, tossed off when the camera was idle, that "you can't afford to be too personal when you're running a business." In that statement, as anyone who has even remotely examined the history of the Hudson's Bay Company fur traders can tell, was distilled a great deal of the essence of the company's history.

MODERN TRADING POST

The Hudson's Bay trading posts of the modern day are quite unlike what most people probably still imagine a trading post to be: there is nothing colorful or exotic about them, except that just inside the entrance stands a small knot of Indian people watching life flow by. The stores are just like self-service stores everywhere, with their rows of goods—very often highly priced—their checkout counters, and in the back behind a glass partition, sitting above the store and looking out over it, the manager, turning over his ledgers. In Mistassini the credit files were kept on a counter just outside the manager's office, and there was always a girl clerk or two there, turning the cards over with an Indian customer. This Hudson's Bay store, though it looked just like any other store, was really not quite as it looked, just as Speers was rather more complex than his sober appearance suggested. The company store in Mistassini still outfitted the hunting families of this culture for their season in the bush, advancing the hunters credit against the likelihood of their returning with a good supply of furs, and Speers was the man who personally flew around all the bush camps twice every winter to collect the furs they had caught and prepared, and to take them some more supplies, usually flour, tea, sugar and occasionally a few packages of biscuits, to provide a bit of variety for the later months of the hunting season.

Speers, though a great talker, is also a reserved, intensely private man. On this first meeting he was polite to me and fairly informative, but he gave me no indication of the remarkable depth of his knowledge of the James Bay area and its people. He had spent the greater part of twenty-five years in James Bay in the service of the company, and it was not much of an exaggeration to say that he knew every Indian family in James Bay. But I would have to wait a year or two, and establish my credentials by showing what I could write and by returning again to learn more, before he would begin to talk freely .

He now outlined to me how the hunters would outfit themselves from his store for their winter in the bush and he would arrange their flights to the bush. Twice a year, in January and March, he would fly around the camps, picking up the beaver fur, on behalf of the Quebec Department of Fish and Game, which offered it for auction in the city, and buying all the other fur directly from the Indians for the company. He said Mistassini hunters caught and prepared between 7,000 and 8,000 beaver skins a year, bringing in about $160,000 in income. There were that year fifty hunting camps (a number which rose to seventy-four three years later) in which betwen 200 and 250 people worked as trappers. The area of the band, he said, was some 100,000 square miles. But there was more to it than just getting beaver fur. "They go trapping," said Speers in his clipped, understated way, "because they live better in the bush. They are free."

A VILLAGE UNLIKE OTHERS

The Reverend took me up the hill to the little white Anglican church and showed me proudly its simple interior, mostly provided by the work of its Indian parishioners. From the church one looked across a small valley, over which ran an old board sidewalk to a cluster of houses straggling along the hill on the other side. Mistassini was slowly becoming urbanized: the government had begun to build houses for the Indians a few years before. They were small wooden houses, painted a sort of dirty maroon, and lined up by some Indian Affairs bureaucrat in neat rows like filing cabinets. Inside they were extremely simple, with partitions delineating the rooms, and an oil stove. There were no cupboards, so clothing and bedding had to be pushed against walls, stored in cardboard boxes or trunks, or shoved along the top of the partitions below the roof, a necessity which contributed largely to the ramshackle, overcrowded appearance of most Indian homes.

The village still had a good many tents, fairly sturdily built, and it retained the improvised air of a place that most people figured on living in for only a couple of months between hunting seasons. The government wanted the place to be more permanent, which is to say that it wanted people to stay there longer, and in the next few years a lot of money was spent on building more houses, a large primary school and some apartments with all modern conveniences for the teachers. At this time the only electricity was generated by the Hudson's Bay Company, but a year or two later the village was joined to the provincial electricity grid. Then only the government establishments were hooked on to electricity, the Indians having to content themselves with unneeded street lighting.

The Mistassini reserve established by the Quebec government was a point sticking out into the lake, eight miles by five, of which a large part was water. The province retained the ownership of the land and the right to reclaim any part that it should want for something else. (Indeed, on the far end of the point, right in the middle of the reserve, it soon built a provincial tourist lodge.) The act of creating the reserve established a curious constitutional position, because the British North America Act says that the federal government is responsible for "Indians and lands reserved for Indians," and yet the land was owned by the province, not by the Indians or their protector, the federal government. The province had also created in the very wide area around Mistassini Lake a provincial game park and had forbidden hunting in it. Not far away from the reserve the Ikon mine had been sunk close to the lake on land that the Indians had always used for trapping and hunting. In all of these ways the Indians were being crabbed and restricted in their use of the land; and as this occurred, the provincial authorities were congratulating themselves on having looked after the Indian needs by "giving" them a reserve of forty square miles!

AN OUTPOST, FAR AWAY

Now that Mistassini is only an eighty-minute flight from Montreal, followed by a fifty-five-mile drive north from Chibougamau, it is hard to realize how utterly remote it was well within living memory. One of Speers' predecessors, J. W. Anderson, in his autobiography records how he was recruited for the Hudson's Bay Company in Scotland in 1910, traveled to Rupert House directly by ship, thence overland by canoe to Mistassini, and did not set foot in southern Canada until 1920. Mistassini was in those days so far from the centers of population that news of the outbreak of the war in August 1914 did not reach the post until March 1915. Anderson was confronted by a curious ethical and practical dilemma: even if he had wanted to enlist, as all red-blooded young men were expected to do, he could not have made his decision, conveyed it to his superiors and had them send him a replacement before the summer of 1916! So he decided to sit it out.

Mistassini was the farthest inland post administered from Rupert House. The river was its highway. The Indians would hunt throughout the winter, and then after the spring flood most able-bodied men would join the canoe brigades down the Rupert River with the winter's catch of fur, returning in August with supplies for the next winter. The men would have only ten days or so of rest after their winter in the bush before leaving on the big trip to the coast. They used four thirty-foot canvas canoes which had been designed by John C. Iserhoff, an Indian canoe builder, one of the four sons of a Russian seaman who had been shipwrecked on a voyage from England and had remained in the James Bay area, marrying an Indian woman and founding a family which is now spread throughout the James Bay villages. Each canoe could carry 4,000 pounds of material, and was powered by six expert paddlers.

Mistassini Lake is 160 miles from the coast as the crow flies, but the river is 400 miles long, winding its way through an incredible maze of lakes and roaring over superb rapids every few miles. The canoes would take ten to fourteen days to go down to the coast and thirty to thirty-five days to return, laden with the supplies for the coming year. There are fifty portages between Rupert House and Mistassini, and many stretches needing strenuous work with pole, paddle and tracking lines. These canoe supply brigades had an almost heroic quality about them. The men would be away on the journey for two months and on their return to Mistassini have only a week or two of rest before going off into the bush for the next season's hunting and trapping. Anderson's head canoeman for many years was Solomon Voyageur, an Indian whose great-grandfather had been in charge of Rupert House and whose grandfather had been in charge of Michiskun post for many years. Solomon was the great-grandfather of Philip, Willie and Solomon Awashish, and was one of many Indians with the blood of former Hudson's Bay factors in them. He it was who would give the signal to his canoemen to stop their paddling as they came within reach of the first rapids on the Rupert River on the downstream journey. He would distribute tobacco to everyone, they would drift quietly toward the rapids, and before they reached them he would drop tobacco in the river to propitiate the Great Spirit.

There were about 250 Indians at Mistassini in those days, though they spent most of the year spread out over the huge territory of their hunting grounds, an area as large as Britain. And this work of shuttling furs and supplies in and out of the vast James Bay interior had been going on ever since the Hudson's Bay Company began to establish inland posts around 1787. In fact, it appears that the French beat the Hudson's Bay Company to the James Bay interior: it was in response to the arrival of the company on the coast in 1670 that Father Charles Albanel was sent overland from Montreal the following year to check out the effects of these godless people on the Indians.

O'REILLY: Have you studied the history of the Waswanipi band?

HARVEY FEIT: Yes, of Waswanipi and Mistassini. There is more history for Mistassini. It was on the major trading route for Lac St. Jean area and James Bay. There are records of a group of people called the Mistassini living on Mistassini lake as early as the 1640s. The first record of anyone visiting Lake Mistassini was Father Albanel in 1672. He was stopped when he approached Lake Mistassini by a number of Indians who said he should not continue until he had spoken to their chief. He was welcomed and wished well on his journey to Rupert House. He went by Nemaska down the Rupert River, and then went to the bay where he met another group, who may well have been the Eastmain people.

A NETWORK OF POSTS FOR TRADE

On his return to Montreal, Albanel related how he had advised the Indians on the coast to look away from the godless ones on the North Sea toward the Christian people farther south, and the French wasted no time in backing up his advice with practical measures. They are believed to have opened a post on Mistassini Lake in 1674, a post that was operated continuously during the next century and was believed still to be functioning as late as 1802, when it was leased for ten years by the Northwest Company of Montreal. This post was established more than a century before the Hudson's Bay Company penetrated inland. Their first post, apparently, was at Neoskweskau, a place on the Eastmain River that is still marked on the map, though it has been abandoned as a trading post for many years. This post was moved to the north part of Mistassini Lake in 1800, and then south to the site of the present Indian reserve in 1835, where it has been ever since. It was after the Hudson's Bay Company and the Northwest Company merged in 1820 that the Rupert River became the main highway to the coast. The chief factor at Rupert House at the time, James Clouston, an Orkney schoolmaster turned trader, made the most remarkable journey ever made by a white man across the interior of the peninsula, the object being to survey the company's operations in the area. With two white companions and four Indian guides, he traveled up the Eastmain River to Nichicun, north through the interior to Caniapiscau Lake and then (like Stephen Tapiatic) north along the Caniapiscau River to its confluence with the Larch, where he was compelled to halt because his guides were said to be in fear of meeting Eskimos. Instead of going back the way he had come, he struck across the tundra toward the coast halfway up Hudson Bay.

His decision was that the cheapest and most practical way to service the inland posts was along the Rupert through Nemaska, 100 miles upstream from Rupert House. So for generations the yearly summer canoe brigades would make their way downstream through the wilderness from the faraway posts where the only white people were the Hudson's Bay Company factor and his family—from Neoskweskau and Nichicun on the Eastmain, from Nemaska and Mistassini on the Rupert, from Waswanipi and Michiskun on the Nottaway—and then they would make their way upstream again carrying tea, sugar, salt and tobacco, the staples that the Indians took into the bush with them, and flour, salt pork, baking powder and raisins, the luxuries that they allowed themselves for times of emergency or feast. "The Cree are the most progressive and intelligent Indians I have ever encountered," wrote Anderson; and among the settlements he knew, it was the Mistassini people who were the best Indians he had ever dealt with. "They were for the most part honorable and upright, paying their way, providing for their families and discharging the obligations of life." Anderson was a Scotsman. His category of virtues is very Presbyterian.

LE BEL: Are you aware, Mr. Tanner, of the successive closing of many posts in the interior, such as Neoskweskau, Nichicun, Kanaaupscow, Nemaska, Indian River, are you aware of that?

ADRIAN TANNER [University of Toronto anthropologist, who had studied the Mistassini Cree]: Yes, I'm aware.

LE BEL: How would you account for the successive closing of the posts ?

TANNER: The trading policies of the Hudson's Bay Company. Posts were opened in the interior mainly to offset competition from traders from the St. Lawrence, and when a monopoly was established many of these posts were no longer needed. The Indians could then be encouraged to go to the coast.

O'REILLY [re-examining]: In what way did the way of life change with the coming of the Hudson's Bay Company?

TANNER: Mainly the annual trek from the bush to these Hudson's Bay posts, which became organized in what are known as canoe brigades; the small period in residence, sometimes only a few days; in the earliest days at the settlement where the Hudson's Bay Company post was, there was the introduction of guns and much later repeating rifles and steel traps.

A SYSTEM OF CREDIT

Throughout the middle Canadian north, the Hudson's Bay Company used the same system with Indians: the post manager would advance the hunter credit in the fall, to be repaid with the winter catch in the spring. Anderson describes the method he used in Mistassini in 1911:

> In September the Indian received his winter outfit or grub stake from the trader. This was important business for the Indian and the whole family—husband, wife and children—participated in it. Each family took its turn at the trader's store, often taking up as much as a half a day of his time. The extent of his advance or debt would depend on the individual trapper and his record of accomplishment. When it had been decided by negotiation with the trader, the amount would be measured out in tokens. Trade tokens—HBC beaver to-

kens—were still in use when I started trading. The beaver tokens would be set out in rows on the store counter. If the debt was, say, $200, the tokens would be set out in piles of ten. . . . With time and deliberation and the family council, the Indian would finally set about his shopping. So many piles of coins would be set aside for food, clothing, ammunition, fish nets and twines, not forgetting the all-important tobacco. If there was a surplus after the heavy items had been taken care of he would generously give it to his wife to buy needles, thread and other housewifely necessities.

Glen Speers retains a great deal of influence with the Mistassini Indians, but he would be the last to compare his situation with Anderson's complacent description of how he saw his own position sixty years ago:

My own experience as a trader, coupled with my observation and supervision of others, has led me to liken a successful trader to a happy mother with a brood of children. The successful mother has everything under control. Her children know just how far they can go and no more. Mother's yea is yea and her nay is nay. She is happy with her children and they with her. The weak mother, on the other hand, has no pleasure in her children, for they are always knowingly disobeying her. She threatens to tell Daddy when he comes home, but the children very well know that nothing will come of it. And the result is a disordered and unhappy family. It used to be very much the same with the Indians or the Eskimos for that matter. Being relatively primitive people, with a childlike naiveté, coupled with not a little cunning, they were quite adept in probing the weak spots in the trader's armor. . . . I believe that your old-time trader developed something of an extra sense in dealing with his native customers. Or it might be applied psychology.

O'REILLY: Would you describe the Indian society at the time of contact with the white man as a primitive society?

EDWARD S. ROGERS [Royal Ontario Museum, Toronto]: We don't speak of them as primitive. It is true they didn't have an elaborate political system, as we know it, but . . . it fitted. It had to be flexible and fluid in order to exploit and utilize the resources in that area. You couldn't segregate in one spot for a long period of time. You had to be able to break up, to move where the game was, either as a body, or if game was very scarce, to be able to move out in small groups to survive. There was an adaptation to the environment. It isn't a matter that was primitive, it was very skilfully done to allow these men and women to survive, raise families in this subarctic environment.

O'REILLY: What was the system of land use by the Cree people in the 1600s?

ROGERS: You had these named groups existing by hunting big game and fishing. They remained together as an economic and social unit for the greater part of a year if not the whole of the year. They moved through a large amount of territory, it's not the richest in the world for game, to exploit the big game and to use the fisheries. . . . I don't know the earliest date of their mentioning the sturgeon fishing [at Nemaska] but the supply of sturgeon was right in that area, and they would come in there for the sturgeon. These groups, again, we have no population studies . . . they would probably number in the neighborhood of 75 to 100 people, each unit occupying a particular segment of a drainage basin. . . . The boundary line tended to be the heights of land separating group from group. There would be in addition water fowl, other small game, rabbits which were important for their furs for rabbit skin blankets, beaver both for fur and meat. Also the need for particular floral elements, the birch tree for birch-bark for the canoe. Cedar for the ribs of the canoes. Cedar only extends to the southern end of Lake Mistassini, it's not found north of there. A variety of resources, not evenly distributed but found in different particular habitats, were used by the people and they moved throughout the year to take advantage of these at the appropriate seasons of the year.

O'REILLY: Individual hunting or mostly group hunting?

ROGERS: It was an adaptation to the environment in a very beautiful, as far as we can tell, balance with the environment, the utilization of it so that everyone received what was needed to survive, and through the mechanism of sharing they saw that everyone had. If one was lucky in getting caribou, the meat would be shared or if it was group hunting and one hunter happened to be the one that killed the animals, still shared.

THE PROSPECTORS ARRIVE

While the Indians were providing the furs on which the Hudson's Bay Company grew rich and were doing so without too much interruption in their traditional way of life in the bush, other people were from time to time eyeing their country with a view to using it for other purposes. White men would occasionally penetrate from the south during the nineteenth century, though for the most part it was an unmapped, forgotten part of the world, wrapped in snow and silence. A fur trader named Peter McKenzie returned from a trip north in 1904 and showed some rocks to Joseph Obalski, Quebec inspector of mines, who wrote a report to his minister: "I cannot too earnestly call your attention to this new district and to the important discoveries made there, for I consider it as destined to play a great role in the industrial development of our province." (His words were almost exactly echoed by the premier of Quebec when he announced the James Bay project in 1971.) Years later the son of Peter McKenzie described how in the winter of 1903–04 he and another man had left his father's party in Chibougamau and traveled north to Lake Mistassini, "where we obtained a valuable cargo of furs in exchange for dry goods and trinkets.

"The Mistassini Indians were totally different from the Lac St. Jean tribe. Their complexion was much fairer, and most of them spoke English, although they had never left the district and knew nothing of the outside world. They had learned the language from the HBC traders. They were a happy and congenial people, hospitable and gay and during the festive season spent weeks visiting one another. Dances were held every night with a violin and accordion as orchestra, and the scene was as lively as a New York night club." Before they left Mistassini the traders were given a banquet of bannock, bear fat and three pounds of meat and fish, served on the only two plates in Mistassini.

This is rather a colorful account, and seems hardly likely to be accurate since seventy years later most Mistassini Cree still could not speak English. McKenzie may have been misled by the fact that the key posts in the village at the time of his visit were occupied by two of the Iserhoffs, the sons of the canoe designer and grandsons of the original Russian shipwrecked sailor. Joseph Iserhoff was the Hudson's Bay manager at Mistassini before Anderson arrived and his brother Charles was the Anglican priest. Another brother, Samuel, brought up by his mother in Waswanipi, later became manager of the Hudson's Bay Company post at Ogoki, on the other side of James Bay, and of him the story is told that after having been south to North Bay to be fitted with new dentures and spectacles he wrote to his boss on his return to ask for a raise "because as a result of recent dental, medical and ocular treatments, I am a new and more valuable man."

As a result of McKenzie's find, the Chibougamau Mining Commission was set up in 1910 and mapped geologically 1,000 square miles of the Indian lands (though of course no one thought of them as Indian lands). Anderson records that probably the first tourist ever to hit Mistassini came in 1916, a judge from New York; but by the 1920s there was a small invasion of prospectors, engineers and promoters moving through the country searching for gold, silver, copper, iron, lead and zinc. This moderate boom collapsed in 1929, was revived somewhat in the 1930s, when two shafts were sunk and a town of 1,000 people was created, and was finally consummated after the Second World War with the construction of an all-weather road to Chibougamau from the Lac St. Jean area in 1945. Quebec was finally beginning to claim its own northland. And Jimmy Mianscum had to move aside to make way for the white man.

SMALLY, A LARGE YOUNG CHIEF

We went to the little green building that served as post office and band office to talk to the chief, Smally Petawabano, the large young man I had met in the bar the previous day. He was the elected chief of the Mistassini Indian Band, made up of every Indian from that settlement with a government registration number. Every registered Indian, as they are known, is attached to one of the 2,000 odd bands (the word is used officially) in Canada, and as a rule each band has a reserve, or lives in a particular location. An Indian does not lose his membership in the band by moving away. Each registered Indian living in a city, for example, continues to be a member of his band, with the right to vote in elections, to receive treaty money (if any), medical care and other special rights. (A curious anomaly of the Canadian system is that a registered Indian woman who marries a white man or a non-registered person loses her Indian status; in the words of the country's Indian Act, she is "deemed to be no longer an Indian" and is struck off the band list; whereas a white woman who marries a registered Indian is granted complete Indian status. Thus it is possible for an Englishwoman born in Yorkshire to become a registered Canadian Indian, as many white women have done; while an Indian woman from Mistassini, for example, who speaks nothing but Cree and has always lived in the bush in a hunting family may be declared to be no longer an Indian if the man she marries happens to have "lost" his Indian status in previous generations through the accident of his having had a white male ancestor.)

Smally had been one of the first group of Mistassini children ever to leave the village to go to school, when in 1946 twenty-six boys were sent away to Chapleau, a little town in northern Ontario, to be educated in the white man's way. Before that the priests were in the village to conduct some classes in summer when the people returned from the bush. But few people except those working with the company had learned any English. Smally's 1946 group stayed away for two years and was never allowed home during that time. They made it through grade four, and when they came back Smally became a leader because of his knowledge of English. When in 1952 a group of men was recruited to go south to cut lines for the Canadian International Paper Company, Smally became the gang boss. This was the first involvement of Mistassini men with the wage economy. But their experience was such that by the end of the decade most had returned to Mistassini to try to re-establish themselves as hunters and trappers.

By 1964 the old chief, Isaac Schecapio, had resigned to make way for the young man who knew English and could represent his people better to the outside world, now moving in so much more threateningly than ever before. In Smally was exhibited the ambivalence which the James Bay Corporation lawyers three years later found difficult to understand: although he was no longer a hunter himself, but an administrator, always dressed in city shoes, slacks and sports coat, his concerns as chief were nevertheless almost exclusively with the hunting and trapping life. For in spite of any appearances to the contrary, the culture of Mistassini remained that of a hunting people, and the yearly rhythm of their lives was dictated by the search for animals, birds and fish. Now Smally told me that his own children would get "a proper education" so that they would be able to live away from the reserve if necessary, since he did not believe there was a future for them in trapping. Yet Smally was not as far removed from the bush as he may have seemed, sitting in his office beside the radio-telephone to Chibougamau. It was only thirty years before, within his own memory, that the first government agent had arrived in Mistassini. He came in from the south by canoe, and his first job was to register the people. At that time, still, people had only one name, so for two summers this agent tried to discover the last names of everyone's grandfathers and give these names to people as surnames. The federal government then sent in nurses in the summer (the first health service available to the Mistassini people) and occasionally helped people going into the bush by providing them with some rations.

In those days the Indians went off to their traplines and hunting territories (the word in Cree is the same for both) by canoe, most of them paddling up the lake fifty miles or so to the incredible maze of streams and small lakes which is the headwaters of the Rupert River. They would move west and north along the Rupert, and those who were going farther would strike north, portaging time after time from lake to lake as they laboriously made their way deeper into the wilderness. Those going farthest north would take a month or more, undertaking 100 portages before reaching their territories.

Indian hunters may look disheveled and haphazard when they appear in white communities, but in the bush they are businesslike and efficient, forced to follow strict schedules dictated by the seasons. They had to arrive at their trapping grounds in time not only to set up their camps and catch enough fish and game to live on until the trapping season proper began in November and the beaver became available for food, but also in time to make a detailed inventory of the animals on their traplines. They had to see the lodges of the beaver that they would be trapping in the coming months, before these became half hidden by the snow. Depending on the area—some parts of the huge Mistassini territory are better for moose, some for caribou—they would try to find big game toward the end of the winter to supplement the beaver diet that would by this time be coming to an end. They would trap otter also in late winter, and by early spring as the ice broke up they would be after fish. They had to plan so that they would have enough bush food to take back to the village to keep their families going for some weeks. For no sooner would the hunters arrive in Mistassini in mid-June after a month's journey south than they would have to leave at the beginning of July for Oskelaneo in the south to pick up supplies for their next season in the bush. Their trip south by canoe would take six weeks: they would return by August 15, stay for only a week in Mistassini and take off north again for their hunting territories late in August.

This pattern of life through the 1940s and 1950s was not very much different from that described by Anderson in the early years of the century, except that the old canoe-brigade voyages west down the Rupert River to Rupert House had been replaced by a trip south to Oskelaneo at the south end of Gouin Reservoir, and above the headwaters of the Gatineau River. This pattern—a year-round pattern, as Smally observed —lasted until the introduction of the planes, which were only gradually adopted in the late 1950s by those Indians who had enough money to afford the charter costs. Well into the 1960s many people went to their territories by canoe, making every year journeys that for white men would have been regarded as heroic.

Until the planes came into general use, there was hardly need for employment in Mistassini, for the men were occupied all the year round with their trapping life. The planes freed them from the journey to Oskelaneo and gave them more time to spend in Mistassini in summer. Then they would look for work, to get some money to help them buy their outfits for the winter. The only work available in Mistassini was the building of houses that were being provided by the government. "But even if we had jobs available," said Smally, "there isn't one hunter who would work all year round. When the trapper knows it is time to return, when it is time for the water to freeze, he quits work because he prefers to return to the bush."

Smally was fighting a losing battle. All efforts by the Cree to defend their life in the 1960s had proved futile. They had to watch helplessly as everything that had been dear to them was trampled over by white men. Not only were the government's game wardens literally seizing the meat from out of the mouths of the hunters' families, but after the Indians had obtained verbal permission to start a commercial fishing operation on Lake Mistassini, the game wardens had arrived by aircraft from Quebec, carrying rifles and revolvers, and had seized the twenty-eight nets the fishermen were using. "We tried to fight back," said the chief resignedly, "but there was nothing we could do."

A SAWMILL THAT DIDN'T SAW

A couple of years before, a group of experts investigating the economic conditions of Indian reserves had made a study of Mistassini and thirty or forty other reserves in Canada. By the indices of white economic activity Mistassini was nothing: indeed, it was almost worse than nothing. Only 2 percent of the workers were skilled. Only 40 percent had work and that for less than six months. Only 2 percent had got past grade eleven. Only 3 percent of those over sixteen were in school. Eighty-four percent of households were receiving welfare. There were no community organizations. The band had only one cent per capita in capital funds. Statistically, Mistassini was a picture of unrelieved gloom. Because of that kind of reasoning Indian Affairs occasionally would try to get some community development project going. To date they have not been very successful.

In 1969 a sawmill by the banks of the stream that runs into the lake alongside the reserve was lying idle. It had been built by Indian Affairs to provide timber for the housing program that was just getting under way. "We could sell timber to people around this part of the world, definitely," said Smally, with vague optimism, when I asked him why they didn't run the mill and take up some of the economic slack, as it were. As it turned out, the mill never did get going: the local lumbermen did not welcome the competition, apart from anything else. But it is doubtful that the Indians would ever have provided much competition. Their hearts weren't in that sort of enterprise. The sawmill kept breaking down and no one in Mistassini could fix it. A year or two later it closed entirely.

For a while it just sat there. The people asked the chief if they could take the boards from the building and use them in their houses, but Indian Affairs said no, the mill was government property and must on no account be touched. It sat there, "rotting away quietly," as someone later said, until one night someone put a match to it. The next day everyone in the village was happy. They were laughing about having pulled off such a coup. That would teach Indian Affairs to be so mean with their wood. And besides, the mill had been spoiling the view down by the river. When people looked down the hill they could no longer see the river for the white man's monstrous industrial improvement project. So much for that!

6

1971: MISTASSINI

The masculinity of Chibougamau is especially marked when the town is resting. On a normal weekday the main street is full of cars, large cars for the most part, bearing signs of the rough roads and frontier environment. But on a Sunday morning when not many people are around, the very tranquility has a slightly raw feeling. The desultory groups of young men who gather in places like the Pekinois Café, where the waitresses are pert and long-legged in the shortest possible hot-pants, carry an assurance in their idleness that suggests a week of hard work well performed. They wear purple jackets with the successes of last year's baseball league blazoned across the backs, or the name of a motor oil or a brand of spark plug decorating their chests. Their language is Quebecois, harsh and rapid, incomprehensible to the outsider, the private tongue of a fairly private people. They toy with their glasses of Coke for a long time, slouch slowly along the half-deserted street, mooch over the girlie magazines across the road in the drugstore. If Jimmy Mianscum has been disinherited, these are the inheritors. But they spare no thought for him. Social awareness is not the mark of a frontier town in Canada. The working class is strong, as in other countries, but its ethos is that of the whole society, and few among its members challenge its assumptions.

It was in this Chibougamau, on a weekend, that Philip Awashish picked up a day-old Montreal newspaper in the drugstore and read about the James Bay hydro-electric project. Two evenings before, the premier of Quebec, Robert Bourassa, had announced to a cheering crowd of his party's supporters in Quebec city that the province would build "the project of the century" in the north.

Philip read that all of the rivers running into James Bay would be dammed and diverted and reservoirs built all along their length. The Nottaway, the Broadback, the Rupert, the Eastmain, the La Grande would all be controlled, the waters to be transferred back and forth from one river to another. Seven power stations would be built on the Rupert alone. Mistassini Lake would be blocked at its outlet on the Rupert and would become forty feet higher and half again as big in area. Dozens of lakes intimately known and used by the Cree people would disappear. Extravagant claims were made for the wonders this would work for Quebec, the many jobs it would create, the prosperity it would bring, the energy it would supply to meet the ever escalating demand of the cities in the south. The Cree lands were to be invaded and their riches carted away for use by other people; and with not even, so far as he could see from the newspaper, a tip of the hat to the people who had always depended on these lands. "The project of the century" would henceforth be the cornerstone on which the whole future of Quebec as a French-speaking survivor in North America would be built.

Philip was flabbergasted: his father and his grandfather and everyone who had come before him had fed themselves and survived on what this very land provided. The Awashish hunting territory was on the headwaters of the Rupert River. Philip knew that his people had rights in the land, for they had never been conquered, they had never signed a treaty and they had never surrendered, ceded or yielded up, as the treaties say, all their rights, titles and privileges to Her Majesty forever.

A MARGINAL INDIAN

This outright attack on the rights of his people also represented a great challenge to him. For he knew more about the white world than most of his people, and that knowledge would be desperately needed if this project ever became a reality. Unfortunately, in spite of his return north to be among his people, he had remained a somewhat marginal figure (a "paper Indian," as he once remarked self-deprecatingly). He had never been able to establish with his father anything like the intimate relationship that Willie had as a result of having hunted and trapped with him for so long, and the hunting culture did not place great worth on the intermediary role that Philip could play. Most of the older men of Mistassini, men like his father, had never seriously tried to relate to anybody outside their own culture, and they did not understand the difficult position of their alienated sons. The Cree Indian Friendship Center in Chibougamau, started by Philip and his wife, Anne-Marie (a Metis— mixed-blood—from northern Ontario), with meager federal funding, was important for all the Mistassini Indians, because it provided a place for Indians from the reserve to stay overnight when they were stranded in Chibougamau. But it was not particularly well supported by the band council in the reserve.

Philip was on his way to Rouyn for a meeting about the Friendship Center when he read of the project. Edna Neeposh and Anne-Marie went with him. "I guess we were pretty shocked," said Edna. "We talked about it all the time on the way there. Everything flashed through my mind, what would happen. I kept thinking about the people at home, how they could find another way of living if their land was to be underwater. We were really upset as we kept talking about it, and we wanted to see if we could do something."

As they talked they decided that the leaders of the James Bay Cree should hold a meeting to discuss the threat to their ancestral lands. Philip returned to Mistassini, told a meeting of the people there about the project and persuaded the band council to invite the various villages in the area each to send two people to a meeting. "When we first told people about the scheme," Philip later recalled, "they just laughed. They couldn't really believe that anybody would do such a stupid thing as to flood the land. And they couldn't understand the scale of the flooding that was proposed."

THE FIRST MEETING

The meeting in Mistassini was held at the end of June 1971, and was the beginning of a political process such as the Cree people had never known. It was the first meeting ever held by the James Bay Cree in the 5,000 years of their history. The elders of the culture, whose most sophisticated traditional political structure had never extended to more than 100 or so people, had never felt the need for a meeting at a higher level. They had spent their time subsisting in the forest, and in recent years had accepted uncomplainingly the many derogations from that life which were occurring more and more as outsiders moved into their lands.

Now that they had decided to get their people together, these younger Indians tried to avoid the Indian political movement that had developed in the southern cities, financed by the federal government, now claiming to speak in the name of all Indians in the province. The Quebec arm of this growing bureaucracy was the Indians of Quebec

Association, dominated by the leaders of the two most urbanized groups of Indians in the province—the Caughnawaga Mohawk band, whose reserve is, in effect, a suburb of Montreal, and the Loretteville Huron band, whose reserve is on the outskirts of Quebec city. The southern Indians had a range of concerns different from those of the culturally coherent and self-confident (though politically retiring) hunters of the north, and Philip particularly wanted to get his people together without the distraction of political involvement from the outside. The association, however, had a great deal of money from the federal government, while the Crees had nothing. And whatever reservations the northern Indians may have felt, there was, in practice, no chance of avoiding an involvement with the association if anything effective was to be done. The Mistassini band council tried to get money from other sources to pay for this first meeting. But once the ebullient secretary-treasurer of the association, Max Gros-Louis, got wind of the meeting, he ran up and down the James Bay coast in a chartered plane picking up delegates, transporting them to Mistassini and grandiloquently announcing on his arrival that all the costs for the meeting, including a *per diem* payment to the delegates, would be paid by the association.

"The meeting was intended to be," said Ignatius La Rusic, who happened to be in Mistassini at the time, "one at which the hunters could say what they felt. But gradually, with the association representative taking over, it turned into the usual Indian Affairs meeting, with people speaking English most of the time, and occasionally translating into Cree for the older men."

The meeting lasted for three days. No one there had any details about the project, so all they could do was express their fear that their hunting grounds would be destroyed. The association's legal adviser, Jacques Beaudouin, told the hunters that since they had never surrendered their territory they could consider it a reserve, which meant that the Quebec government could not take it over without their consent. Since it was a reserve, however, Quebec could develop it with the permission of the federal government, so the only way to stop the project would be to ask the Minister of Indian Affairs to refuse his permission for it.

"The chances of stopping it through the federal government are really remote," said Philip Awashish. "There are Liberals in power in Ottawa, and there are Liberals in power in Quebec."

"This is the only means to oppose the project," said the lawyer.

But, argued Philip, native people have never been involved in northern development. "I cannot see Indian Affairs stopping the project." His sense of political reality was sharper than the lawyer's, whose advice was questionable on several counts. The area was not a reserve, federal authority was not needed to proceed with the project, Quebec was able to take over the territory without the consent of the Indians and the Indians had many other means of opposition open to them, including legal means, much more powerful than a mere request to one federal minister.

ONLY THE BEAVERS HAVE THE RIGHT

Eventually, however, the Indians took the lawyer's advice and decided to appeal to the minister to defend their interests. They drew up a remarkable petition:

> To the Minister of Indian Affairs:
> We, the representatives of the Cree bands that will be affected by the James Bay hydro project or any other project, oppose to these projects because we believe that only the beavers had the right to build dams in our territory, and we request the Minister of Indian Affairs Northern Development, to use his legal jurisdiction to stop any attempt of intrusion of our rightful owned territory by the government of province of Quebec, or any other authority.
> And we have signed this document on the first day of July, 1971.

Gros-Louis' interest in the meeting was to establish the association as the spokesman for the Cree to the federal government. And he, too, obtained his wish, for the meeting decided to ask the association to forward the petition to the minister. For months nothing happened. The Cree received no answer from the minister, and they were confused, because they were under the impression that it was his job to defend them. Several times the minister told Parliament in Ottawa that he had never been asked to intervene by the Indians of James Bay and would not do so until asked. Many months later, fed up with this lack of response, Philip gave me a copy of the petition and I mailed it to opposition leaders in Ottawa. An emergency debate about the federal responsibility for defending the Indians of James Bay was arranged. The minister stuck to his story: he said he had never heard about this petition until that very day.

It was not until Philip and Chief Billy Diamond of Rupert House joined the association as communications workers early in 1972 that the mystery was cleared up. Philip found the original petition in the association's files. It had never been forwarded to Ottawa. The Cree had made their decision about the project within three months of its announcement. But they had lost almost a year of valuable organizing time through the negligence of others.

7

1972: MISTASSINI

PHILIP AWASHISH [in evidence]: . . . When Billy Diamond and I were up north from February to March 1972 the people were still anxious to hear from the federal government through the department of Indian Affairs to respond to our petition we had signed July 1, 1971. . . . In Waswanipi the people were greatly alarmed by the fact that the . . . scheme would be flooding a good large percentage of their trapping territories. The Mistassini were alarmed by the fact that Mistassini lake being a reservoir system would flood nearby trapping territories and the community of Mistassini. . . . The possibility of the village of Mistassini being flooded was verified by a meeting with an engineer from Hydro-Quebec. We found out that about 40 buildings would be flooded. So with all this in mind the Indian people urged the Indians of Quebec association to proceed with the court injunction against the James Bay project.

The two communications workers went north accompanied by four journalists, of whom I was one, and an anthropologist, Ignatius La Rusic. Mistassini had grown considerably since my visit two and a half years before, the outside world having been brought closer by the completion of the road from Chibougamau. A new school had become the biggest building in the community. Some small apartments had been built for the teachers. One of Edna Neeposh's brothers, Philip, had quit hunting to become a taxi driver, the second in the community. The people could now come and go more easily between the reserve and Chibougamau. More liquor was brought into the reserve every year.

There was still nowhere for strangers to stay in Mistassini. On the first night Smally invited us to sleep in sleeping bags on the floor of the new band office, just moved from the little old green post-office building to one of a new line of Indian Affairs houses. Philip Awashish's family were out in the bush, so he opened the empty house, started the oil stove and the next night invited us to sleep there.

La Rusic, a persistent observer of the response that the James Bay Cree were making to the challenge before them, now analyzed for me their developing political leadership. Their leaders were divided between the white-oriented and the traditional. The first spoke English, had been to school, had worked for the Hudson's Bay Company, the church and the state. This group had begun to emerge even before the children were sent out of the reserve for their schooling after the Second World War. The second group regarded a good hunter as a successful man. These spoke only Cree, did not undertake much, if any, wage employment, and most had never been to school. This group needed a hunting ground, the other an administrative center.

The highest level of political organization traditionally had been the getting together of groups to secure trade goods. There had been a loose arrangement of clans, families, kin-based groups. There had been no chief, but for a time a man was recognized as "a trading chief," a man chosen by the hunters to provide the contact with the outside traders. Now, as white influence increased, Cree society was moving away from these kin-based groups to a recognition that politics was about policies. That is why, he said, all the chiefs on the James Bay coast were young men, English-speaking, with an orientation into the white world rather than back into the community.

"The band government has been set up by Indian Affairs," said La Rusic. "That is of critical importance in understanding the situation. So people feel the need to have spokesmen who are articulate in terms of getting resources out of the government bureaucracy rather than in terms of backing up the desires of the people." Thus the young chiefs could ignore the old Indian habit of consensus politics because they knew what was important for Indian Affairs and the people did not. In this set-up the white man—whether representing church, state or commerce—knew that by giving the leaders goods and services to distribute to the community, they were consolidating the power of a certain group.

By now La Rusic found that the influence of the "new *auchimaus*," the new brokers mediating between the Indians and the white society—the timber companies, the tourist-camp operators, the mining companies— had begun to decline as the Indians became convinced they had not delivered the goods. In Mistassini the balance had begun to swing back a little toward the traditional people. If the meaning of hunting could be re-established in some way, great pressures would be put on the present leadership of the band. "The critical thing for the hunting," said La Rusic, trotting out his most endearing cliché, "is that the young people do not lack the physical skills, but the theological ones. Hunting is really religion, not ideology, that's the critical thing. Once education has broken down the religion, it becomes extremely difficult for younger people to keep it up." A heavy responsibility lay on the new generation of leaders, such as Philip and Edna, better educated than any of the Mistassini before them, now returning to the reserve to bring about the New Jerusalem. La Rusic was deeply committed by ties of friendship to both of them: they depended on him when they were in the city and he was generous with his help. "Philip is the key figure," he said. "He is trying to come back, re-establish his roots. If he succeeds, he could, in my judgment, become one of the significant leaders of Indians in Canada."

Getting to know the elders

At long last I had begun to realize that the usual journalist's perspective was useless in getting to understand a complex Indian community like Mistassini. The key lay with the older people, guardians of the value system and the traditional skills. I asked George Matoush, a young Mistassini Cree who had just returned from the Indian occupation of the island of Alcatraz in San Francisco harbor and was now living a marginal sort of existence hanging around the reserve, to take me to meet some of the older men. He took me to a little house occupied by an elder of his family, William Matoush, an old man with a thin, beaten face, now retired, and prevented by poor health from accompanying the younger men into the bush. He was one of those who in his younger days had worked on the canoe brigades hired by the Hudson's Bay Company to freight the furs from Mistassini to Rupert House, and travel back with the season's supplies. He did it every summer for ten or eleven years, just as Anderson describes in his book. He told me they did not take tents in those days. For shelter they would cut small trees, lay them over as the roof and place moss between them.

He began to warm up as I spoke to him: it had never happened before that a white man visiting the reserve had gone to talk to him, since his experience had never been regarded as of any significance outside the Cree culture. He told me how they would in those days search out big rocks and build their houses around them: they would build a fire around the rocks, which would retain the heat. He talked on for a long time, explaining how they went about killing the caribou, trapping the beaver, how they survived in the intense cold as they moved across the country. He opened up and began to talk of the respect which the Indians had for the animals, their responsibility to respect the animals they killed, a subject which I had been told the older Indians were reluctant to discuss with strangers (this was not true: usually they were not asked about it, and they would not offer their opinions without being asked). They had to keep the dogs away from the bones of otter and beaver, from the head of the moose and rabbits. Sometimes they would tie a little cloth on part of the skeleton and put some tobacco in it: "In return he knows what you are doing to him. If you show respect to him, you will get something from him. But if you allow a dog to get hold of those bones, then you will have a hard time to kill that animal in future, because he makes himself scarce."

O'REILLY: Is there any religious content in respect to animals?

ADRIAN TANNER [anthropologist]: All killing of animals has a strong religious significance, governed by religious beliefs and values. Usually whenever a beaver is brought in, it will be drawn by a ceremonial string. When eaten, particular parts of the animal are put into the fire for the spirits, the bones are carefully preserved so the dogs will not eat them; particular bones are put into a tree, others are put onto a platform or thrown back into the lake. The caribou antlers are erected on a tree facing east, and decorated with streamers for religious reasons. This was the major part of my study.

O'REILLY: In a nutshell, what is the significance?

TANNER: The animals are seen as persons, or as being controlled by spiritual persons and the hunter is engaged in a religious occupation when he hunts and is exercising spiritual powers by hunting. The older a man gets and the more animals he kills the greater he achieves this spiritual power.

"You didn't go hunting," said old William Matoush to young George, "you went drinking. Your father was a good hunter, but the young people now, they are getting worse. All they do is drink." As for the damming of the rivers that was proposed, that sure as hell would not help the hunters. The money from the development would go to other people's pockets. But there were many things the white man could learn from the Indians if he wished. "The white man is going to come and flood the land. That's teaching Indians how to flood land and build dams. But the Indians can teach the white man about nature. We have lived with nature. We were born with it, we have got to look after it, not to destroy it. It's like our mother. To us it is like putting something in the bank when we do not kill all the beaver we could. We leave the beaver to allow them to multiply.

"If they are going to flood the land, they are taking all our savings away. We talk about it among ourselves, this flooding. We discuss the old days and if we can do something to stop them, but we are getting too old. We have to look to our grandchildren to see what they can do about it. We cannot do anything. Unless maybe we have to use medicine."

George laughed in rather an embarrassed way as he translated this last touch. The younger people from the Cree communities do not like to talk about the powers that some of the older men have. They now only half believe in these powers themselves, and think that if they speak about them, outsiders will laugh at them. But there are still shamans living in the Indian communities who are capable of performing the shaking-tent ceremony, that strange exercise in the supernatural, and there are people who curse other people, and whose curses have led to their deaths.

DEMON WORSHIP, SORT OF

Larry Linton, the Baptist missionary who lives in a small house in Mistassini and uses his mission's bush plane to take his Christian message around the hunting camps, had seen a shaking-tent ceremony a few years before in Rupert House.

"There was an anthropologist there who asked them to shake the tent. I was of the opinion that it was a demon worship, sort of, that there was something to it, but I thought, Well, I'll go and see just this once, I'll satisfy my mind, and then if it is real I won't bother with it again. As a Christian minister I don't think I should play with demon worship, and if it is a farce, well then, I'll explain it away as a farce.

"So I went to the tent shaking, and we took a tape recorder, and we took cameras with 500 ASA film, but they wouldn't allow any light, it didn't come out, but on the tape you can definitely hear the rustling of the tent.

"That tent was twelve feet high, three feet around at the top, six or eight feet in diameter at the bottom. This old man went into it alone. He had a bad cough. He was old and one of the reasons that he wanted to shake the tent was, he said, If I have power by my spirits again, then they will give me a few more years of life. I am an old man, and I am sick, and I am afraid this is the end, and if they don't give me any power I am going to be finished this winter, I don't think I will see another spring.

"He shook the tent. He kept it going for two or three hours. I can't remember the time exactly, it went by rather quickly, because he was talking to his spirits and they were answering him back. We had all this on tape. I remember one time in particular the tent stopped. I was feeling kind of uneasy as a Christian minister in this demon worship, satanic worship. I had realized by this time that it was such because the old man couldn't be shaking that tent for that length of time. I shook the tent before he started, and I shook it again after, and it was solid, no man was going to get in there with a bad cough and shake that tent three feet off center on each side. All of a sudden, bang, the tent stopped. It really was quite an experience. But then it started again.

"I calmed down. At least it was not me causing this trouble, so I went on watching. Finally he stopped and he said, Well, that is all for tonight. The people started asking him questions, to find out things. When they are shaking the tent they have real power to foretell the future to a certain degree, and real power to tell what is in the past. They have been talking with their spirits, and these spirits do have the power, there's no doubt about it. One man came and he said, Where am I going to kill my moose's fawn? The old man said, Well, I can't talk about that today, because my spirit that would tell me about the moose isn't here tonight. This is really just a thing for this white man that is here with us tonight, and we are not going into real things.

"But they do have these powers. Originally if you had an enemy and you could shake the tent, presumably he would drop dead miles away. There's lots of stories that this has happened."

He knew for sure, from his brother-in-law, of one case that occurred in Nemaska a few years before. The people there had not liked a teacher who had arrived to conduct a summer school, and had decided to get rid of him. They shook the tent at him and he became so terrified that he was almost driven crazy. Only the intervention of the priest stopped them. "If he had died," said Linton, "definitely they would have claimed it as a victory for the tent shaking. It is a real thing, but it has died out completely."

A few months later someone in Mistassini tried to shake a tent for the delegates to a provincial Indian convention. It didn't work.

A PLETHORA OF PRIESTS

At this time the solidly Anglican village of Mistassini had priests of four denominations competing as missionaries for the souls of the Indians, and two of them were active in the meetings held in the next few days. The general opinion at the public meeting of those Indians who remained in the village—most were out on their land trapping and hunting—was that the white man had gravely offended against Indian customs by not having asked the permission of the people before entering their lands. This was regarded as one of the worst crimes a man could commit. Father Alain, the Roman Catholic, and the Rev. Tom de Hoop, the Anglican, were active at this meeting, offering to translate documents and hold follow-up meetings to discuss the project further. "Very interesting to watch how the priests behaved," remarked La Rusic later. "The critical thing for them would appear to be to establish themselves in some broker relationship between the band and the James Bay Corporation, so that they become indispensable to both sides."

Father Alain was one of three or four French-speaking Roman Catholic missionaries who had been trying for ten years to shake the long adherence to the Anglican Church of the Cree and Inuit people in the villages along the coasts of James and Hudson bays. They had been notably unsuccessful in winning converts (indeed, none of them had ever made any converts), but some of them had built up commercial and educational establishments which dominated their villages.

Tom de Hoop was quite a different type of character. Like the Rev. Muller, he was a Dutchman, a tall, blond, amiable and slightly maladroit young man who seemed to be out of his depth among the Indians. He was the inheritor of a powerful position, for the Anglicans had been a major influence among the Cree for more than 120 years, but he had little personal authority, and Smally did not hesitate to speak disrespectfully to him before strangers when he came into the band office to use the phone while we were waiting for a meeting of the band council to begin. De Hoop put a brave face on it, but it was obvious to us that as he gathered up his black robes and tramped off into the muddy village he commanded only the tattered remnants of the majesty and dignity of his church's long service in James Bay. De Hoop, too, was moved out of the reserve within a couple of years, leaving the Anglican living vacant.

A DEMAND FOR ACTION

The band council confirmed La Rusic's analysis of the changed balance of political power. Every member was young and could speak English, except Jimmy Mianscum, who had finally abandoned Doré Lake to take up the government's offer of a new house in Mistassini. The meeting was conducted almost entirely in English. The councillors listened patiently as Billy and Philip described the runaround they had been given in Ottawa and Montreal when they had tried to find out more details about the proposed project.

Billy said: "A distinguished lawyer in Quebec city, Henri Brun, says the Indians have clear rights, and if the corporation carries out work north of the Eastmain River they will be in all sorts of trouble. The Indians of Quebec Association is sitting on the injunction which has been prepared. They are using it as political leverage to try to get a better deal from the government on certain other matters affecting Indians elsewhere. I told Max now is the time to push it, but they have replied that I am only one Indian. There are seven thousand others. So we are here to find out what the seven thousand others think. I can tell you that my father's trapline will be affected by the Nottaway diversion. He is thinking of taking court action himself. He says that even if he cannot speak any English he will go to court and speak Cree."

The council passed a resolution demanding that the Indian Association proceed immediately with the court injunction. "We understood that the object of the injunction was to stop all construction until agreement has been reached between the affected Indians and the government of the province of Quebec."

FINALLY, A HUNTING CAMP

In a small plane piloted by an old bush pilot, Jimmy Kershaw, we flew the next day the 100 miles or so northwest to the hunting camp of Philip's father. We passed over apparently featureless country, a maze of small lakes that were sheer white under the snow and ice, dotted over a black-and-white landscape of snow and evergreens. Kershaw had no trouble finding the place (he knew every camp, since he serviced them for Speers, who was at this time of year making his round of visits to pick up furs and take the people some new supplies). We circled as we approached the camp, a mere dot in the wilderness, and picked out two tiny figures trudging over the land several miles from the camp. They were Isaiah and Willie on their way back from a trip out to check their traps.

It was my first visit to a hunting camp. I had by this time talked to a lot of hunters and their children, and knew how attached they were to the bush life. But nothing had prepared me for the shock of walking into Isaiah Awashish's hunting lodge. The tent was almost hidden under a pile of snow, and was identifiable only as a mound of snow with a door in front. The canvas door hung free, a pole slung across it about two feet from the ground, heavy enough to hold it in place against all but the strongest winds. At the same time the lower part of the door was sucked in, providing an ideal drafting mechanism for the interior fire which all of these camps originally had. When I lifted this door and walked into the tent, I was astonished by the warmth and glorious smell. Beaver was cooking beside the tin stove, hanging by a string from one of the boughs bent over to support the tent roof, spinning round throughout the day, the fat dripping into a pan on the ground as the flesh cooked. But the predominant smell came from the spruce boughs on the floor. Every Indian tent has them. The boughs are picked about a foot long from a large number of trees, never enough to spoil any one tree, and are laid down on the floor in a fanlike fashion. They make a soft, springy floor, comfortable to sleep on. And they are ideally practical in other ways, for anything dropped just falls through the boughs and doesn't clutter up the floor or make a mess. But their primary function is aesthetic, for they fill every Indian home in the bush with one of the most delicious of natural scents.

The lower part of the lodge was made of split logs insulated with moss and with snow piled outside. The upper part was a network of boughs bent to take the strain of the roof. Clothing was hung from poles strung across the roof to dry over the two or three tin stoves which blazed away in the center. Philip's younger brother, Kenny, twelve, had been taken out of school to join the hunting camp. Isaiah's brother Sam had arrived only the day before with his wife, their new baby and some of their older children. Mary, only two weeks of age, was wrapped in a papoose laced with moccasin hides. Along the rafters was stored dried moss which Sam's wife used for the baby's diapers. The wood for a new pair of snowshoes was being cured and bent.

Everywhere one looked, someone had been at work fashioning something or other from the materials of the forest—moccasins, snowshoes, shovels, beaver stretchers, sheets, even the stoves were hand-made. I sank down on the spruce boughs, my back propped against the wall, and absorbed the atmosphere in wonderment. So this was what Edna was talking about when she said it was a good life in the bush! Little Sandy, two, was tied to the wall to prevent him from stumbling into the hot sides of the stoves. The baby slept peacefully. The children went racing outside into the cold air. I went with them. There were two or three other little tents for storing things. Beaver skins were hanging to dry on tamarack stretchers. A platform held the bones of the animals that had been eaten, and at the entrance to the camp, in a neat parcel of birchbark, the arms of a bear had been wrapped and tied to a tree. The camp had been carefully placed away from the winds, though close to the shore, and the month's supply of firewood that a trapper always keeps by him had been piled like a wall to cut down winds blowing toward the door. I began to realize what Philip meant when he told me how degenerate Mistassini was as an Indian environment, and I understood for the first time how much he must have felt his divorce from the qualities of this life. We waited for an hour or so, but Sam figured that Willie and his dad would take three hours to get back to camp. We could not wait that long, so Philip had to miss his father this time around. They gave us some meat to take back to Mistassini, and we took off.

Adapted to the environment

o'reilly: Now during your field trips did you have occasion to observe any of the places where Indian people camped or did specific types of fishing or hunted?

edward rogers [Royal Ontario Museum]: Well, yes, both the location of archeological sites, plus the old abandoned camps of more recent vintage, and when we were travelling with the people themselves. They have their concept of the geography, of how it is organized, where the resources are to be found, how to deal with it to make life as easy as possible. For instance, where rivers come together is an area that appears from the basis of their archeological sites, old camp-sites, and where we were with them where they camped, to have two things: one, it's good in resources, that is, for food, pots and floral material; also, of course, it's the point at which you depart for other, well, they're like highways along the rivers. They would always follow the rivers, camping near water, camping where resources tended to be more productive than elsewhere. They had to be cautious in terms of winter camps, for instance it's not a matter of just going out and putting up a camp. You will ideally wish to get your back to the northwest wind, which can be pretty cold in January and February.

You'd put your camp ideally, for instance, on the west side of the lake, hopefully with a little hill in back of it for protection from the wind, but you also would want to have a supply of firewood nearby, spruce boughs for the flooring for the lodges. All of these environmental factors had to be taken into consideration, and, as I say, the native people had tremendous knowledge of the environment and how these would sort of organize and fit together. This also, as you can readily imagine, meant that they would move quite frequently because of running out of spruce boughs or firewood, or having utilized the game resources in the area, move on to another location. Again, it was this adaptation to a somewhat varied environment that allowed them to survive.

O'REILLY: Now, did you have any occasion to notice yourself whether the animals were in greater abundance in these areas that you've just been describing or in other areas?

ROGERS: I can only give impressions of this based on the fact that these were named areas, used over and over again, as we have evidence from the archeological material, so that we have three, four, five, six hundred years of utilization of these particular spots and from this point of view, what the people have said, that these were the areas where most game will be found at particular times of the year. You wouldn't go and camp on top of some barren rocky hill. It wouldn't have the resources that you would need.

HE STOPPED A MONSTER

We gathered in Philip's house with a few cases of beer after a meeting one night, and began to drink and talk. From time to time the door would open and someone would come in—a group of children, some women, an old man or two—and often they would just stand and listen for a while and take off as suddenly and wordlessly as they had arrived. It is exactly like this in every Indian community I've visited in Canada. It takes only a few minutes to get a party going, and soon the evening became quite cheerful. Two old men who had already been drinking a bit appeared in the doorway and stood there expectantly. They were John Loon and Philip Petawabano. They each had a son living in Montreal and working around the periphery of the media and documentary-film worlds. John Loon was a tough old bird with a long, heavy face. He had been a hunter all his life, couldn't speak a word of English and had never seriously tried to communicate outside his own Cree world. He happened not to be hunting this year, and appeared to be doing considerable drinking on his

off year. Philip Petawabano was a smaller man with white hair. It was one
of those ironies to which I have never become accustomed that this
extremely private and spiritual old hunter had a son, Buckley, who was
a television star with a considerable following among the young girls of
West Germany.

Of Philip Petawabano I had been told this story: Five years before
when a tornado swept through the area, blocked the road to Mistassini
and appeared to be headed for the reserve, he went out with another man
to confront it. The tornado evidently was a monster out to demolish the
reserve and its people, but the two old men stopped it dead in its tracks
by sprinkling tobacco in its path and presumably uttering some sacred
imprecations. The incident had been interpreted as a warning that other
things were headed for Mistassini—a warning that seemed now to be
justified as the monster of technological progress threatened to devour
half the village. Five years later the huge area of flattened and twisted
trees could still be seen, the place where the two old men stopped the
monster.

PHILIP PETAWABANO

Philip Petawabano tells what he knows:

"I am an old man. I caught a last glimpse of what it was like before
the coming of the white man. My father raised me, and when I was still
young, seventeen, he died and I was on my own. There weren't that many
animals to survive on at that time, but after a time it got better and I
learned what I would need in order to survive from the land. Long ago
in Mistassini you couldn't even really get anything from the white men.
For the whole year in the bush a man would probably take only ten
pounds of flour, so that when he arrived in the bush from the trading
post, a man had to start hunting right away to catch some food to eat. I
would put out my gill net in the lake when it was still high noon or before
morning, and in the evening check the net before the sun set. That would
be the first thing we would catch to eat, a few fish. We would have half
a cup of shells for our guns from the Bay, or perhaps some powder to
make our own shells with. And if we didn't catch fish, since at that time
all the game was scarce, we would go off with our guns for partridges or
rabbits, and bring them back so that our children could eat.

"At that time fish eggs were a delicacy. I would hang them by the fire
to thaw out, take off the hard little heads and crumple them up and then
cook them. They were just like pancakes. They were very good. This was
one of my daily diets. This is how I lived. Life was good and hard. I don't

think too many people are still living here now who saw those times when life was very hard.

"Here is what I think about the bush. The wilderness is just like a store where you can get all of what you need. Everything I needed to survive on came from the land. Of course, I could never part with my land. If somebody lost his land, it would be just like shooting him. All the animals on this land, the moose, the deer, the beaver, the lynx, the fish, all these animals have gone down in the past few years. I hope this doesn't go too far, for this is the only way we know to survive, this store that is put before us to feed our children from. All this that I've been talking about will be gone soon when it's flooded. When I first heard about this James Bay proposal, I wondered about it for a very long time. It almost came to the point where I could not really relate to what the other society has done. I don't know what will happen to our land. I couldn't really tell you."

A DRINKING PROBLEM

The old man had asked if he might talk to the journalists. And while listening to him talk (interpreted by Philip Awashish), we did a lot of drinking. A number of people went to sleep on the floor, and it was early in the morning before we began to run out of steam. The two chiefs, Smally and Billy, came into the house for a few minutes during the evening, but left abruptly. And we were just settling down to go to sleep when Philip and I were brusquely summoned to Smally's house nearby, to be upbraided by the two chiefs for having purposely got the old Indians drunk so that we might interview them. It was one of those tense occasions that white men experience from time to time in Indian communities, when they find that, without having realized it, they have offended their hosts. At such times a white man is liable to discover that a deep bitterness and defensiveness underlie the placid face that many Indians present to the world. The edge was somewhat taken off this conflict by the fact that the chiefs themselves were sleepy, quickly accepted our explanations and became smiling and friendly. "I think the critical thing," said La Rusic, when we told him about the incident the next day, "was the surfacing of that tension, that rivalry, between the threatened chief and the young man back from college. I think that's the critical thing there."

IKON MINE

The next day I talked to a young man, Philip Schecapio, who told me what had happened to his family's hunting territory when the white men arrived to sink the Ikon mine, not far from Mistassini. They did their work without permission and told the Indians to go away. Where there had been beautiful rapids, where the Indians had caught pike, whitefish and walleye, the white men had turned the rapids around to send the water over where it was needed to produce copper. The otter used to play there. The people would sometimes catch 100 partridges at a time under their large nets. Now they could catch nothing and if they went near they were told to go away by the people running the mine. "The mine does not employ Indians. I asked for a job, but even though I told them it was my hunting ground, they would not give me one." After working three years as a truck driver hauling logs, he was now taking a training course as a carpenter. Would he not benefit from the work that would be generated by the James Bay project? In the background his young wife, hearing the question, just laughed. They had seen it all from close up. They had been simply kicked aside.

JIMMY MIANSCUM AGAIN

Jimmy Mianscum goes into more detail about the effects of the white man on his territory around Chibougamau:

"Even when we do not touch our land for a year so that the animals can come back, when we return the following year it looks as if it had already been hunted—because of the coming civilization. The animal hears the noises of civilization. And he also dies from eating leftovers, left by the white man who goes into the bush. He throws away old tin cans and other stuff, and if the animals get hold of them, they probably lick the inside of the cans, which are dirty. The white people throw the cans into the river, which has the same effect as throwing away gas or oil, the same way of killing the fish, the animals, the beaver, who have never tasted anything like these sorts of things. This will probably happen to us, too. Before the white man came, I had plenty of beaver, plenty for another family, too, and it always made me feel good to know that people could survive from my land.

"The fish are poisoned by the mines.

"The sewage from the town flows right into the lakes, and in the spring you can see the debris floating in the lakes and rivers. The water is a murky color, you cannot drink it any more, and the animals die from that, too.

"There used to be many ducklings born in the spring on those rivers and lakes, but now hardly any ducks have ducklings there.

"It's been over five years now since I've seen the last loon in that area. He is to be found now only on the most remote lake of my territory, farthest from the town.

"You cannot eat the fish any more.

"On your way to the most remote lakes, you cannot stop off any more to camp on the shores of the rivers, or have a cup of coffee or tea, unless you happen to find a clean stream coming in from the ground.

"We are right in between two roads that have been built across our territory, so wherever we try to go the white man's presence is always there. If we set our traps out for a day or two, when we return to check them they are gone. We used to leave our belongings out in the open. But wherever the white man puts his footprints, you are bound to lose your property. So I have to bring my things back to Mistassini, where I can see them and make sure they are all there."

PULLED TWO WAYS

Charlie Bosum, a young man in his twenties, who works with the white man in the mines and also hunts and traps on his father's land, speaks of the confusions and uncertainties of someone caught between the two worlds. His father's trapline is not far from Chibougamau, and for six years he has tried to get out onto the land on every possible occasion when he is not working in the mine. Minerals had been found on his father's land, and a road was being built to take out lumber. "When I first started hunting, there was an abundant supply of fur-bearing animals and fish and fowl, but you can feel the white man is coming closer and closer. Even if there are hardly any animals left on my territory, I would still feel much better to be there than where I am right now. I think everybody who loves and respects the land would feel the same way."

He admits to a feeling of great confusion about his two houses, one in Mistassini and one in Chibougamau. For neither of them is really home to him. "I do not have that true feeling about them. Home is when I'm in the bush.

"It made me wonder and wonder a great deal when I heard they were going to flood and destroy the land. . . . It seems like I'm somebody cornered, an animal so desperate to get back to my land, because I have that feeling that I might lose it any time. I guess I will feel desperate in the future, when I do actually see the land going underwater, which would be the end of it. . . . The white man, you can see, is already here. But still I have that love and respect for my land."

1972: RUPERT HOUSE

Everything was laid on for us when we landed in Rupert House. We were met at the plane by a flotilla of Skidoos which whisked us over to the settlement, one of the oldest in Canada, there in that same spot at the mouth of the Rupert River ever since it was founded by the company in 1670 on their very first voyage. The village still had a rudimentary look, as if most of the buildings had been put up in the last ten years. There were fewer houses and many more tents than in Mistassini. And by this time of year more hunters had already returned from their traplines and the village was full of beaver skins put up on tall poles to dry. There was the usual large institutional network, the compound dominated by the company, very handsome here, with white storehouses down by the shore, and a new school and a big sign announcing the impending construction of a health center.

The groups of Cree people centered on each small village on the east coast of James Bay differ in many ways. The coastal people speak a different dialect from those inland. When Jimmy Mianscum of Mistassini was giving evidence in the case, he complained that he could not understand the Cree spoken by the interpreter, Ted Moses, who came from the coastal village of Eastmain, and Philip Awashish was hurriedly sworn in as an interpreter for the afternoon so that the evidence could continue. The people also follow a different hunting cycle from those farther inland, for on the coast the spring migrations of ducks and geese have a primordial importance. The inland people go into the bush earlier and return later, but on the coast, as Billy Diamond told the court, they take part in the fall and spring goose hunts and for that purpose cut down the time they stay in the bush trapping beaver.

Rupert Bay is at the foot of James Bay (which is at the foot of Hudson Bay). Rupert Bay is formed by the three rivers, the Nottaway, Broadback and Rupert, which feed into the bottom of the bay within fifteen miles of each other. The three rivers drain a crescent of 50,000 square miles of land (an area half the size of Britain) to the southeast.

They are strange rivers, sluggish and slow-flowing in the late summer and through the winter, then suddenly released into a wild, tempestuous flood as soon as the snow starts to melt in May. This change is so great that within a week or two these rivers can swell to fifteen times their winter size. All life along the James Bay coast is adjusted to this annual flood, including that of the plants on which the animals depend, and of the animals on which the people depend. It is one of the rare remaining examples of a food chain which reaches locally uninterrupted up from the lowliest micro-organism in the river to man himself.

Over the centuries the rivers have created in Rupert Bay a swampy marshland of *Zostera marina*, or eelgrass, which the geese and ducks like to eat on their annual migrations from and to the Arctic in fall and spring. Rupert Bay is a stopping-off point on the Atlantic and Mississippi flyways for the migratory birds, and Indian life has always been adjusted to this pattern. Now the government had announced its intention to disrupt completely the flow of these rivers.

All of the people in Rupert House knew that no good could come of such a scheme. As in Mistassini, these experts in the biological processes of the land whose behavior they understand so well were opposed to the plan, and they had good reason for their opposition. But at this time they were having difficulty translating this firm conviction into political or legal action.

A WARM FAMILY LIFE

Billy Diamond had not been home for many weeks, and was happy to disappear into the tiny cottage where he lived with his wife and baby, his parents, brother, sister and their families. He had arranged for the other journalists to stay with a single teacher who lived alone in a good-sized house on a hill overlooking the village and the bay. Philip and Smally went down the hill to stay with an Indian family. And since there was not enough room for all of us with the teacher, I trudged along to the other end of the village, where Billy put up a stretcher for me in the band office.

I went for a walk and Billy came out and invited me into the tiny house to meet his family. Inside, the atmosphere was like something out of the boyhood of one's father, when so many people lived a rural life dominated by family and with strong interaction—a life that preceded the development of the isolated, fearful, suspicious urban nuclear family remote from its neighbors. The Diamonds had controlled Indian life in Rupert House for many years. Billy's father, Malcolm, a squat and smiling figure exuding that air of physical and emotional strength that seemed common to many of the older Cree, had been chief for years and it was in the nature of things that Billy should have succeeded him as chief almost as soon as he returned to the village from high school. His mother, Hilda, was a strong woman with a full, handsome face under a circle of graying hair and the heavy figure of the Indian woman, which so often belies the hard life, the solid physical work they have always done uncomplainingly. She was the sort of woman, kindly and perhaps a little stern, that one thinks of when one hears the word "grandmother."

Billy's sister Annie had just returned from the trapline. His younger brother Albert was not long back from having been thrown out, he said, of Trent University in southern Ontario for displeasing the professors who ran the native-studies institute there. They were all jammed into the little house with spouses and children. Every room was overcrowded. One could not imagine where they would all sleep. But that seemed the last thing on anyone's mind. There was great hilarity, a keen sense of pleasure among them all as the old man produced a magnificent pair of snowshoes he had made for Billy during the young chief's absence. While Mrs. Diamond singed the hairs off the nostrils of a moose, preparing for a family feast to welcome the young man home, the old man clowned as we snapped pictures of the snowshoes, peering and grimacing comically through the webbing at the camera.

BILLY STIRS THEM UP

PHILIP AWASHISH [in evidence]: In February the James Bay Development Corporation released their environmental report, the federal-provincial task force report on the impact on the environment by the James Bay project. By the time Chief Billy Diamond and myself . . .

O'REILLY: What did it say, that report?

GEORGES EMERY [lawyer for the James Bay Development Corporation]: I object to that, My Lord.

MALOUF: Objection maintained.

AWASHISH: By the time Chief Billy Diamond and I went up north to begin
    informing the people about the James Bay project, the report itself
    had been distributed among the Indian bands. The Cree people
    rejected the environmental report on grounds that they were not
    able to understand the report, although it was written in Cree. The
    James Bay coast people burned the report in Rupert House because
    they were not able to understand the language, the Cree written. The
    Waswanipi band mailed the reports back to the James Bay Develop-
    ment Corporation politely explaining to the Corporation that they
    could not understand this.

Billy Diamond turned out to be a most effective orator, in Cree, at
the band meeting that night. Though I could not understand it, I could
tell that he was injecting such an emotional tone into the recital of the
same old facts that soon everyone was caught up in the terror and injustice of it all. Many more hunters were in the village than had been present
at the Mistassini meeting, and they reacted strongly. "Throughout the
country," said Billy, "the Indian people have been the social casualties
of development projects. I have seen it out in the west, towns booming,
but the Indians poor, gone in prostitution and booze."

Old Malcolm then told how in Ontario, to his own knowledge, Indians had been ignored by the developers of hydro schemes, and since he
could remember every generation from his own great-great-grandfather
down to his grandson, he was concerned about what would happen to the
land.

"The people are concerned about how the fish will get up to Lake
Evans and Lake Nemaska if the Rupert is to be dammed," said Billy. "I
told the people in the corporation that, and they said they would make
ladders for the fish to climb up."

That caused a stir of amazement, that anyone could entertain such
an idea. Then a man said he had been working with Hydro-Quebec
but he knew the work would not last, and Johnny Whiskychan, a well-
known Pentecostal, warned against the attractions of money. "What is
money?" he asked. "I have seen people get money who do not know how
to spend it. We should not consider money as being more important than
the main thing, which is what we now obtain from the land." Philip said
that in Val d'Or at a meeting the chiefs of James Bay had authorized an
injunction proceeding to try to stop the project. "If the association is
going to use it as political leverage, that is not going to stop the project,
and that is what our people are concerned about."

A young man, Lawrence Jimiken, his shy, sensitive face half hidden by glasses and a shock of long black hair, said in English: "I have had the misfortune of seeing both the old and new worlds. I spent four years down south, and I can tell you, those four years were pure misery. Now I have returned and I have a trapline of my own. And I can tell you I do not want anyone destroying that trapline."

Billy Diamond handed out to the meeting the thin, orange-colored Cree version of the government's ecological report, which had, in its first English draft, contained 300 pages, reduced to fifty-five pages in its second draft, and in this Cree version, including the English translation, contained only seventeen pages. He held up the English version and compared the thickness of the two reports. "What have they left out of the Cree version?" he cried. His sister Annie took a copy, and as Billy read the English translation provided, Annie tried to read the Cree version. She stumbled a lot, finding the reading difficult. "She doesn't understand it," cried Billy. "Does anybody understand it?" The people in the hall, trying to read along with Annie, muttered and cried, "No!" Billy said that in the English version on page forty-five, the government experts said that the Indians were no longer economically dependent on fishing and hunting and had become strongly dependent on the white man's society. That conclusion had been omitted from the Cree version of the report.

He took his copy and waved it around his head. "What shall we do with this report? It is translated into Cree and we cannot even understand it." Some people started to throw their copies toward the front. "We may as well burn it," shouted Billy, and a man in the crowd took up the idea, shouting, "It should go into the fire!" Suddenly the air was thick with flying orange booklets, like a migration of geese in the fall. One old man who'd been sitting in the front row, drunk and half asleep, suddenly woke up, picked up copies as they fell at his feet and pounded them at members of the official party at the head table. The people were yelling, and the younger ones grabbed up armfuls of the booklets and started stuffing them into the stove at the side of the hall. "We burn them, we burn the bloody things!" they shouted, and scooped them up and started to carry them outside.

It was probably the first book-burning in Canada since some German tomes were put to the flames at the beginning of the Second World War. I raced into the band office behind the hall to get a flashbulb from one of the other journalists, who was asleep on my stretcher, but the flash wouldn't work. Outside, I used the light of the merrily burning pile of books to take some pictures by. In a few minutes the paper had burned away. Only a few people stood around kicking the curling pile of burnt paper on the snow. The meeting was over: the Cree people had given some sort of answer to the James Bay Corporation. But it was typical of them that they had not burned the report in anger. They had done it in fun, and instead of filing back into the hall and passing some stiff, angry resolutions, they drifted back in to take part in the next order of the evening's business, the weekly film show. I went back with them and sat down. I was excited by what had happened, but there was no sense of drama or excitement left among the Indians. They waited quietly until the first images flickered onto the screen. The movie was called *Nashville Revels*. It had something to do with the music business in Nashville, a country singer, a girl and a rival. The film was, to be kind to it, complete drivel. I would like to be able to report that the angered crowd, stirred to great passion by their intense desire to defend their culture, rioted and burned down the cinema. But that was not in the nature of the Crees. All around me they were watching with total absorption.

Two experts on geese

John Blackned, seventy-eight years of age, sitting on his bed in a dark, cluttered little house, told me how his father's death had left him at the age of sixteen to look after his mother, sister and three younger brothers. "When a young man's father dies and he is alone, he turns to the other trappers. He looks to the other men, and he follows them. That's what I did." He said the rules of respect for the animals that the hunters had explained to me inland also applied to the birds that were hunted along the coast. There was no one in Rupert House, young or old, who did not have respect for the birds. I asked him a lot of questions about the cycle of the birds' visits, the grasses they ate, the places they landed, the sort of land they needed. He explained how the geese would behave when the water was deep and when it was shallow; the times when the Indians would kill all they could shoot, and the years they would restrain themselves because they could tell that the numbers were low. He told me about the rituals for eating the geese, the parts that must be eaten first, the ornamental beadwork done on the skulls of the dead geese, the sense of outrage the Cree felt at the disrespect shown by white hunters, who would eat what they wanted and throw the rest away for the dogs. He knew what would happen when the machines came in to do their work. Already the planes had disturbed the birds along their flyways on the coastal strip. If the white men cared about the animals and the Indians, why could they not take their planes along different routes, farther away from the coast, so that they did not disturb the birds? "My goose camp is on the coast. This year the planes came buzzing and flying overhead, and the geese did not feed along the shore, they went into the swamps. The geese can be protected, if only the white man will listen to the Indian and take suggestions we make to them. But if the white man does not really care about the geese or the Indian, he cannot do it."

HAROLD HANSON [a white man who does care about the geese, a wildlife biologist from Illinois, testifies]: First of all, across the continent, I can account for over 30 varieties of geese, and the further north you go, the smaller goose you get, because the time they require to go through the nesting cycle is less than the bigger geese. And the smallest geese go further south, while the Giant Canada goose, which I rediscovered a few years ago and wrote a book on, moves the least. These geese winter primarily in the States, and the bigger honkers that we're talking about in northern Ontario and Hudson Bay, Ungava, they start northwards in early March, and they spend about a month in what we call staging areas. One of these is in southwestern Saskatchewan (not for these geese, but as an example). The snow geese use the area around Winnipeg, our geese use the area around Oregon-Wisconsin, and the geese in question here, I was interested to learn this fall, their great spring staging area where they put on this tremendous layer of fat is western New York in the Montezuma and Iroquois refuge area. They stay in that area for a month, and then, in a matter of just, if the weather is good, in a day or two, they're up on the breeding grounds. In some cases, where you'll get recurrent snowstorms, they may be forced back two or three times before they finally arrive. Now, when they do arrive, they've been able to do practically no feeding along the way because much of the country is covered in varying degrees with snow. They're still extremely fat, and when these birds are shot by the Indians, and I've examined a very large number of them, their eggs are extremely well-developed. In some cases you'll get eggs in the oviduct with a soft shell. Time is critical at this period and in the late spring in northern Minnesota they found eggs just laid on the ice. They must get there and get to their nest site in a very short space of time, otherwise they'll lose those eggs. Now these geese, the veteran nesters, return right to the same spot in this vast look-alike muskeg apparently year after year. If you look at a nest scrape, of say, in the spring of the previous year, and you dig down, you'll find as many as two, three layers of egg shells below this one scrape, apparently made by the same female goose. We know they do this in more confined areas in the south, where they've been able to be studied intensively, they return to the identical nesting site, in many cases, and when they can.

Now, I suspect that if the birds return and find the area under water, many of them would have to be relocated sometimes under conflict pressure from other pairs, and for many of them it may be too late to get relocated and nest that year, because of this critical period for egg-laying. If they lose their eggs in the north they don't recycle for a second clutch. Now in the south, in extended seasons, the Giant Canada goose can do this. But as I've heard now from Churchill lake [the reservoir formed behind the Churchill Falls development in Labrador] if you've had this winter drawdown and you have these exposed islands, the geese will nest on these and then the waters rise rapidly in the spring and of course they're all flooded out. Now, that nesting for the year is through for those birds. It's only been a year or two since that reservoir was filled. I suspect that that situation won't last at Churchill or any of these other reservoirs, because once the water has been upon the land long enough, it will kill all vegetation, and the goose or any other bird is not going to nest on a mud flat or a mucky dead situation. They like to nest under a little spruce tree, or either they settle beside a rock. With that gone, of course, they won't use that area at all.

JACQUES BEAUDOUIN: Can you tell us what kind of food is needed for the geese?

HANSON: It's a question of what the goslings have to have. Now you do have moulting adults that have failed to breed and aren't old enough to nest, and their food requisite is very small at that time. . . . But the goslings must have these tender shoots of sedges and grasses. I've done an immense amount of flying through the continent, and in Ungava or northern Manitoba, you can be flying over country that is not fit for geese, you don't see any, then suddenly you come on . . . these rich, vivid green marsh grass areas adjacent to the rivers, and you say, there's ideal goose habitat, and invariably, it's just a matter of minutes and you see broods and adults in the general area. It's quite uncanny.

THE ANGEL OF HUDSON BAY

It was in Rupert House that Maud Watt, later celebrated in a book called *The Angel of Hudson Bay,* had made her name. She was the wife of the Hudson's Bay Company manager, James Watt, who had been in several of the posts of northern Quebec during the famine years of the late 1920s and the early 1930s. Together they had made an epic journey across the entire Ungava Peninsula, from Fort Chimo in the north to Sept Isles on the St. Lawrence—the first time it had been crossed north to south by white people (it must be said, however, that the journey was "epic" only in white terms, for the Indians who accompanied them knew the country well and were used to moving across it, though they had never bothered before to cross from top to bottom).

More importantly, it was James Watt who halted the decline in the almost extinct beaver when he forbade the Indians of Rupert House to kill any more beaver and offered to pay them instead for any news they might bring about live beaver that they had located. Watt and his wife, a woman of great determination and strength of character, then pressured the Quebec government to set up a beaver preserve within which the killing of beaver would be restricted to Indian hunters, who themselves would be given quotas. This preserve was set up in 1932, the first of a series which eventually covered almost the whole province; and until this day the Indians are given quotas for their beaver kill each year. Since the quota is normally based on the number of occupied beaver lodges observed by the hunters in their territory the previous year, it can be said that they arrive at their quotas themselves. (In the huge Mistassini region, there will normally be a total of between 7,000 and 8,000 beaver lodges: only one beaver per lodge may be killed, so the band's quota is normally set at around that figure.)

After the death of her husband, the strong-willed Mrs. Watt quarreled with the Hudson's Bay Company, which would not even give her the time of day when she returned to Rupert House independently and set up a small store. She gathered money in the village and built the community hall in memory of her husband. Billy Diamond one day took us into the hall, pointed out with contempt the inscription honoring the name of James Watt and said that it would come down. But older people remember him more generously. In the many years in which they remember white people mixing into the affairs of their hunting lands, he probably did more than any other white man to help the Indians maintain their way of life.

Daniel Wapachee, aged sixty-four, remembers: "Mr. James Watt was the one who brought back the beaver, to make sure it did not go extinct. After that time he opened up the territory for trappers to hunt the beaver, and then times were very good again. What Mr. Watt had done was very good. But the government did not come in to help us until about six years later. Now it has been twenty years since the white man has been on my trapping territory and he has destroyed quite a bit. They've been drilling for minerals on my land for ten years, and a road has now been built through it, but still I don't receive anything from these people who come and destroy my way of life. When we were given these territories by the government we were told to make sure that no intruders came, and if they did come uninvited, we were told to report it and we would be given compensation for what the intruders were getting from our land. But since these people arrived, we haven't received anything at all in return. So I guess it is understandable that I have a little bit of dislike for the white man."

WHAT KILLED OFF THE BEAVER?

Given this more recent history, the memory of James Watt and what he did has faded. But the events in which he played so important a role remain among the most interesting in the twentieth-century history of James Bay. Among the people who have investigated this near-extinction of the beaver in James Bay, there is no agreement as to its cause. Some hunters such as Isaiah Awashish, who lived through the lean times, explain that the decline came from the natural cause that every species of animal took a downturn in its regular cycle at one time. Other hunters whom I have asked have no explanation. One hunter in Fort George, Samuel Pash, later told me that the decline came because the land was tired and needed forest fires to bring on the regenerative growth that would be good for the animals. When the forest fires arrived, he said, the land and the beaver recovered. Still others said the decline was caused by a disease which spread through the beaver and other animals, wiping them out. An exchange of letters written in the 1930s, found in 1974 by a research team working for the James Bay Cree, indicated strongly that white trappers who had moved into the area used poisons that killed off the beaver.

Anthropologist Harvey Feit, who has tried to identify the most likely story, is puzzled by three things. One is that for the first and only time the ethno-ecological system of the Cree, in which their religious beliefs and their general world view enjoin them to maintain themselves in a suitable balance with the animals, broke down in the 1930s for reasons which they themselves do not seem able to explain satisfactorily. Secondly, if disease had been the cause, Feit feels sure that the finding of thousands of dead animals throughout the area would have been handed on in the oral tradition of the hunters and would have occurred in their explanations, but no such reports have ever been made. Thirdly, though the explanation given by J. W. Anderson in his book seems the most likely one—that the area was opened to intensive white trapping with the construction of the railroad to Moosonee on the Ontario side of James Bay —the hunters themselves never put this forward as an explanation. It remains, however, the explanation which Feit favors.

Anderson who, after all, lived in the area throughout these hard years, has no doubts on the matter: " . . . deleterious for the long-term welfare of the Indians was the advent of the white trappers, some of them erstwhile construction workers, while others were professionals from other areas. . . . The territory was excellent fur country, little disturbed by the natives, and therefore a bonanza for the newly-arrived white trappers. They reaped a rich harvest, many of them building up sufficient capital to start fur trading ventures for themselves; but by the time the stock market crash of 1929 came around they had practically cleaned out the country and were pulling out. The fur resources being exhausted, the Indian could have his country back again!"

Anderson goes into detailed reasons for the behavior of the white trapper who depleted the fur resources: for the whites trapping was a money-making business, whereas for the Indian it was a way of life. The white man was never satisfied, would reach for $4,000 if he had made $2,000 the previous year. The white trapper required $1,000 to outfit him for the winter compared with the $300 or $400 with which the Indian was content. The white would need furs to cover this $1,000, plus another $1,000 or even $2,000 to keep him during the summer. "The land and its fur resources could not stand this drain, so this fine fur country was exhausted in a comparatively few years."

NEMASKA ABANDONED

Lawrence Jimiken said that Nemaska, where he had been born, was a quiet, idyllic spot on a lake 100 miles inland on the Rupert river system. When he had been sent away from Nemaska to go to school in the south he had cried a lot from homesickness. But now he had his own trapline, thirty miles by ten miles, near Nemaska.

"I was still in school when I heard that the company post in Nemaska was to be closed. Someone spread a rumor that there was no food, and the chief and two councillors panicked and were the first to leave in July 1970, just after they first started to talk about the James Bay project. I came back from school and found some people still living there, but then everyone left. It was quite unnecessary. We could have organized a co-operative food store. But the chief abandoned his people." Like the

Waswanipi people before them, the Nemaska people were now scattered, some in Rupert House and about half in Mistassini, far from their hunting grounds, living on welfare, not quite accepted by the people in the reserve, living in improvised shacks and unable to find a way of making a living.

Former Nemaska chief Bertie Wapachee, a fat old man, round-faced, owl-like behind his spectacles, hopelessly confused by the years of change, now living on welfare in Mistassini, does not feel that he has failed in leadership. The Hudson's Bay Company says that the Nemaska people were given two years' warning that the store was to be closed, that the decision was taken in 1967, long before the James Bay project was thought about, that the Indians were always fully informed. But the story most Indians believe is that they returned one year from their camps to find the stocks of food low and they had to leave so that they could get something to eat. Bertie Wapachee, in short, seems not to have been an effective chief. He remembers the coming of the first white men (leaving aside the company managers) forty years before, the geologists and surveyors who arrived to put up claim posts and test the water depths, and who never explained satisfactorily what they had in mind. The Indians were surviving well before the white man arrived with the many things he had invented for surviving on, far more things than any Indian really needed. Only when Bertie became chief did they receive help from the government. "When an Indian sees a white man in his territory he must find out what the white person is doing and what he's up to and keep an eye on him, to make sure nothing happens like what happened in the past," Bertie concludes.

A DISASTROUS IDEA

Three months after we arrived in Rupert House, the focus of the James Bay scheme was shifted off the Rupert River onto the larger river farther north, the La Grande. But at this time people still thought that their river was going to disappear, that the yearly rhythm of their lives would be irrevocably shattered, that the animals and birds would be destroyed, and they were deeply uneasy. For many outsiders, too, the fact that the Quebec government was willing to disrupt the delicate natural balance of Rupert Bay without bothering to devote even a moment of study to how it functioned came as a depressing affirmation of the insensitivity with which modern governments can act, in spite of all the scientific information that is supposed to be available to them.

Under the project, two of the three rivers that form Rupert Bay (the Nottaway and the Broadback) would be diverted into the third (the Rupert). The bay is used by millions of migrating waterfowl every spring and fall. The Indians were not allowed to bring evidence in court to indicate the folly of this proposal, because it belongs to Phase Two of the James Bay scheme, for construction, presumably, in the 1980s. But the dismay of wildlife biologists was briefly indicated:

HAROLD HANSON [Illinois wildlife biologist]: Now, I can't think of a project that would cause more mischief in terms of the welfare and survival of geese than if you were to go after the Broadback and Harricana and these southern rivers. This would be a disaster. Everybody believes it, and certainly I do. A little off the subject at the moment, but part of the future plans. . . .

He was stopped from developing the argument by an objection from the corporation lawyers.

FORT GEORGE BRIEFLY

We flew north to Fort George by a scheduled flight of Austin Airways, whose little plane landed on the river beside Rupert House the next day and stopped just long enough for us to clamber aboard. The James Bay coastline beneath us looked like a cold black-and-white landscape painting, stripped in the modern fashion of all emotion and drama. It appeared to be utterly empty except for the tiny villages of Paint Hills and Eastmain, fifty miles apart along the coast, and the occasional track of a snowmobile or a pair of snowshoes. The Fort George airport was manned by white men employed by the southern airlines, and there was none of that intimate, family feeling that we experienced at Rupert House. The James Bay Development Corporation had already begun preliminary work in Fort George, which sits on an island in the mouth of the La Grande River (which is known as the Fort George River in English, Chissibi in Cree).

We stayed only twenty-four hours, but the divisions and tensions of the place were evident even to a casual visitor. Fort George is the only Indian village of James Bay with more than a token white establishment, some 200 out of the total population of 1,600. The village had, over the years, attained a large institutional network which provided a considerable amount of work for the Indians, who had to a much greater extent than in the other villages abandoned hunting as a full-time occupation, though they hunted regularly in their spare time and were still very dependent on the food they could kill in the countryside.

Fort George was within the boundaries of Nouveau Quebec, that huge principality formed by the Ungava Peninsula, which was administered by a special department of the Quebec government set up in the early 1960s, the Directorate-Générale de Nouveau Quebec. The foolish struggle for power that had been taking place in this principality between the Canadian and Quebec governments throughout the 1960s, with the Indians and Eskimos being used as a football to be kicked around among the territorial ambitions of the two governments, was here evident on the ground. Both the provincial and federal governments had schools in the village. The federal government maintained a large student residence and a health center, while the province administered the big new hospital, an adult-education center, the water supply, garbage collection and some municipal services.

That division between the two European cultures which made itself evident in the courtroom was also evident here. The division was not only between two levels of government, but between English- and French-language administrations, Catholic and Anglican religions. The Catholics had a huge establishment, occupying one end of the village: here, as in other villages, they were short on adherents, since the people were 100 percent Anglican, but the priests had poured their energies into commercial enterprise and now had a large, successful private school and residence, a laundry, garage, machine shop, bakery, poolroom, cinema and a small plant for the making of concrete blocks. Strangely enough, the Indian parents preferred to send their children to the Catholic private school rather than to the provincial school, for throughout Nouveau Quebec the native people were resisting the efforts of the Quebec government to assert its right—surely reasonable enough—to establish an administration in the northern part of its province, just as other provincial governments had done. The roots of this resistance were historical. The Quebec government had traditionally shown little interest in the Indians and Eskimos of northern Quebec; in fact in 1939 they appealed to the Supreme Court of Canada for a declaration that Eskimos were Indians within the meaning of the British North America Act and that therefore the federal government was responsible for the administration of Eskimos in northern Quebec. Only with the rise of a more aggressive nationalism in the early 1960s did the Quebec government decide to oust the federal administration from its north lands. An agreement to this effect was made by the two governments in 1964, but its implementation was delayed; and when in 1969 officials of the two governments jointly toured the region to explain the new deal to the Eskimos, they met powerful resistance to the very idea. The Eskimos and Indians had always depended on the federal government, the only outside level of power which had ever concerned itself with their welfare, and they were apprehensive that this dependency link should be broken. (Termination psychosis, the phrase coined in the U.S by Alvin M. Josephy, Jr., now manifests itself throughout North America every time any move is made to end or even to modify the trust relationship of the federal authorities with the native people.) In Fort George this burgeoning institutional structure, divided and competitive, had brought in 200 white people as administrators, and the hunters had become janitors, repairmen, student supervisors and the like.

We were installed in Tom Webb's lodge, the only hotel in any of the Indian communities of James Bay, a concrete rectangular block in the middle of the village, run by a friendly white man who had arrived in Fort George on a construction job ten years before, married an Indian girl and decided to stay. He didn't have much space, for the preliminary work on the project was bringing many people through the town, but he put three of us up in the lounge downstairs and charged us $20 a day each. We went into the long dining room, which reflected the lodge's status as a way-camp for transient workers. Tom went around with a huge tray of steaks, piling three or four on each plate before the hearty trenchermen who filed in after their day's work. The men ate with solid determination, got up and left.

Josie Sam, the sharp little man who was the Shell agent in Fort George as well as the regional chief for the Indians of Quebec Association, had accompanied us north from Rupert House, and he now organized some interviews for us. Instead of taking us around to homes in the village, as we had wanted, he ushered people into our hotel room, one after the other, to tell us what they thought of the James Bay project. Young and old, they opposed it. There was a curious ambivalence about the younger men, those working around the town in menial jobs in institutions, yet protesting their love of the land and their determination to return to it eventually. Most of them had excuses of some kind for not being out on the land—their mother-in-law's illness, family troubles and so on—and it was not being unduly skeptical to feel, as one of the journalists with me said, that "if the time comes, I doubt if we will find them flocking back to the trapline rather than living on welfare."

Some of the younger people had lived a long time in the south and had returned in search of that very small-town peace, that proximity to the land and the wilderness, that was now about to be destroyed. All of them were still hunting regularly, after work and on weekends. And in their alarm over the prospects, they were expressing a real perception, common to native people all over the Canadian north, that only the land is lasting—that, whatever happens, the land will be there and the people will be able to return to it. "The people here cannot live as the white man lives," said one young man who had worked for years in Ottawa laying tiles before returning to the village. "They like to know that, whatever happens, they can go back to the land. What would happen if this whole area was filled with cities? Where would *we* go then?"

A REMARKABLE DOOR

If the flood does back up, said one young man, the moose and caribou will go back to where they came from. And where was that? Fifty or sixty years ago, he said, there had been many caribou and moose, but they had then gone away. Ten years ago, 500 miles inland, around a certain spot by the side of a mountain, many caribou had gathered. To judge by their footprints, they appeared to have come from inside the mountain: it was as if a big door had opened and they had come through the door. That is where they had disappeared to. Now they were returning. Every year people noticed that the caribou tracks were moving farther west. Last year they had been seen along the bay, ten miles north of the village. But if the flood came, the caribou would go back inside the door.

THEY WILL NOT GIVE US A CENT

Two older men shuffled into the room, big, tall men with long, strong faces, dressed in waterproof jackets that bore evidence of long wear, their feet clad in soft moosehide moccasins. George Pachano had just come off his trapline, which lay athwart the river and would certainly be flooded when the dams were built. "I have never seen a man-made lake," he said, "but when a beaver dams a river or a creek, upstream everything is spoiled, the earth is spoiled, and the trees will never grow again. A man from Hydro-Quebec has told me that only the tops of the hills on my trapping ground will not be flooded. They have told us that even if they produce power from this river, they will not give us a cent. They have not asked for permission to invade our land in this way."

The other man, Job Bearskin, did not talk much. He smiled and nodded and said that long ago, before his father even, the old people had already heard about this thing, that someday the white man would come and dam the river. And now today it was happening. "I can tell you I am not interested in this thing at all. I am not interested in the money. We will still survive even if this project is closed down. Right now the only way many people can survive is to go into the bush to do some hunting. They go there to get the clean air. They do not want to make money. It is hard to know what we can do, but at any time I would be glad to hear that this is stopped."

I asked if I could take their picture; when we had finished talking, I went outside with them and quickly got a picture. As George Pachano and Job Bearskin stood erect in the freezing air, Job, the tall one, suddenly smiled at me in a most charming and simple way, his long, handsome face opening as if to try to cross the barrier of language and culture, of history and background, that lay between us. I am not much given to this kind of instinctive reaction, but I was moved by his expression. And months later when I returned to Fort George, his smile stayed with me, and I made an important decision based on my intuition about it.

9

JOB'S GARDEN

A year later Job Bearskin gave evidence before the court. He was called at the stage of counterproof to refute the evidence of a government witness that the Indians of Fort George were making an average of $10,000 a year. He had not, of course, ever laid hands on any such sum, or even on anything near it, and his counterproof convinced the judge that the government witness's figures were not authentic.

By this time I knew Job Bearskin well, for in the intervening summer I had built a one-hour film, called *Job's Garden,* around him. He was a man of fifty-nine who had never in his life missed spending the winter hunting and trapping around the headwaters of the La Grande River. He had a profound but simple perception of the meaning of life which ill-prepared him for his ordeal in court. As he groped to understand the confusing questions put to him, the corporation lawyers constantly interrupted with a barrage of objections, for they did not want the counterproof to cover ground that had already been covered by the Indians' evidence-in-chief. The transcript of his evidence reveals, perhaps more brutally than any other single document, the unbridgeable gap between the two sides in their attitudes to the purpose and meaning of life. The difficulties began with the very first question, when the old man was asked his address.

INTERPRETER [Albert Diamond]: Your Honor, I think the witness did not understand the question.

MALOUF: What did he say?

INTERPRETER: He said, I have come from what I have survived on. You see, there is not one word, one word exactly, meaning the word address.

O'REILLY: No. Where does he live? Where does he stay, partly?

BEARSKIN: I have lived in the bush and I am from the bush.

MALOUF: What part of this province does he live in?

INTERPRETER: I think the witness is confused, Your Honor. He asked, I am not sure what I have to say.

O'REILLY: No.

MALOUF: Just a minute, Maître O'Reilly. [To interpreter] If the witness says something, you will repeat exactly what he says.

INTERPRETER: Yes, sir.

MALOUF: Notwithstanding the fact that you may or may not consider it important. What did the witness say?

INTERPRETER: He said, he's from this land. Then he went on to say, I don't know exactly what they want to know.

O'REILLY: May I ask him where you will go back to, when you go home?

INTERPRETER: He asked the question to be repeated. He said, say again, please.

MALOUF: Are you married?

BEARSKIN: Yes.

MALOUF: Where is your wife at the present time?

BEARSKIN: At Fort George.

MALOUF: Is that where you live?

BEARSKIN: Yes.

O'REILLY: Perhaps the witness can be told that he doesn't have to keep his hand on the Bible the whole time. Mr. Bearskin, when was the last time you were trapping in the bush?

BEARSKIN: This past winter.

O'REILLY: Mr. Bearskin, you know about the James Bay project, do you not?

BEARSKIN: Yes, I know.

O'REILLY: Mr. Bearskin, how much money will it cost you if the James Bay project goes through as planned?

BEARSKIN: I will lose a lot because a lot of what I have will be destroyed.

LE BEL: I object, Your Lordship, to this testimony which is only repeating testimony already given by the Indians and Inuit about the consequences of the project. . . . In counterproof one cannot confirm a proof already made, and . . . one . . . cannot . . . contradict proof already made by the defense.

O'Reilly argued that the new proof was to quantify the losses that would be suffered, something that had not been done previously.

MALOUF: I will permit you to continue but I wish to point out to you that if the evidence you intend to produce is simply confirmatory of plaintiffs' petition which it has been up to now, I will intervene and put an end to the testimony.

O'REILLY: Mr. Bearskin, if the project continues for two or three more years, and I am referring particularly to the roads and to the camps, and to some 5,000 people mentioned [as coming in] by 1974, what will be the effect upon you or the damage to you?

BEARSKIN: I know I will lose a lot. Already I have seen some of the land being destroyed and already I can feel the effect this has on my hunting and trapping.

LE BEL: I object to this reply, Your Lordship, which merely confirms proof already given.

O'REILLY: Your Lordship, on the surface . . .

MALOUF: Under reserve . . .

O'REILLY: How much money would you take, Mr. Bearskin, for what the James Bay project is doing and will do for the next two or three years?

EMERY: I think I will have to object, that question has been put, My Lord.

MALOUF: The objection is maintained.

O'REILLY: Mr. Bearskin, why do you say that the James Bay project will continue to affect you and continue to hurt you for the next three years?

BEARSKIN: I say that because of what they are destroying. The trees that they have destroyed will never be able to grow back again. They have taken the roots right out of the ground and they will never grow again.

LE BEL: That's exactly the testimony of Chief Billy Diamond, Your Lordship.

O'REILLY: Do I respond to that, My Lord?

MALOUF: No, it's not necessary, he didn't make an objection.

O'REILLY [to interpreter]: Is that his answer, is that the witness's answer?

BEARSKIN: I will say, I will talk more, if I am asked to say more.

LE BEL: That's what we want to verify.

O'REILLY: I will ask you more. Mr. Bearskin, do you receive $10,000 each year?

BEARSKIN: I have never had $10,000 in one year. I really don't have, ever have that much money. I don't think I ever had $10,000 in my whole life. . . .

O'REILLY: For the area around Fort George which has been affected, how much in terms of money has that been affected now and will be affected over the next three years?

BEARSKIN: It's the same thing again. The game will go away and what I have survived on will be gone. What we have lived on will move away and there will be nothing there for us.

O'REILLY: But how much money damage will be caused to you? . . .

BEARSKIN: When you talk about the money, it means nothing. There will never be enough money to pay for the damage that has been done. I'd rather think about the land and when I think about the land, I think about the children: what will they have when that land is destroyed? The money means nothing.

LE BEL: I object to the contents of this reply, Your Lordship.

MALOUF: Under reserve.

LE BEL: For the same reasons.

INTERPRETER: The witness just asked me if they're going to ask more questions.

MALOUF: Ask the witness to be patient.

O'REILLY: Mr. Bearskin, how much money would it take for you to buy in the store at Fort George what you say you will lose over the next three years if the project continues? . . .

BEARSKIN: It can never be that there will be enough money to help pay for what I get from trapping. I do not think in terms of money. I think more often of the land because the land is something you will have for a long time. That is why we call our traplines, our land, a garden.

A PERSONAL VENTURE

It was the great effort being made by the Cree people to defend their way of life which unexpectedly took me into James Bay in the summer of 1972 to make a film, something I had never done before. Philip Awashish and Billy Diamond had reported to the Indians of Quebec Association after their tour around the villages that the people wanted to try to halt the project, for they were in fear of it. Consequently the chiefs approved an application for a permanent injunction to halt the project, and this was entered into court in May. The Indian Association had been negotiating with the Quebec government about Indian rights in general for some time, unsuccessfully, and now wished to use the leverage given them by the James Bay issue to increase the scope and tempo of these negotiations. Rather than proceed with the legal attempt to stop the project, as requested by the Cree people, they decided to try to negotiate the issue.

(Philip Awashish quit the association in disgust, went off fishing with his
father for the summer and did not return to the battle until the negotiations broke down some months later.)

At first the Quebec government and the James Bay Development
Corporation refused even to discuss the James Bay project with the Indians: it was "non-negotiable," they said. They were persuaded by the
federal Minister for Indian Affairs that they must discuss the project, but
they did so reluctantly and their concept of negotiation embraced no
more than the most perfunctory discussion of the likely effects of the
project on Indian life. Soon the negotiations were again on the point of
breakdown because the two sides could not agree on what effects the
project would have. The Indians decided to send a team of scientists into
James Bay to try to make an evaluation of likely effects that would mean
something to the white men with whom they were negotiating, and it was
when he was putting this team together that Professor John Spence of
McGill University suggested to me that I might like to accompany the
scientific team and make a film. Working as a freelance journalist, I wasn't
overburdened with work, nor was my friend Jean-Pierre Fournier,
another freelance journalist, so we plunged in and decided to try something we knew next to nothing about.

The Indian Association had decided to spend on the James Bay issue
$250,000 it had received from the federal government for research into
land claims. But they could spare only $8,000 to make a film, about a fifth
of what was needed. It enabled us to buy some film and hire some
equipment and a cameraman, Guy Borremans. When we flew north with
the scientists in the world's oldest DC-3, chartered from St. Felicien
Airways, Borremans was the only man in our crew who had ever made
a film. We still didn't know what the subject of our film was to be, since
we had dispensed with the normal months of planning, had no scenario
and had not done any preliminary research.

The scientists were rather an interesting lot, who, like most wildlife
or biological scientists, had spent much of their lives in remote places
immersing themselves in the minutiae of animal and vegetable life. One
of them would be going up the river to examine the vegetation; one
would be going inland to estimate beaver populations and pin down the
sort of food the beavers were eating; one would fly up and down the coast
counting white whales and seals; and others would be investigating the
food habits of the Indians of James Bay, checking on fish spawning places
and on the migratory habits of geese. There were fifteen scientists. They
would be scattering over 100,000 square miles of wilderness and I soon

realized we would need half a dozen camera crews if we were to keep up with them. In any case, scientists holding up fish samples do not make the most fascinating of subjects; quickly we decided we would concentrate on the Indians, try to get them to tell their story.

MUSKEG COUNTRY

It was a superb cloudless day as we flew north up the James Bay coast. The landscape was a hazy blue and purple, the coastline below us had no definition: over a mile or so the land gradually gave way to water, the firm edges of solid land seeming from our height to disappear reluctantly into a kind of half-submerged nondescript, neither sea nor land. This was the coastline that was in such delicate equilibrium between the sea-water and the fresh-water inflow, that responded every spring to the huge run-off of melted snow with the growth of new grasses and sedges among the swamps. The islands just off the coast which supported seals and polar bears were a part of this equilibrium. The ducks and geese depended on it. The Indians moved to its ancient rhythms.

KENNETH HARE, Director-General of Research, Canadian Department of Environment; Professor of Geography, University of Toronto; former Master of Birkbeck College, University of London; former President, University of British Columbia; one of the world's leading northern climatologists; testifies:

HARE: At the coast James Bay is very marshy with extensive muskegs. These muskegs extend perhaps fifty miles inland, they are not unbroken, there are some islands of higher ground with poor forest. Then, coming east, the ground rises continuously and the forested landscape becomes dominant from fifty miles inland across to Mistassini with very few outcrops of rocks and very many lakes.

O'REILLY: Is it firm ground?

HARE: The further up the slope you go, the firmer it becomes.

O'REILLY: What kind of sediment is there in that muskeg area?

HARE: The muskeg consists of peat, several feet thick, partially flooded with water, with here and there trees, mainly black spruce. They are able to grow, but they are deformed, they are unable to develop properly, but they do survive. . . . If the water level [in the muskeg area] is raised, the trees and shrubs tend to die. If it is lowered, the tendency is for the shrub areas to extend. These muskegs have existed for thousands of years. They are in a state of equilibrium. The animals are used to the water levels which exist in nature. If you change them, you change the habitat of the animals.

AN OLD FRIEND

We had no sooner arrived in Fort George than I met an old friend, Gilbert Herodier, a startlingly handsome young Cree of mixed blood whom I had known during the three or four years that he had spent in Montreal going to a variety of schools and colleges. He was the descendant of a Frenchman who had worked as a trader for Revillon Frères, a company that had established trading posts through northern Quebec half a century before in opposition to the Hudson's Bay Company, but had finally sold out to Hudson's Bay and withdrawn to France. Even in the city Gilbert stood out as a remarkable boy, not only for his good looks, but for his unusual gentleness, his unfailing good humor and his totally erratic behavior. I had first met him when he was working as a counselor in a summer camp attended by my own children: they all worshipped the ground he walked on, and wanted to be with him whether he was doing anything or not. In Montreal he had drifted from one school to another without ever graduating from any of them: he had worked through the National Theatre School, the Indian film crew at the National Film Board and a school of photography, before deciding to go back north to live. It was not unusual for southern institutions to give up on young Indians, but Gilbert always had the air of having given up on them first.

I asked him if he would help us on the film: he said he would. Could he suggest an Indian trapper around whom we could build a film which would tell the story of how the project was being imposed on the Cree? "It did occur to me," I said, remembering my previous visit, "that Job Bearskin would make a good subject."

"I was thinking about him," said Gilbert. "He was the first one who came into my mind. Because, you know, Job Bearskin is a beautiful man. I could go and ask him. I could talk to him about it."

The next day he told us Job would be willing to help us in any way. Gilbert also found a friend, Billy Bearskin, with a canoe which he would put at our disposal, and they suggested that we should spend the next day, Saturday, going up the river in Billy's canoe to the first rapids, where many people went at this time of year to catch whitefish, now making their annual run upriver from the sea.

A TRADITIONAL FISHING PLACE

Billy, a strong, cheerful fellow with an infectious giggle, played the clown around town, but I realized within minutes of going on the river that he had great personal strength and qualities of leadership and command that one would not have suspected.

The river was broad and swift and remarkably impressive: the tall banks were soft and sandy and every few hundred yards had slipped into the river, carrying hundreds of trees with them. The three of us huddled under our waterproof coverings, but the masterful way that Billy handled the canoe and the contentment with which he and Gilbert faced the cold wind added a dimension that had been lacking before: they seemed happy being in touch with these natural forces of water, wind and land, and as I huddled there the theme of the film struck me: the river. It was their river, they knew every inch and rock of it, all its channels, tributaries, bays and moods. They knew where to fish and where to find game. They knew the passages to avoid, they knew where the traps lay hidden under the water. To them, the river was life. But the white man was not content with it. The white man had come to claim their river, and to improve it. The theme seemed so obvious I wondered I had not thought of it before.

We ran upstream for about an hour, the river narrowing slightly as we traveled. And finally we could see far in the distance the white cloud of mist where the rapids boiled down over the rocks. Several times we passed the abandoned skeleton of a teepee, the crossed poles left standing for the next man who might come along. We stopped at one of these: the spruce boughs around the ring of rocks in which the fire had been built were still quite fresh. The teepee had not long ago been used.

As the river narrowed, the waters flowed faster. But we were still about a mile or so short of the rapids when we pulled across to the left bank and ran the canoe up on a gravelly shore. Inside a tent pitched on the beach a woman sat cleaning a huge pile of fish, with a stew made up of the innards of the fish bubbling on the little tin stove before her. We sat around the tent with the Dick and Pachano families while they poured us tea. (The Indians of James Bay drink their tea lukewarm rather than hot, perhaps because as they move around the country in the winter they try to maintain an even body temperature so that they will not become either overheated or chilled.) The men of the families were fishing at the rapids, so after a few minutes' rest we headed upstream, stepping up onto a platform of striated red rocks worn down to a polished smoothness by centuries of the rush and hammering of these wild waters. The rocks went under the flood every spring as the great river gathered in the melted snow from 37,000 square miles: it all came pouring down here, beating these rocks on its bed and shore until now they were one of the most perfect works of nature that one could imagine. About a mile long and 100 yards wide, the rock platform ran right up to where the rapids, even in August when the water was low, boiled through the narrow channel with an unremitting roar. Below the rapids the water raced downstream and then, sucked back by the vacuum at the foot of the rocks, swirled upstream in a great circle before heading off toward the mouth of the river. There were thirty or forty people there, women and children sitting idly on the rocks around little piles of burning twigs over which kettles were warming, as they waited for the men to return from the fishing. Down by the racing waters, the men were clambering over the rocks with their nets: a team of eight or nine of them holding one net, marching along the rocks, the net rising and falling when caught by the wind as the men moved along. They would throw the net into the swirling water at one end of a broad bay between rock outcrops. At the other end one man, standing as far out in the stream as the rocks went, would haul on the rope, pulling the net across the mouth of the bay. The water would bring the fish into the little bay as it swirled back upstream, and they were caught in the net the next time it receded. The Indians had been fishing here for centuries: it was much more to them than just a day's outing by the river.

O'REILLY: Have you ever fished at the first rapids?

STEPHEN TAPIATIC: Yes, I have fished at the first rapids, and this is where I have been always fishing, and this is the place where most of the people of Fort George take their fish in the summer time. All the people I have seen fishing at the first rapids were from Fort George.

O'REILLY: How many people have you seen from Fort George at any one time?

TAPIATIC: I have seen about 1,000 people fish there.

O'REILLY: How many members are there in the band?

TAPIATIC: I cannot tell you.

O'REILLY: When you say 1,000 people, do you mean most of the band members fish at the first rapids?

TAPIATIC: Yes.

O'REILLY: When you were last at the first rapids, did you see any fish being caught?

TAPIATIC: Yes, I did see about 200 fish being hauled in from the net.

O'REILLY: To your knowledge how many fish are there in the first rapids that can be caught during the summer?

TAPIATIC: I cannot answer your question.

Gilbert asked an old man who was sitting on the rocks, Willie Rupert, to explain what was happening. "The fish are plentiful when the water is normal, as now," he said. "When it rains a lot inland the water rises, and then the fish wander in different parts of the river and don't come into these coves. I have fished here since I was a young boy, and I have noticed that the fish are decreasing. Whenever I am not working I come here to fish."

Later, with their questionnaires, the scientists discovered that the Indian people of Fort George obtained 12 percent of their total food supply from the fishing beneath these rapids. Not only was it an essential part of their culture, the yearly rhythm of life, but it was also economically important to them.

We wandered back along the rocks to the Dick family tent. Clifford Dick was a laborer working for the Quebec government, but as long as he could remember he had been fishing below these rapids. Norman Pachano, a large, shambling twelve-year-old, came out of the bush carrying a porcupine and a rifle, followed by two smaller boys, seven and eight. They were fulfilling the traditional function of small boys, hunting for small game.

EINAR SKINNARLAND, Norwegian-born engineer, a commando sent back into Norway to destroy the heavy-water installations in that country, now a member of the Canadian and United States committees on large dams, the Canadian Nuclear Association, Fellow of the American Society of Civil Engineers, testifies as an expert witness for the Indians:

A dam and a powerhouse to be known as LG-1 will be built across the river at the site of the first rapids. It is an earth and rock filled dam, with the power station built right into the river. There will be a channel excavated on the south side where the river will flow during construction. It will be a spillway when the project is completed. The level will be 105 feet on the upstream side of the dam, falling to 12 feet, almost tidal level, on the downstream side. The dam and powerhouse will be concrete, and on the north side of the river it will be a fill-dam, a combination of earth, sand, gravels and some rock. There are rapids in this whole region now. By the time you build a dam, they will be under the lake. You will find the forebay above LG-1 will come right up to LG-2, so there will be no more rapids in the river. The river will disappear.

CARPETED PLATEAU

Later that afternoon we wandered through the densely wooded land along the banks of the river to the uplands, where the vegetation was less thick and the trees stood in an open landscape covered with the light green lichen that lies over much of this country. Rose Sam, Billy Bearskin's sister-in-law, led the way. She was a tiny girl with a finely etched Oriental face, a student counselor in the federal school residence. She had watched our clumsy efforts to pick spruce boughs for the floor of our tent, and having had a bit of a laugh at us, promised to show us "the easy way." We walked along in a glorious open plateau, springing up and down on the soft carpet of lichen. She moved from tree to tree, snipping off spruce branches about a foot in length, not taking many off any one tree, before placing them carefully in a pile on a regular fanlike pattern. Like all young people from Fort George, Rose knew this land around the first rapids intimately. She had been going there for fishing weekends all her life, and had many happy memories of going upriver and walking up to the plateau in the moonlight or early dawn after a dance in the village. Under the James Bay project, this place was earmarked for a "borrow area" from which would be gathered materials for use in the dams—a vast gravel pit, in other words.

O'REILLY: What kind of vegetation is found between 48 and 60 degrees parallel in James Bay and the Ungava and Labrador peninsula?

KENNETH HARE: That's the whole peninsula. There are four basic types. Tundra is very roughly north of 59 degrees [marks a line on the map some 350 miles north of Fort George]. It is a Russian word which means treeless waste. It is a treeless landscape, the ground is covered with vegetation made up chiefly of small shrubs, with a considerable amount of sedges which are quite like grass, with lichens, and with a certain amount of mosses, which form a carpet which is called the tundra.

Then there is forest-tundra, whose northern limit on the average is around 59 degrees and southern edge around 55 degrees [draws a line whose southern edge runs just inland from Great Whale River, north of Fort George] though the lines are complicated. It is a land which is still mainly treeless, it's mostly tundra, but there are isolated groves of trees, mainly on low ground, scattered within the tundra.

The third zone is woodland: it is an open kind of forest land, like a park. The trees are scattered in the landscape with in between them a carpet of other vegetation, mainly lichens and shrubs. This goes down to about 52 degrees [draws line running north of Rupert House, south of Eastmain, including the La Grande valley.] The fourth type is the true forest, south of 52 degrees, where the trees completely shade the ground. The usual way of defining it is that a squirrel can make its way without touching the ground.

o'REILLY: What about the areas between 52 and 55 degrees?

HARE: The coast becomes a little more hilly, the land more rocky. There is less muskeg [than farther south] and near the coast in for 25 miles there is quite a lot of good forest scattered among the muskeg. There are extensive rocky areas, and as you go inland the land rises. Inland you encounter woodland, open to the sky, the trees do not form a continuous carpet, they are present only in clumps, the floor is brightly lit, and it extends as you go up the slope all the way through the centre of the peninsula. There is an area of this woodland about 100 miles wide east to west, beginning 25 miles inland, where most of the ground between the trees appears to be covered by fairly dense shrubs. When you get beyond this, the shrubs become less common, and most of the floor between the trees is lichen.

o'REILLY: What is lichen?

HARE: It is a complicated plant which is made up of two plants living in union, but they never occur separately, so they are called one plant. There are many, many different kinds. They form carpets on the floor, on the land. They are photosynthetic, able to manufacture their own tissues from sunlight, but they have no roots, they simply lie as an inert carpet on the landscape. They take very little from the soil. They depend on the sunlight and the air. They form this carpet by slowly building up these tissues.

o'REILLY: What is the relationship to the animal population?

HARE: For the small and large animals, foraging in winter, lichen is the staple source of diet when the ground is snow-covered. The caribou, most of the mammals, including lemmings, nibble at it. But this is only from observation. I was concerned mainly with the populations rather than the behavior of the animals.

o'REILLY: In terms of its quality as vegetation, is lichen a vegetation which subsists near dead areas?

HARE: It is a very slow-growing ground cover of very low efficiency. By efficiency I mean the fraction of solar radiation that is converted to energy. It takes decades to develop.

O'REILLY: How far in area, in square miles, does this lichen extend in the centre of the peninsula?

HARE: Of the order of 160,000 square miles. The whole peninsula is 500,000 square miles.

BUMBLING WHITE MEN

Rose got a lot of amusement from our stumbling efforts to perform simple tasks. We tried to carry the loads of spruce boughs down to the river. They are suspended from a strap around the forehead, and require a strong neck: carrying the spruce boughs is women's work, but we collapsed under the strain several times, Rose doubling up with laughter at our feeble efforts.

She checked with Mrs. Dick to make sure which way the spruce boughs should face when placed on the ground. "It's different in a tent from a teepee," she said, "and it's bad luck if you face them the wrong way." She laid them methodically in a fan shape, following the path of the sun, east to west, jabbing the stick end of the bough under the previous row, like laying tiles on a roof. She started a fire for us, watched as we hacked our fish to pieces while trying to clean it, and then with a few deft flicks of the knife cleaned one for herself. "You cook yours first," she said, and watched as one of us burned his fish, the other undercooked it. "*Omstegushu,*" she said. "You white men." She laughed at us till the tears ran down her cheeks, then cooked her fish, showing us the easy way. Darkness fell. We sat around the small fire among the rocks. The rapids roared away to our left. We were all deeply contented.

The wilderness river was a lovely sight when I awoke early the next morning. In the distance, over the rapids, the mist hung in a white cloud. The day was clear, but chilly down by the river, and overnight everything had been soaked by the mist from the river. The bush around the banks, where I walked in search of dry firewood, was slightly eerie in the heavy dawn air. Along the path, the Dick and Pachano children had set rabbit snares and had caught three rabbits, two of which were still alive and were leaping about and squealing furiously in an effort to escape. The three small boys walked up through the bush as the sun rose. The sun filtered through the trees in thousands of silver shafts, and where it penetrated, the forest floor was aglow with millions of tiny water particles hanging from every stem of the delicate furry plants that covered the ground. The spectacle was utterly enchanting. The boys moved swiftly to their snares, zeroed in on the rabbits, dispatching them quickly and efficiently, young hunters already absorbing the lessons of killing and respect for the prey.

SCIENTIST IN ACTION

While we were watching the people fish that afternoon, a couple of anthropologists came through by canoe and left again, and John Spence arrived, his Polaroid camera at the ready, snapping fish tails and fins and dead hawks and geese and rabbits and anything that might conceivably be of scientific interest. He would peel off the instant picture for an amazed group of watching Indians. Spence was an intense, crew-cut fellow, a fisheries biologist who had recently been working in Czechoslovakia. His main immediate task was to check out the reports by the Indians that the whitefish they caught under the rapids had come in there to spawn.

O'REILLY: How long did you spend at the first rapids?

SPENCE: I myself spent two days there, other members of my team spent up to three days working at that site.

O'REILLY: And did you see any Indian people fishing at the LG-1 rapids?

SPENCE: Yes, at the time I was there I counted somewhere between 30 and 40 Fort George Indians.

O'REILLY: Did people camp there overnight?

SPENCE: Yes, certainly.

O'REILLY: And what were they doing?

SPENCE: The main object of going to the first rapids was, if you want to put it that way, to intercept the whitefish that were coming into the river at that time from James Bay, and fish them in a very traditional method that these people appear to have used for a very long time.

O'REILLY: And were these fish being caught?

SPENCE: These fish were being caught in the tailwaters of the rapids. There are several little bays and inlets where these fish accumulate at certain times, and the net is thrown over the mouth of the bay and hauled in, in a standard seining, partly gill-netting technique.

O'REILLY: Were many fish caught when you were there?

SPENCE: Yes, I observed personally anything from three fish per catch up to 15 per throw of the net, and we estimated on the day we were there that some 300 fish were taken, and of that we examined, did meristic counts of some 70 of those fish.

O'REILLY: And were those fish edible?

SPENCE: Very edible. The fish there were all sexually mature, and the gonads well developed, and for this reason their food value was very high.

O'REILLY: Did you take any samples of the fish?

SPENCE: Yes, one thing I did sample was the status of the ovaries of all the females that we caught, and I've brought one sample to the court, it may be of interest.

There followed a somewhat comic argument as to whether these ovaries should be admissible as evidence or not. The judge didn't want them unnecessarily cluttering up the record, the corporation lawyers argued that they were totally unnecessary and irrelevant, but O'Reilly insisted that he needed them, because he knew that in the preliminary negotiations the corporation had denied the existence of a whitefish-spawning area at the first rapids. The judge eventually permitted them to be produced. Spence's further testimony made plain that excellent conditions for spawning not only existed at the first rapids, but that these grounds were critically necessary to the survival of a particular, large population of whitefish.

An Outraged American

While we were at the first rapids, the American beaver expert Garrett C. Clough had flown farther up the river to the second rapids, the point seventy miles inland at which the first and biggest dam, LG-2, was to be built. He returned indignant. He couldn't believe what he had seen and heard. Engineers who had been working there for a year had never seen a biologist before. Millions of dollars were being spent on construction work, yet no solid biological knowledge about the area had ever been collected. The engineers were taking no care about what they did to the environment, though they were publicly claiming that the project would prove that Quebec could build a huge development and take the most sophisticated measures for protection of the environment, and be a world leader in both aspects.

"Without any basic biological knowledge," said Clough, "the idea of measuring the environmental impact of the scheme is a farce." Yet the federal and provincial governments claimed to have done it already in that cursory report which the Cree of Rupert House had so sensibly burned.

"I flew around in a helicopter with this ex-Vietnam pilot who is being paid $220 an hour and is up here on a 140-day contract. He kept telling me there is nothing here, and yet when I looked out of the window I could count plenty of beaver lodges as we went along. I did some work on the Alaska pipeline study: even though a host of biologists have been working for two years, they do not think they have collected the basic information to judge the effect of the pipeline on the environment. Yet here no work has been done at all.

"The reports that these people are sending back that this land is desolate are utter foolishness. I just have this feeling of hopelessness before some sort of monster. The Indians I have spoken to have the same sort of feeling. It's a crime for these people to say that these things up here have no value to the people who depend on them."

Another American on the task force, though one who had transplanted himself to Montreal, was anthropologist Tom Shiveley, and while Clough had been looking at beaver on the river, he had been talking to people in Fort George. He had found that the village was traditionally divided between the coastal people, who depended on the seasonal migrations of geese, seals and polar bears and were oriented to salt water, and the inlanders, who had come to the coast to trade furs and stayed to live. Until the Department of Indian Affairs had started building the new homes the year before, there had been two separate communities, with a space between them—a marsh that now was drained and covered with sand and new houses. The list for the new houses was alphabetical, and people moved in without retaining any of this traditional identity.

Shiveley had noticed that the people, when they cleared the settlement, had used a light touch and produced a type of vegetation which was full of soft mosses, was beautiful, easy to walk on and didn't cause all sorts of erosion difficulties. Now white men had arrived with their huge machines, roaring around and clearing the bushes out, and creating the sort of environment around the new houses in which sandstorms actually occurred when a strong wind blew. These new houses were designed and sited with no concern for the fierce elements of Fort George: some doors opened east, some north, some south, and there were no air-lock vestibules designed to keep in the heat when the doors were opened. "It seems that in general the people who come up here to work with the government are misfits who can't get jobs doing anything else, incredibly stupid people who come up and perpetuate their stupidity on the Indians, and so this is what white men are for the Indians. Stupid."

THINKING ABOUT THEIR GRANDSONS

The next day I went to talk to Job Bearskin. His little house stood on the far side of the inlanders' village, surrounded by the waving grass Shiveley talked about. A large teepee stood beside it, and alongside that a tall pile of logs on end ready for use as firewood. The house had only two rooms, a large central room with a wood stove in the middle and some beds around the walls, and a smaller room almost filled by a double bed. As in most Indian houses, there were no cupboards to speak of. The walls were covered by hanging clothes, linen and other things.

Job remembered me from my previous visit and greeted me with painstaking politeness. He was taller even than I remembered him, his long, strong face quickly lighting into a warm smile when he was making contact with a stranger. He sat on the double bed in the little room with his wife, Mary, standing behind him, her eyes mostly on his coat, picking off pieces of dust nervously and stroking his back as they talked to us through Billy and Gilbert. "They are anxious to help," said Gilbert. "They want to do everything they can to help because they are worried about what is going to happen."

Their voices were extremely soft, the personalities they projected were gentle in the extreme. Job would say, "Ah-ha, ah-ha" and something that sounded like "Gee-wa, gee-wa" (that's right, that's right) as Gilbert explained what we had in mind.

"They want me to tell you that they are not thinking about themselves," said Gilbert. "They are thinking about their grandsons and all the young people who are yet to be born. That is why they want to help." Job started to talk again. Billy said: "He keeps thinking about the beaver at Kanaaupscow, where he has his hunting ground, and wondering what is going to happen to them."

The old man kept talking, tears springing into his eyes. "There is so much involved in it," said Gilbert. "He has been telling me how he lost his stepson, the father of the boy that he has brought up. He was up the river at the time and he had to bring a pregnant woman down to Fort George, a long journey over all those portages. He managed to do it: but he is thinking of all these things that have happened to him in the past. There is so much involved for him.

"He says he will go up the river," said Gilbert. "We will take two canoes, we will go up past the rapids, make a portage and then travel for another hour or so. He will do everything for you the Indian way, as he always does it. He will build a teepee and set nets in the river and snares, everything to show you how he lives."

"Tell him I am grateful," I said. "I hope we can tell people outside how he feels and they will take notice." The old man and his wife nodded. We got up and shook hands. They took us to the door. The whole thing seemed to have taken on a new dimension. As we walked away Gilbert giggled, his handsome copper face glowing. It was often hard to figure out how he was thinking, how he was really reacting to what one did or said. But now he seemed pleased. "Since I came back home," he said, "I have gone around talking to a lot of the old men. Job has been teaching me."

JOSIE SAM, CAPITALIST

To get two canoes full of fifteen people onto the river early the following morning became my immediate responsibility, and Josie Sam, band manager, regional chief, gas supplier, taxi driver, the sharpest business-man among the Indians, seemed to be the key man, because we needed gasoline.

LE BEL: You mentioned that you're a gas station owner?
JOSIE SAM: Yes.
LE BEL: For how long have you been a gas station owner?
SAM: For about two years.
LE BEL: Two years. You sell gas and oil to the Fort George band?
SAM: Well, no, to everybody.
LE BEL: To the members of the band?
SAM: Yes, to everybody.
LE BEL: To everybody.
SAM: Not only the members of the band.
LE BEL: How many gallons do you sell in a year?
SAM: Just the gas is 100,000 gallons.
LE BEL: Just the gas?
SAM: Just the gas.
LE BEL: And what else?
SAM: Oil, fuel, fuel heating oil.
LE BEL: How much, what is the quantity of fuel oil that you . . .
SAM: About 350,000 gallons a year.
LE BEL: A year. Can you tell us how gas and oil is delivered to you or to
    your gas station?

SAM: Well, it's delivered by boat from Montreal.

LE BEL: By boat?

SAM: By boat.

LE BEL: The gas and oil is contained in barrels, I suppose?

SAM: Well, part of it comes in bulk, gas and oil.

LE BEL: But part is also delivered in barrels?

SAM: In barrels too.

LE BEL: And what do you do with empty barrels, Mr. Sam?

SAM: Well, they're stored, I store them away till next spring and then I ship them back again to Montreal. I don't leave them in the river.

LE BEL: You don't leave them, you don't put them in the river. Ah, but you will notice that I didn't ask you the question. . . .

SAM: I know you're going to ask me.

LE BEL: No, no. Do you have any other commercial interests in Fort George?

SAM: Yes, I'm a Skidoo dealer and also I'm a . . .

LE BEL: You're a Skidoo dealer?

SAM: Yes, and also an outboard motor and chain saws.

LE BEL: You sell chain saws?

SAM: Yes.

LE BEL: You sell outboard motors?

SAM: Um-hum.

LE BEL: And you also sell Skidoos?

SAM: Yes.

I had arranged with Josie Sam—who was also Fort George's only taxi driver—to come with his truck next morning at 7:00. But when he didn't arrive I went and knocked on his door. There was no answer for quite a time. I kept knocking, and finally an arm came through the window and dropped a key in my hand. "Take it," said the taxi driver, "and leave it somewhere."

It had taken us almost five hours the previous evening to secure the gasoline we needed for the canoes, running back and forth from one side of the village to the other. And in the morning it took us another two hours of hard work to gather Billy's and Gilbert's gear and all our camera equipment, to carry Billy's heavy canoe 100 yards across the mud flats that had been exposed by the low tide, and to get out on the river. Job and his family, ready at the appointed time of 7:30, waited impatiently by their loaded canoe on the riverbank.

The family in the canoe we were following was an extraordinary spectacle. George, fifteen, the boy who had been brought up since infancy by Job and Mary, wore a pink tuque and sat right in the front, leaning forward into the wind and naturally striking a pose like that of the figure-head of a ship. Job and Mary sat in the middle with a couple of young men who had come along to help, and at the back, working the motor, looking for all the world like an illustration for a James Fenimore Cooper novel, was Samuel Cox, a teenage boy with a long, sensitive face, black hair streaming below his shoulders and a beaded headband bearing the emblem of the nuclear-disarmament movement. The symbolism was almost too strong: the symbol awoke very personal echoes in me. I had been in London in the early sixties when the huge Aldermaston marches had taken place, at a time when all political argument seemed to be concentrated on the morality and ethics of the hydrogen bomb, the most impersonal of all weapons, as it was the most powerful and deadly result of the white man's devotion to technology. And here we were ten years later on a superb river with the inheritors of a hunting culture, the oldest type of culture known to mankind, whose existence now seemed to be threatened by that same impersonal technology that we had protested against in London. And here, unconsciously but suitably, was the symbol again: the two protests joined accidentally by this strikingly handsome boy who appeared to have resisted all the efforts of our technological society to turn him into a white man.

A DANGEROUS EVENT

It was a magnificent day on the river: one could feel the deep contentment of the Indians as the canoes plowed steadily upstream. Between the two boys at front and back, Mary Bearskin crouched low in the canoe under a warming tarpaulin; but Job never relaxed his attention for a moment, gazing around him at the surface of the water and across to the banks as if every yard held objects of extraordinary interest for him, which indeed it did. Every now and again he would raise himself in the canoe, his large head under his cloth cap keenly alert as he peered forward. He would make a gesture which Samuel would follow unquestioningly. Job had been traveling the river all his life and he knew every rock of it. Though his canoe was smaller and slower than ours, and sometimes we would maneuver ahead of him so that Guy could shoot back at his group, he was the leader. No one had to state it: it was just assumed by everyone that the old man with the experience knew best. We were traveling behind them when they pulled into the shore. Something had gone wrong with their motor. We seemed a long way from anywhere, not the ideal place to have a mechanical breakdown. But no one was worried. Mary sat patiently in the canoe. Job got out and walked along the shore, stopped every few steps to examine some rock or plant minutely. The two young boys stripped the motor.

"What the hell's wrong?" I asked Billy.

"Nothing. They've got some motor trouble."

"Is that all?"

"That's nothing. If you're worried you should be thinking about them rapids." He guffawed with his inimitable laugh. "We've got to run right up into them rapids. You make one mistake and that's it, boy."

"How close do we go?"

"We go right up as close as man can get. We go right up to the edge, where the water's rushing back and forth. You make one mistake there, boy, and that's it. There's nothing more."

"How many times have you done it?"

"Me, I've never done it."

"Christ!"

"One mistake," repeated Billy, with relish, "and that's it. Job knows the way, though. He knows what to do."

"That's some consolation, with him traveling in the other canoe. I hope I have time to wave to him as we're going under."

"He knows what to do," repeated Billy. "But you can't afford to make mistakes on a river like this. It don't give you no second chance." Gilbert, listening as he loaded a magazine with fresh film for Guy, using a big rock as a table, looked inquiringly at me to see how seriously I was taking all this and laughed. "That's right," he said enigmatically. By this time the boys had fixed the motor and Job had finished his walk of inspection along the shore. They paddled their canoe out from the shore, pulled the engine a couple of times and we were away again.

I did not appreciate it yet, but this constant repetition of the dangers that lay ahead was part of the enjoyment that these people took from the challenge of the river. Every aspect of the experience, every moment of it, had to be savored fully. Each one of them enjoyed every minute. That is a behavioral trait which came from the hunting culture. A subsistence hunter does not hoard things against the future, he does not accumulate goods so that when he is old he will be able to enjoy the fruits of his hard work. His every fiber is strained to make the most of the present. His hunt is for food for today and the next day; his enjoyment is now, in the act of living. He knows that when he is old, those he is caring for now will take care of him when their turn comes. I was beginning to feel something of the perceptual need that these people had for contact with the land

and the river, the plants and the animals. I had not spent a great deal of time among them, but enough to revise any opinion that I might have formed on my first trip to Mistassini, when so many of them had seemed so poor, so rough, so stoic and silent.

Sure enough, we ran very close up against the rapids. The keen anticipation everyone had enjoyed was rewarded as we ran dramatically into the rocks along the southern shore, the water swirling and boiling angrily around us. Job led the way, sitting high in the canoe so that he could guide Samuel through the hidden rocks. Billy followed in the same path. We ran the canoes up on the rocky shore: the men jumped out quickly and held them secure until they could be pulled to a safe position. "Come on, you guys, pull!" cried Billy, so I jumped out, got one boot full of water, tried clumsily to get in position to help pull.

"How'd you like that?" cried Billy, laughing happily over the roaring of the rapids. We were on the opposite shore from where we had camped a few days before. The canoes were safe now, and the unloading started. I was anxious to show the Indians that we, too, could help with the heavy work: there seemed no other way of establishing credit with them. We had to accept that we were clumsy and fairly useless, but they should know that our spirit, at least, was willing. I started to carry stuff over the hill: the packages were heavy, it was steep and I wasn't an accomplished packer.

This was a portaging place that the Indians had used for as long as their collective memories went back. We found that logs had been cut and laid from the shore all the way over the hill and down the other side, to the point probably 250 yards away where we were to put the canoes back into the water. The men rolled the canoes up the hill on these logs. When they reached the top of the hill with the smallest canoe, they turned it over, picked it up and carried it the rest of the way on their shoulders. But the big canoe was too heavy for that, and they rolled it all the way. The women, after having carried much of the stuff over the hill, went around gathering firewood, and when all the carrying was done we sat down on the huge rocks to rest and drink the tea they had boiled in their pails.

Ahead of us, upstream from the rapids, the water was smooth, moving in a deadly sort of way toward the place on our right where it plunged over the rocks in a spectacular swirl of white water and spray. I was wet with perspiration. "Jesus," I said, "do you do this often?"

"That's nothing, " laughed Billy. "When Job went up to his trapline every year, before the planes, he would make fifteen portages like this.

And this is about the shortest one. Most of them are long, some of them five miles, and he would carry everything himself, including his canoe. This is just nothing, this one."

The journey upstream would take them about two weeks in those days: and they were by no means at the end of the Fort George trapping grounds. Other Indians would go much farther, a month's journey into the interior, and it was to serve them that the Kanaaupscow post, now abandoned, had been opened up by the Hudson's Bay Company. Glen Speers had been the manager who built that post, and this 100-mile, fifteen-portage, two-week journey upriver from Fort George was the one he made every fall with the supplies needed for the winter. Speers had told me that the La Grande was one of the most beautiful rivers in James Bay—indeed, one of the most beautiful rivers anywhere.

We loaded the canoes again and headed off into the smooth, fast-flowing water above the rapids: the waters here raced downstream at a tremendous rate, a very high proportion of them having been concentrated, because of the contours of the stream bed and the banks, between a rather large island and the southern bank of the river.

O'REILLY: Between the first and second rapids, have you ever hunted, fished or trapped yourself?

STEPHEN TAPIATIC: Between the first and second rapids, I hunt and fish, and in the summer the people from Fort George fish there, because all summer long I see fish nets there on both sides of the Fort George river.

O'REILLY: And between the first and second rapids, what kind of animals are there?

TAPIATIC: It's exactly the same as down the river. [Rabbit, ptarmigan, porcupine, bear, beaver, lynx, squirrel, mink.]

During his solitary walk along the shore while the boys were fixing their motor, Job had not been idle: he had collected rocks with which to anchor his nets and on the way upstream he had been busy tying them to the nets. Now, twice, we stopped while Job took his canoe over to some weed beds and set the nets in the river with expert hands. When he had finished the second of these jobs, we ran downstream for half a mile and pulled the canoes onto a sand spit. Job looked over the banks of the river, gesticulating to the young men who were clustered around him, and then, the decision made, we ran the canoes onto a sheltered shore about thirty yards wide from bush to water.

Before our stuff was unloaded, Mary was already cleaning out the floor-bed of an abandoned teepee site. Job disappeared into the woods with his ax, returning in about fifteen minutes carrying four long poles already trimmed and ready for use in building the teepee. Mary and Billy's wife, Martha, went into the bush to cut spruce boughs for the floor. The Indians had swung into action like a machine, everyone doing his appointed job as if at the flick of a switch. Expertly and without any fuss they put the teepee together. Before the canvas was on the poles a rock fireplace had been assembled and a fire lit in it right in the center of the wooden frame. Old sheets of canvas they had brought with them were wound around some sticks and expertly fixed across the top of the teepee poles, leaving a hole at the top for the smoke to escape from. Job fixed a door similar to the one I had seen in the Mistassini camps, and before the teepee was finished the smoke was already issuing from the pole at the top, the pail of water was boiling for tea and Mary was squatted on the spruce boughs cleaning fish. Billy cleared a space for his tent just behind the teepee: the two were joined to make one big canvas home, plenty big enough for the fifteen people in our party.

We took our own little tents, the orange-and-blue sleeping tents that are so familiar over all the camping grounds of Europe, and put them up along the shore a bit.

"You're not going to sleep in those things, surely," said Billy. "You better come and sleep with us, everybody together."

"Okay, if you're sure it's all right, we will."

"Sure it's all right." His face broke as he emitted his high, mocking giggle. "You can't sleep in those things, that's for sure."

INSIDE THE TEEPEE

I poked my head through the teepee door after half an hour or so—shyly, I must admit, for I had a sense of intruding on something private—and was astonished and immediately captivated by the atmosphere. Everything seemed to have both an aesthetic and a utilitarian purpose. The fire in the ring of stones was cheering and delightful, but it also provided as much light as would be needed, and was cooking food and boiling water which were already available for whoever needed them. Mary had pots and pans spread out between herself and the door, and between her and Job sat some small boys ("grandchildren," as the Indians call all small boys who are with older people) unconsciously absorbing the lessons to be learned.

Mary was never still. Job already had a supply of long sticks by his side; whenever his wife needed one, he would whittle it to a point in a matter of seconds and she would use it to suspend some food over the fire. She had already mixed some bannock, that Indian delicacy which was brought to James Bay by the Scotsmen who worked for the Hudson's Bay Company and is nothing more nor less than Scottish scones; she had rolled the bannock round some sticks and it was cooking over the fire, being turned every few minutes by herself or Job. One job followed another, and to do them all she scarcely had to move more than a foot to the left or right. She cleaned some fish dexterously, chopped them into large chunks and placed them, too, over the fire on sticks. Coffee, tea, water for drinking and washing, fish, bannock—one after the other these things effortlessly appeared as, outside, the sun sank and night lowered over the river.

That evening we all sat around the flickering flames, relaxing and enjoying the quiet time after the heavy day on the river. The teepee was pervaded by the scent of the spruce boughs; the copper-colored faces were illuminated by the fire which also kept the place warm; Mary never stopped putting food on sticks or in pails over the fire; the children sat by their grandparents, watching their every move, listening to their quiet chat. Fournier, Borremans and I watched for about three hours. The others would move up to Mary's collection of pots and sticks periodically and eat something. Finally Billy said, "Don't you guys want something to eat?"

We said yes, that would be good.

"Well, what do you want? You want some fish?"

"That would be lovely."

"I thought you guys weren't gonna eat. I don't know how you can wait this long to put something in your stomach."

We rolled over onto the spruce boughs, laughing. "We didn't know it was mealtime."

"We don't have mealtime. We don't run by the clock up here, like white men. We eat when we're hungry. I never knew anybody eat as late as you guys."

I awoke at 5:30 the next morning. It was already daylight, and Job and Mary had the fire burning low and were moving methodically as they rolled up their bedrolls, washed, prepared tea and got the crowded tent in order for the day. When everyone was up, we went off in Job's canoe to check his nets. The day was glorious. The river was unlike any river I had ever seen before, smooth like glass, the clouds and the surrounding trees reflected as if in a mirror. Its beauty on this morning was truly transcendent; everyone was moved by it, perhaps the more so because we knew that a decision had been made in the south to destroy it.

Job went about his fishing with what I can only describe as that innate sense of propriety that characterized all his actions. He killed the fish with quick blows on the head, commenting quietly to the younger men on their various qualities. The river was performing its traditional function, that of helping the Indians to survive; the fish, too, were playing their allotted role, and one had only to see Job's face, the quick and expert weaving of his hands over the net, the way he bore himself as he dipped the paddle into the water as a signal that the work was finished and it was time to return to other tasks, to feel something of the deep satisfaction he took from the work.

ROBERT LITVACK: Could you describe the fish species which are known to exist in the La Grande River?

GEOFFREY POWER [Professor of Biology, University of Waterloo, Ontario]: The principal species, the more conspicuous ones are brook trout and lake whitefish. This latter species occurs in two forms, a form which goes to sea and a form which stays all the time in fresh water. The form which goes to sea is called technically anadromous. Pike, sturgeon, there's a small species of whitefish called round whitefish, as distinct from lake whitefish. There is a group of fishes called suckers, two species. And a number of small species, small forage fish, which certainly from the point of view of eating them, they're of no significance.

LITVACK: Now if the mean annual flow of the La Grande river is increased by 70 percent and if hydro-electric dams are built . . . and if a series of power pools are formed behind these four dams . . . could you describe the possible effect of these changes on the various species which are now found in the La Grande river?

POWER: Building of dams, and the turning of a river into a series of lakes like this is equivalent to a natural catastrophe of major proportions. In terms of natural events it's the equivalent of five major landslides blocking portions of the river system. This will have a disruptive effect on the fish populations that inhabit this river. . . . I think to properly explain this we should perhaps consider the fish species by species. If we start out with the trout. The trout in a river system such as this are divided into a large number of sub-populations and these populations are more or less discrete in terms of the section of the river they inhabit, the places in which they feed and of great importance to us, the places where they reproduce. There is of course some mixing of populations but for all practical purposes fish like trout return to the same places that their parents did to spawn.

These places have to be of a very particular nature in order for the spawning to be successful. Trout spawn in flowing water and except in exceptional circumstances, lakes are unsuitable spawning sites.

LITVACK: Doctor, would the bed of the lake in any way affect the suitability of the spawning site, that is, if it were gravelly as against silty?

POWER: Well, the only places in lakes that trout ever spawn are where there are underground springs or wells which produce a flow of water through gravel on the bottom of the lake, and in the Quebec-Labrador area such places are rare. The result is that the trout spawn in rivers and streams. The female requires gravel of a certain consistency in which to dig her nest. She also requires gravel through which water is flowing and this kind of gravel is generally found at the tail end of gravel pools in rivers and streams in which there is a series of rapids and pools, because in this gravel the water percolates through the gravel and provides oxygen for the developing eggs. The trout spawn in the autumn and the eggs are incubating during the winter months. The whole life cycle is adjusted to the natural rhythm of events in the river, and everything has to be precisely timed. . . . The method by which fish find their spawning area is not completely understood. We do know that the homing is precise and accurate.

LITVACK: When you say homing, does that mean that they return to a home?

POWER: The homing is going back to the spawning area. The cues that are used by fish in finding their way back include a number of things and probably things that we don't understand. But they use the sun, they use the stars. This is probably for long-distance movement. They use smell and taste and this is probably very important in allowing them to discriminate between, for example, one tributary stream and another tributary stream, so they have in their memory the smell or taste of their home stream. . . .

LITVACK: Would the damming of a stream cause any difficulty to a fish in swimming upstream or in homing?

POWER: Well, that depends on the damming. If the obstruction is such that the fish cannot get around it or over it, it'll stop the migration.

MALOUF: What then happens?

POWER: Then that fish is lost. It may not spawn, it just may sit at the bottom of the dam, and hammer its head against the dam, trying to get up to where it belongs.

A HUNTER TALKS ABOUT THE RIVER

When we returned from checking the nets, we gathered again in the teepee for an on-camera interview with Job. Mary began to clean some fish, Job sat on his heels, his back straight, the young men gathered around. We asked him to tell us about the river.

"I want to tell you a story about this river, and this spot where we are," he said without any nervous tiptoeing round the theme. "My father was very old and he used to paddle his canoe here and throughout his life hunted all along the river. Where my nets are now, as far back as I can remember, my father used to put his nets there too. Exactly where my nets are, that's where he put his. Before I started hunting, he brought me here. All along the river he used to catch fish. Then he couldn't hunt any more because he was too old. He stopped hunting when he couldn't get around any more. When we were children he provided for us with his hunting, and when he became old, we did the same for him. He was 108 when he died."

Mary, leaning over and scraping the fish, kept up a low prompting as her husband told what he felt. Some people, he said, would travel 200 miles up the river before they reached their hunting grounds. Some would stay away in the bush for seven months, and others never returned to Fort George at all. "Some of the old people who are no longer able to hunt love to talk about the days when they hunted, but many who still hunt are thinking about their children and how someday they will want to hunt, too. Now they tell us what they are trying to do to this river." He flung his arm up over his opposite shoulder, gesturing indignantly as he spoke. His voice was soft, his manner dignified, he had an air of immense authority. He gestured toward the small children. "Look how happy they are when they are trying to see what has been caught. These two boys are starting to hunt. They've already killed ptarmigan with nets. Where they killed them down the river it will all be flooded and there will be no more ptarmigan. A child is very happy when he kills something. You must have been like that when you first killed game." He motioned in the direction of Brian, Billy's little boy. "Look at this youngster, he will be happy when he first kills something."

"You're so right," said Billy. "He wants to come every time I go out."

"I've seen it myself," said Job. "When we go to raise the nets, Brian always wants to come with us."

EVERYONE HAS THE SAME SOUL

Prompted all the time by Mary, Job talked of the destruction that was planned for the river, the fish, the beaver, the rabbits, even the birds that depend upon the branches on the trees along the river. Those down south who had said that the animals would not be much affected by the project did not know. "I know, I'm the one who's telling the truth," he said. "I'm over fifty years old and I've been hunting over this territory since I was a young boy." He spoke of the relatives buried on the most beautiful spots along the river. "The person who gave the word to go ahead with this project," said Job, "the Indians think he has no consideration at all for the graves of our ancestors. I have to say this about him: he has no consideration for the Indian at all. I would tell him exactly this if he were sitting right in front of me."

"*Gee-wa*," said Mary, "*gee-wa*"—right, right.

Job gestured at a rabbit lying along the wall of the teepee on the spruce boughs. "Look at that little rabbit there, he probably has a brother and a sister. That's how everything goes on this earth. Look at the earth, things keep growing on it, human beings and animals keep multiplying. We Indians worry about the things that grow. The animals were given to the Indians so that we could survive. I think the white man doesn't care about this."

When the water rises in the river, he said, it washes the shores and the water tastes of earth. You cannot drink the water in some of the inland lakes for the same reason. But this project would destroy the taste of the river completely. All the lakes whose water is undrinkable would merge with the river and spoil its water. "We love our river because we were brought up on the food that came from it," he said. "We were given this river for our use. I could talk a lot more, because there is a lot more to be said."

His wife added that even the man who proposed the project must like planting things in his garden. "It is the same with us. We love our garden. We love the animals in it and everything that grows in it."

Now Job said that he had heard that the man who launched this project believed this was just a wasteland, that there was nothing here. How could he believe such a thing when so many people were surviving here? That man should come north and see for himself. "He's made the same way we are," said Job. "He would still survive if he lived with Indians. He has the same soul as we have. Everybody was given the same. But he is not using his soul properly, he's using it only for his own gain.

He is trying to destroy lots of men. The Creator did not intend that this land should be destroyed."

"I have read the Bible many times," said Mary, "and I have never seen anywhere that the Creator says the land should be destroyed before he is ready."

"When he's ready," said Job, "then it will be destroyed. Then there will be no life on this earth for a while. I have been thinking about this since I first heard that the river would be dammed."

ROBERT LITVACK: Now would you have any comments with respect to the changes in the morphology of the middle reaches of the La Grande river pursuant to this project?

ROLF KELLERHALS [civil engineer, of Edmonton, Alberta]: Well, the middle reaches, upstream of LG-1, are essentially being turned into a staircase of reservoirs. There'll be one dam, a lake, a dam, a lake, so the river channels are being drowned and the bottom sediments of these river channels which, from what I can see on the basis of present information, is probably coarse gravel in many places, these bottom sediments will be gradually covered with lake sediments, with a much finer stuff. In parts this may be very thin, in other parts, this may be quite thick.

LITVACK: The build-up of this fine sediment, will this continue or is it at short-term?

KELLERHALS: No, this is a continuing process. These reservoirs will fill up and the river will start by depositing very fine sediments over a good part of the reservoir area.

LITVACK: Will this have any effect on the plant life or on the animal life in the so-called lakes or reservoirs?

KELLERHALS: Well, again, this is it. I would say, in my expertise, that I do know that the types of fishes, for instance, that can spawn in a lake of strong sediments are quite different from the fish that can spawn in river channels. Tremendous difference between those two.

LITVACK: Now these lakes, or power reservoirs or forebays or whatever term is the most appropriate, perhaps you could indicate, will they have the appearance generally of a lake? Beside the fact that they will have a silt bottom?

KELLERHALS: No, one shouldn't really call those lakes, one should call them forebays, or I like to call them power pools. I think the most important aspect of a lake is the shoreline, aesthetically, economically, biologically, it's the shorelines of a lake that make the lake, and the near shore zones of a lake. And these power pools, there will be tremendous fluctuations in level, and the fluctuations will be at a different time than they are now in natural lakes. The high water will persist for longer, the high water will occur in possibly August, rather than in June and maybe early July when it's occurring now, and because there's no provision for clearing those reservoirs, the shore zone, the zone that's being exposed each year will be a maze of dead trees initially, debris and dead trees.

LITVACK: Is it current practice in the construction of hydro-electric power pools or power reservoirs to leave standing timber?

KELLERHALS: No, this is going out. I think five years ago one could have said, it's current practice in Canada. Right now, it is not any more. The similar big project under way in British Columbia, the Mica dam, was originally not supposed to be cleared. It has now been cleared. The Beacon reservoir in Alberta, which has just been filled, has been cleared. The Brazeau reservoir in Alberta was not cleared a few years ago, and it is now being cleared at the considerable expense that the level in the reservoir is artificially being kept low to permit the clearing.

LITVACK: Does the fact of not cutting and clearing affect the, well, it's no longer called a river, I believe, but the body of water that is there?

KELLERHALS: Yes, there will be all this rotting wood in it, which affects the water chemistry and, as I mentioned, it affects the shorelines tremendously but it makes some of those reservoirs virtually impossible to navigate because of the driftwood. On the reservoir behind the Bennet dam in British Columbia they have to use good-size tugs because the normal boats are much too dangerous for use.

LITVACK: Is this a rather short-term effect, or an extended term, a year, two years, three years?

KELLERHALS: No, I have seen reservoirs that are forty and fifty years old in British Columbia, and it seems that they look just as bad as the new ones. It takes amazingly long for this to clear by itself. Longer than one would expect, and I think in the cold climate up there, rotting is naturally not proceeding at a very fast rate.

RED RIVER

That afternoon we ran up the river in the canoe on a fishing trip. It was the first time Gilbert and some of the other young men had ever seen this part of the river, between the first and second rapids (they had, after all, spent most of their youth in the south at school), but Job knew it intimately. Until he had started to use the planes to get to his hunting ground a few years before, he had always traveled the river on his annual journey to and from his land, and he was delighted to see this stretch again after having been away from it for a few years. He took us to a magnificent little stream that ran down into the La Grande in a series of spectacular rapids. The water was deep red, for it ran through country with a lot of iron deposits, and the stream—unnamed, of course, by the white men—was known as Red River in Cree. As we were running up the stream a bird flew across the river and settled on a tree near the bank. George shot it, a kingfisher, and in the evening it found its way onto Mary's sticks over the fire. We stood on the rocks beside the rushing water, fishing or just looking while out in the stream Job drifted about lazily in the canoe with young Brian. No one caught any fish, but when we climbed into the canoe again, it being a superb day, Job decided to go farther upriver and show the young people a little more of the country. The river was broad here between the rapids, and as we went along Job pointed out features of interest: the tracks of the bear, the signs of the otter, the berries and plants that attracted the bear to the water's edge and made him vulnerable to attack by a hunter. We passed two Hydro-Quebec camps, neat rows of tents on a clearing that had been brutally gouged out of the wilderness. It was their practice at this time to inhabit such a camp until it got too dirty and cluttered with garbage to be bearable any longer, when they would move a mile or so along the river and set up a new camp.

But there was no hostility in the attitude of the Indians toward these camps, and at one point Billy almost decided to call at one camp to ask for some steak, for he had heard that the men at these camps had suddenly become very friendly toward the Indians and would give them anything they asked for.

Eventually we pulled in beside a gigantic rock which stood on a promontory on a bend in the river, enabling us to look miles ahead, almost to where the LG-2 construction camp had been built at the second rapids. Almost before I had time to ease myself out of the canoe, the boys had a fire of twigs burning on the rock and were boiling water for tea. I had not even realized they had their pot along, but they never traveled anywhere without their axes, knives, guns and cans. One never knew what might happen, or where one might go, and one had always to be prepared. The boys took out the guns and had pot-shots at stones and other objects, and Job talked to them for a long time in Cree, telling them stories about the river, the animals he had caught here, the experiences of people he knew right at this spot, imbuing them with that same awed sense of the strength and beauty of this place that moved him.

THE SONGS OF A HUNTER

Job agreed to sing for us that evening: the songs of the Cree hunter are
an intimate part of his religious, hunting life. Many will not sing them for
strangers. They are usually sung to celebrate a particular occasion, or as
an integral part of the hunting season. But Job had said he would do
anything we wished, and he was as good as his word. Besides, I had the
feeling that he welcomed the opportunity to sing to these young Cree,
to transmit to them something of the old culture. Before each song he
moved into a more formalized sort of language, heightened in tone and
intensity, not singing exactly, but certainly not conversational. They told
me later it was an ancient Cree, a particular type of formal imagery which
the young people seldom heard and had difficulty in understanding.
(That's so wonderful, they said, if only we could translate it for you, but
we can't.) In fact, it was poetry.

"A man who was trying to catch a beaver sang this song, he enjoyed
catching beaver very much, and I enjoy that very much, too. The man says
he is really happy when he finds a beaver lodge, and this is what he
sang."

> Little beaver that swims at night
>   if we can't see it
>   we can't catch it.
> I like to try and find beaver tunnels.
> I go on the ice with a chisel.
> I feel the ice to find where the den is.
> One night I found many dens.
> I've dragged many beaver out of the dens.
> Sometimes I put sticks around their dens
> And when they come out I grab them and toss them on the ice.
> Sometimes I kill them all like that,
> Just tossing them out of their dens.
> Little beaver that swims at night
>   if we can't see it
>   we can't catch it.

We asked him to sing about the river, and quickly he dropped into
the poetic imagery to tell us he wanted to sing a song that he would sing
while crossing the rapids, something he had been doing ever since he was
a child. He still liked to go paddling through the rapids, even though he

realized it was very dangerous. He used to sing this song as he was about to enter the rapids.

> I paddle to clear my way through the water.
> I depend on my paddle for direction.
> The one that works in the rapids
> will get me and my canoe
> through these rough waters.
> The paddle goes out in the rapids.
> The paddle goes out in the rapids
> and leads the way for my canoe.

After his songs we all filled up our cups with tea and lay back on the spruce boughs, resting. I asked Job about the system of respect for the animals that the old men in Mistassini and Rubert House had told me about: did they have the same beliefs in Fort George? Of course, it was the central belief of his life. The first time he went hunting, his father told him not to kill too many animals. If one did, one would thereafter have a hard time killing these animals. They would go away.

When he had first started to hunt, he said, there were few beaver or porcupine and no caribou; but now just as they were coming back, the white man proposed to destroy everything. I suggested that the white man must seem crazy when looked at from the tradition of the Indian hunter. The suggestion was translated, and for quite a time they discussed its meaning. Then Job gave his answer: "The white man has no feeling of love for all life on the earth. That's the way we understand it. The way the Indian sees him, the white man is like a spoiled child, grabbing everything for himself, never sharing, a destructive child who never matures." Gilbert looked up, his copper face glowing in the light from the logs. "He says the white man has no emotion." Though the conversation had been in Cree, we could tell that everyone present had been stirred to intense feeling. There was a long silence before Gilbert spoke, even more softly than usual: "It is really beautiful what he has been saying. He said this whole place is like a garden, because many things grow here, and the Indians are one of the things that grow here. He says the animals were given to the Indians so they could feed their children and old people, and everyone has always shared the food from this garden. He says everyone here will always share. It's always been like that."

We were white men, and the white man was not being praised. But they didn't mean us to take it personally. Billy guffawed in that raucous way of his, to ease the tension, and told us, "Job says he will never forget you guys, even if he never sees you again."

We asked if perhaps Mary might have something to say. She had been talking away, *sotto voce,* all evening, prompting Job as he spoke, in the way of Cree women, knowing their place but never accepting an inferior position. Now she sat up straight and spoke out loudly with an astonishing flow of non-stop eloquence. "I know how a person feels when he goes out hunting," she said, "because I have been a very good hunter myself. Many times I have been out to check his traps and have been happy to bring back a marten or a mink. I would kill sometimes with my own gun. The only animals I have not shot are bear and caribou, because I have never had a chance at them. But I have killed lots of rabbits and ptarmigan. I have killed otter, I have killed all manner of things."

She gave a great sigh, thought for a moment and continued: "I have loved cleaning everything, including the fish. I love working with beaver when he kills them, and all kinds of food, all the kinds of game that he gets. That's why when I think of what is going to happen, it makes me very unhappy. I am not thinking only of myself, but of all those young kids that are just starting to hunt, and those that have yet to be born. I am thinking more so about the parents who are raising their children who will be using the land in the future. We know that the animals will be destroyed. The men are not the only ones who know about the damage that will be done. The women are good hunters, too: that is how I know how a person feels when he is out hunting."

The next morning we had fish stew for breakfast. Such a dish is full of bones and shapes which are well known to the Indians, each with a particular meaning. Some of these pieces are shaped like airplanes, and if you throw them up against the canvas of the tent and they stick there, it means you will get fish the next time you check your nets. Job threw them up, they stuck, and when he checked his nets that morning he came back with four huge pike, five big sturgeon and a lot of whitefish: he had caught enough, on our brief trip up the river, to feed everybody, with fish to spare.

ACCULTURATIVE STRESS IN FORT GEORGE

JOHN BERRY, thirty-three, a cultural psychologist from Queen's University, Kingston, Ontario, told the court about his work in measuring acculturative stress in New Guinea, and among the Australian aborigines, the Eskimos of Baffin Island, the Timney of Sierra Leone, the Carrier and Tsimshian Indians of British Columbia and the Cree of James Bay.

BERRY: Acculturation is the process which occurs when two cultures come into contact with each other, usually a dominant culture and a less dominant one. The less dominant culture is said to be acculturated when the dominant culture intrudes or overlays its forms, language, religion, values, skills and way of life essentially on the less dominant culture. . . . Acculturative stress is the stress due to coming into contact with the large society which is experienced by people living in these particular settings. . . . In the James Bay area I studied two communities, one [Fort George] which I term as having a high contact with the Euro-Canadian life, and the other [Paint Hills] having a relatively lower contact.

ROBERT LITVACK: So you attempted to measure stress, absolutely or stress from a particular source?

BERRY: . . . If we can compare the stress in communities undergoing acculturation with communities that are not undergoing acculturation, the usual findings in the research literature show that communities undergoing acculturation have much higher stress levels and there's a very common association—not necessarily a cause and effect relationship because we don't have the proper studies anywhere in the literature to show this—but a very high probability, that stress will be associated with acculturation, hence the name acculturative stress. [His sample contained sixty people in each place, divided equally between males and females and three different age groups.]

LITVACK: So among the Cree of Paint Hills and Fort George you measured, you've termed it stress. Were there any other characteristics measured by you and your team?

BERRY: . . . I also measured a concept which is termed marginality, and a set of attitudes towards modes of relating to the larger society. Marginality is a concept that exists in the literature of ethnic relations, intergroup relations, and it's used to refer to the feeling of being caught between two cultural systems. Marginal to both, in a sense having one foot in each of two worlds. The manifestations of marginality are ambivalence, uncertainty, apathy, withdrawal in some extreme cases, generally doubt and uncertainty. . . . The attitudes which I attempted to assess were based on a scale originally developed for use with aborigines undergoing acculturation, and modified, of course, for use with Cree samples. The scale is divided into three sections, one of which measures the degree to which assimilation is desired, the second the degree to which integration is desired, and the third the degree to which rejection is desired.

Assimilation, integration and rejection are terms which indicate a pattern of association with the larger society. There are two essential questions to be asked. One is, do the people wish to retain their cultural characteristics and ethnic identity. The second is, do the people want to relate positively with the larger society, or do they wish to avoid contact. . . . Where people say they want to have some kind of relationship with the larger society and they want no longer to retain their cultural characteristics, this is what we label assimilation. They want to move into the melting pot, to lose their cultural characteristics, to be absorbed into the larger society. Where the answer to the question of retention of cultural characteristics is yes, but the contact and positive relationship with the larger society is also yes, then we have integration. . . . This indeed is the official policy of the federal government, now that we have a multiculturalism policy. The rejection is when you want to retain your cultural characteristics, but want to have nothing to do with the larger society. It's a self-segregation. [His results showed that the three Indian communities, Tsimshian, Carrier and Cree had higher stress levels than Montreal and United States urban samples, and among the three Indian communities, the Cree were the highest.]

LITVACK: Is there any particular reason why you chose these three Indian samples?

BERRY: Yes, the three communities are related along a cultural dimension often referred to as migratory through descendantory. The migratory samples have traditionally been hunting and gathering and have low levels of social organization and social stratification.

LITVACK: Which of your three groups would that be?

BERRY: The Cree sample has traditionally operated in that way. The Tsimshian on the other hand have been traditionally urbanized with chiefs and slaves and villages and a high degree of social structure. The reason for sampling along this dimension is that, in my judgment, Tsimshian sample would have been better prepared for the stresses of acculturation in their traditional life than were the migratory peoples. . . . The distance that they [the migratory peoples] have to travel in accommodating to the overlay of the larger society is much greater, because the traditional preparation has been in a different form of life-style. . . . There are lower levels of marginality among the Tsimshian, moderate levels amongst the Carrier, and highest amongst the Cree, and this is consistent with the stress data. . . . We can see looking down the assimilation column that there is relatively speaking a low interest in assimilation amongst the Cree, roughly neutral in the Carrier, and moving on to a somewhat positive attitude in the Tsimshian. Statistically the difference between the Cree and Tsimshian is reliable. In terms of rejection, the opposite pattern is apparent. Amongst the Cree there is a positive interest in rejecting further contact with the dominant society and amongst the Tsimshian there is a disinterest in rejection which can be interpreted as an interest in further relating with the larger society.

Dr. Berry was interrupted by Mr. Justice Malouf, who wondered out loud whether this testimony was relevant, because it seemed to suggest that a people who suffered a greater amount as a result of the project would have more right to proceed against it than a people who suffered a lesser amount. He wondered if the degree of suffering made any difference: was it not just a question of the rights of the parties in relation to one another? O'Reilly quickly urged upon the judge that there could be all kinds of damage besides property damage: cultural damage, psychological damage, physiological damage, medical damage and even spiritual damage. It was an interesting passage, because gradually the Indians' lawyers who began the case thinking only of property damage (the classic cause of an interlocutory injunction proceeding), were having to take into account the entire spectrum of damage that a powerful society could impose on a less powerful one. They were beginning to realize that the case was about the life or death of a culture. "Carrying your argument to the proper conclusion," said Malouf, "then the matter may be much more important with respect to a Cree than to the Tsimshian?" O'Reilly quickly agreed. "I'll allow the testimony," said Malouf, an interesting indication of the way his mind was moving.

LITVACK: Now, Dr. Berry, could you express a professionally valid opinion as to what would happen to the stress levels among the Cree people if a large scale hydro-electric development were built along the La Grande river and as many as 12,000 white workers were introduced into the area?

BERRY: The usual pattern, as I've indicated, is for stress to increase with increased acculturation. I have typified Paint Hills as a lower contact community and Fort George as a higher contact community already. I think it would be quite safe to say that the levels of stress would increase, depending on a number of factors, such as the direct impact that construction and workers have on these areas depending on the attractiveness of the project for wage employment. Everyone wants money from time to time, and . . . I'm quite sure that the initial impact would last for perhaps half a generation, would be quite serious in terms of community mental health.

LITVACK: In your examination of the two Cree communities in James Bay, did you find a surviving Cree culture, or did you find that it was a culture totally inundated by Euro-Canadian culture?

BERRY: . . . My impression is that particularly in Paint Hills there is a viable Cree-oriented community. Most of the people use Cree, speak Cree, many of the people still maintain strong relations with bush as a way of life. Statistically I don't know the details, but my impression is that fewer people were on welfare and on wage employment than in other communities that I have worked with and in general I had a picture of hope that something could be done. . . . In Fort George, my impression was of a community divided. . . . My view is that the already high level of stress apparent in Fort George is ripe for increasing levels, especially if they do get a fairly major dose of construction workers, alcohol availability, easy transport in and out and so on.

INGENIOUS, IMPERSONAL MACHINES

The day after we returned from the river we headed inland again, but this time in a helicopter, the commercial charge for which, if we had had to pay it, would have been something like $700 for the seventy-mile flight. A young pilot efficiently hauled us into the air, and with bored indifference carted us upriver. Only the day before, we had been on the river with the old man, watching how every channel, island, sand bar, even every submerged rock was known to him and loved by him. Now we were in a machine, a marvelously ingenious machine that could hover, move sideways, climb quickly, drop suddenly, do everything but dance, and to the men running such machines the superb wilderness below was featureless, boring and meaningless. Efficiently they had carved the top off a hillside just past the second rapids. Efficiently they had bulldozed all the trees down the bank. Efficiently they had transported into the wilderness a small township of tents and big huts and great machines. Where there had been trees and rocks and water through 8,000 years, disturbed only occasionally by animals and a hunting man, now there was a restaurant with fluorescent lights, a shower room with constant hot water, the ceaseless crackle and roar of the radio-telephone system, the whine of chain saws and machine shops, the constant coming and going of helicopters and airplanes. Where there had always been hunters, deeply contented with their work and their way of life, now there was a largely bored work force, men who were hostile to the landscape and felt the landscape was hostile to them, men who either had nothing better to do elsewhere in the world, or who had come up here just to make money, lots of it, quickly, so that they could do their living later, at some other time that usually never arrived. When work stopped, there was Ann-Margret simpering across the screen to remind the men that if they could ever get out of this god-forsaken hole with money, it would buy lots of beautiful arse when they got back to the big city. A great big pinned-up nude stood on the wall above the head of the cheerful fat cook in the restaurant where we were fed liver and potatoes. We sat next to Brian Deveney, a young helicopter pilot whose previous job had been flying around Cambodia shooting people in trucks. He worked every day from ten minutes past seven in the morning until nine o'clock at night, flying back and forth over the wilderness. "This is some of the most barren country in the world today," he said. "It has all been scraped clean by glaciers years ago and virtually nothing has grown since." The other day, he said, he had flown inland to a Lake Vincelotte. "Now I know where nowhere is."

Though he probably didn't realize it, he was expressing the official government attitude to this environment: the publications of the James Bay Development Corporation described the region as empty, undeveloped, "a land of tomorrow," and the river as "wasting away in foam and swirls." Between this impersonal attitude of the machine men and the deeply personal human attitudes of the hunters, no meeting ground existed.

JOB SEES THE WORKS OF THE ENGINEERS

Job and Gilbert flew up to LG-2 the next day. The old man fitted uneasily into this technological world with which we now surrounded him. He had moved all his life along and around the river, in total command. But this new place that had suddenly appeared on the top of a hill that he had always known well was a shock to him. Two miles downstream from the camp, the engineers had created a landing base on the river by hacking down the trees that grew on a spit of land right where a splendid little stream tumbled over a half mile of waterfalls into the great river. Job had known this spit of land and this little stream all his life. Every year he had pulled up his canoe here, for this was where his longest portage began, inland to join up with a system of lakes that carried him north before cutting across eventually to the Kanaaupscow River, the great tributary of the La Grande. "When he saw that place and what they'd done to it, his eyes filled with tears," said Gilbert. The old man had closely observed the river from the plane: he noticed every change wrought by the white men, the places where garbage, oil cans and other things had washed up onto the shore, and he was disturbed that already the effects of the interlopers were evident.

For the first and only time during my acquaintance with him, Job looked uneasy and unhappy as we lined up in the restaurant for food: he broke into a sweat as we photographed him sitting with Gilbert at a table among the workers, eating the unfamiliar food. That evening the camp directors, worried by a bear hanging around the garbage dump, asked if he could shoot it for them. As Gilbert and Job crept toward the dump, they caught a glimpse of the bear through the trees. Job could tell that it was not in good shape, he told me afterward. The bear was sick. Job took a shot at it, but he had only a .22 and missed. He disappeared into the forest and returned after half an hour. The bear had got away.

We went to bed that night, all four of us, in the cots in the visitors' hut. Midway through the night we were awakened by a most unholy moaning, a terrible pained cry, some strange words wrung from the old man's throat, culminating in a great involuntary shout. For an hour thereafter, lying in the barren, cheerless hut so different from the warm, sweet-smelling tents he was used to, he talked to Gilbert in a gentle, low tone until we all fell asleep again. "He had a dream," said Gilbert the next day. "I couldn't tell you what it was, it was personal, it wouldn't be right to talk about it. But it had to do with what he had seen, all that he had seen at this place."

JOB ON BEARS

Job Bearskin tells what he knows about bears: "I will tell you what I can about trying to kill the bear. If you see a bear and are alone, trying to walk up to him, and if the bear hears you, he is going to run away, and you will not be able to shoot him. His ears are very sensitive. But he cannot see very far. He has not got sharp eyes. He can only see a person who is close and out in the open. Then he can recognize you. That's what the bear does.

"I have walked up to a bear on the open ground while he is eating berries, and I had no trees to hide behind. He cannot always keep his head down, he looks up from time to time, looking around at anything that is moving. He is always on the lookout for something that might harm him. When he looks at me straight, I just freeze there. While I am still quite a way from him he cannot recognize me. That's how you

A man in his element, summer 1972, Job Bearskin by his tent at Kanaaupscow, where the Kanaaupscow River used to join the LaGrande River, now flooded by the LG2 reservoir. Photo: © *Guy Borremans*

Philip Awashish at his father, Isaiah Awashish's camp in February 1972.
He became the Crees' chief negotiator of the hunting, fishing and trapping regime under the James Bay Agreement. Photo: © *Boyce Richardson*

Though founded in 1669, Rupert House still has an improvised air three hundred years later. It was here that the first ships of the Hudson's Bay Company landed to trade for furs with the Crees. By 1990 the village had been almost completely rebuilt and renamed Waskaganish. Photo: © *Boyce Richardson*

Sam Blacksmith and his fourteen-year-old son Malick return from the bush to check traps. Lac Trefart, March 1973. Photo: © *Boyce Richardson*

Ronnie Jolly at Sam Blacksmith's camp, fashioning a snow shovel, with two-year-old Abel Coonishish and Mrs. Jolly nearby. Like every experienced Cree hunter, he could also make beautiful snowshoes and many other everyday artifacts. March 1973. Photo: © *Boyce Richardson*

Once a Cree hunter's drum has been brought into the lodge, it must be played. And no feast can end until the drum is played. March 1973, Sam Blacksmith plays the drum and sings at a feast to celebrate the killing of four moose. Photo: © *Boyce Richardson*

When Cree storyteller Samson Nahacappo holds fourth inside his
crowded teepee at Fort George, the children do their best to listen in.
Photo: © *Guy Borremans*

Job and Mary Bearskin in their teepee beside the LaGrande River on a fishing trip in the summer of 1972. Photo: © *Guy Borremans*

Job Bearskin at the foot of one of his people's graves at the old Cree burial ground above Kanaaupscow Post, 1972. This land is now buried beneath the LG2 reservoir. Photo: © *Guy Borremans*

approach a bear. I have heard many Indians say that, and also my father has taught me the same way.

"The bear knows that man is out to get him. That's how smart he is. He is one of the smarter ones for a human being to try and kill. But if there are trees around, you can hide, and you have a better chance of walking up to him. If you want to walk really close to him, you have a good chance if there are trees around. If you do not want to get really close, you can still shoot him from far off. I have killed a bear with a .3030 from quite far away, because the bear was in an area that was not good for getting close to him.

"That's the behavior of the bear. That is why you cannot kill bears in great numbers, unless the bear decides that he will be killed by a certain person—then it is very easy to get him. That's what the Indians have said, and that's what I have been told, too.

"Some bears as soon as the snow starts to fall make a cave for themselves and start to hibernate. Sometimes he makes a cave where there is a cliff, and sometimes on high ground. Sometimes you find a hill covered with trees, and the ground is lifted, hollow, and the bear finds a place there. Sometimes the bear finds some sort of cave among the rocks and crawls in and spends the winter there.

"But you can still find signs of the bear. You can sometimes tell when the bear is near by the boughs on the trees. He breaks off the boughs, just as a human does. But there is a difference. A man breaks off boughs downward, so that the broken side is on top; but the bear breaks them upward, so the broken side of the bough is on the bottom. That's how a person knows the bear is around, by the way he breaks the branches.

"When a man spots those branches he knows right away there is a bear, because the bear uses those branches to sit on. He also uses them to cover the cave door—that's the way he hides in the wintertime. In spite of the precautions the bear takes, some people can still find him in the winter. After you find these branches you start digging around in the snow, but of course people who have found bears before know where to start digging. They are digging around looking for the cave door. The first thing they are looking for is the dirt that the bear has dumped out of the cave when he dug his winter quarters. You can tell this dirt that the bear has thrown out because it is frozen solid; that's the way you tell that it is the doing of the bear.

"Sometimes there is snow in the cave door, but the snow is melting, and there are icicles at the door, because of the warmth coming from the bear. This is another way you can tell where the door of a bear's cave is.

"There are no berries around where the LG-2 camp has been built,

or the bears would not be eating the garbage. The bear would rather eat
berries than garbage, if there are berries. He feeds on bearberries, as we
call them; that's what he eats. Usually when there are no berries around,
the bear likes to eat meat. He can kill it for himself, like fish. He fishes
in shallow water in the springtime. But after the berries grow he eats them
instead of fish. Some bears will not eat garbage at all. They wait for the
berries to grow. It doesn't matter if the bear doesn't eat for quite a while,
he can handle it.

"A bear eats the plants that have just started growing in the spring,
he eats flowers that have just grown. You know, some bears will eat
anything. They eat mice and ants. They look for ants where there are
dead trees and under rocks. The bear will turn over a rock and arrange
it to make a home for the ants, and then he will leave it and perhaps not
come back to look for the ants for as long as a year.

"A bear has her young ones during the winter, and in the spring she
takes her cubs with her. The bear stays with her cubs all summer and
when it is time to hibernate again, the cubs live with the mother all winter.
The young cubs usually live with the mother for only one year, and the
next summer the cubs take off on their own. Then the young cubs usually
hibernate on their own, but sometimes the bear will stay with her cubs
for two years before she allows them to go off."

A GIGANTIC SCHEME

O'REILLY: Could you describe the dams?

EINAR SKINNARLAND [engineer]: At LG-2, the river is 1,000 to 2,000 feet
   wide. The sides of the river are very steep, which is one of the things
   that makes it convenient to build a dam there. The first thing that
   needs to be done here is to build a dam so that a lake can be built
   to whatever level is desired. That stops the flow entirely, and certain
   smaller dams and dykes are needed to keep the water within the
   reservoir. The second part is to get the water from the lake into the
   powerhouse. The LG-2 powerhouse will be built underground. The
   water goes through tunnels to the powerhouse, and then back out
   into the river. The third thing is to take care that if you have more
   water than you need, you have a spillway to let the water out of the
   reservoir, and back into the river 10 miles downstream. The water
   behind the dam is called a forebay, but it is so large it will also
   function as a reservoir. You will fill it up, and lower it in the winter.
   The technical difference is that a forebay is where you take the water
   into the powerhouse; it is also combined in this case with a lake. The

area of the river marked in white on the map will remain totally dry.

O'REILLY: What is the distance that will become dry?

SKINNARLAND: Two and a half miles. The highest level of water in the lake will be 575 feet above sea level. It will be drawn down to 545 feet, a variation of 30 feet. But where the water comes out into the river again, the elevation is 108 feet.

With Job and Gilbert, I clambered down the steep hill from the LG-2 camp to the river level, where the superb rapids roared ceaselessly, as they had done for many centuries. The rocks were wide and vast on our side of the river, and we walked along them for a few hundred yards before sitting down on a broad, flat rock around which the waters boiled (it was through these rapids that Stephen Tapiatic told the court that he always navigated his canoe by pole). Job had brought his gun, as usual, and an ax and his pot for boiling water, and within a minute or two of sitting down he had a fire going and water boiling for tea.

"Would you tell him," I said to Gilbert, "that when the project is built these rapids will no longer exist. The dam is to be built about five miles upriver, the water will be taken in tunnels to a powerhouse that is to be built five hundred feet inside the rock, and then it will be brought back into the river farther downstream. So just here, where we are sitting, there will be no river, it will be dry. The rapids will be no more."

Gilbert translated it to the old fellow, who looked out over his beloved river. He replied very quietly in Cree—I could scarcely hear his voice against the roar of the waters, and for several minutes Gilbert did not translate what he said. We all three of us sat on the rock, without moving or speaking. Finally Gilbert said softly to me, "I don't think he can believe it. He says it is unnatural."

SKINNARLAND: The dam will be a tremendously large fill dam, to be built out of rock, with a layer of concrete to seal the water. It will start at elevation 100 feet, and rise to 590 or 600 feet, the equivalent of a 30-story building. To be able to build this dam, it will be necessary to take care of the river during construction. . . . In this case all they need to do is to take the river and get it out of the way. There will be two diversion tunnels. When they have been constructed, two small coffer dams will close the river off.

O'REILLY: How wide will be the empty space in the river during construction?

SKINNARLAND: About 2,000 feet . . . After several years you have to close these diversion tunnels, to close the river off completely, unless provision is made to let water out, and there is no provision for that in the plans. Then, no water will be flowing into the La Grande until it has been filled up. This will depend on snow and rain. But based on the schedules we have been given, it may take a year. It will be closed one fall, and it is scheduled to be full next fall, and during that time there will be no water coming into the downstream part of the La Grande.

O'REILLY: How much concrete would it take to build LG-2?

SKINNARLAND: Two to three million cubic yards, if you build it out of concrete. The amount of fill is estimated at present at 23 to 24 million cubic yards of material in the main dam. In truckloads, that would be 2,500,000 and that is just for the main dam. Also large quantities would be needed for the dykes.

O'REILLY: What kind of material is to be found around LG-2?

SKINNARLAND: There are glacial tills, there is dirt with a lot of fines in it, there is sand and gravel in the area, and you do have an abundance of solid rock, which can be blasted and used.

For about an hour and a half we sat on the rock on the edge of the great river. Job Bearskin talked about this person, this river, which had always helped the Indian so much. About its clean, good taste. About its good-tasting fish. About the many places he could camp along its shores. About the vegetation on the banks where the animals liked to feed and were easy to find and kill for food. About its utility as a highway, helpful to the Indian when he wanted to travel up and down the country. Nothing, in Job's experience, had ever been so helpful to the Indian as this river. Poor river, its millennial power in helpfulness to be cruelly rewarded by a death blow. Poor Indian, to be deprived of his greatest helper.

O'REILLY: Could you describe, after the putting into operation of LG-2, the changes, if any, in the channel flow in the lower La Grande?

COLIN TAYLOR [lecturer at Trent University, Peterborough, Ontario, hydro-meteorologist and geo-morphologist]: First of all, the annual fluctuations will be evened out considerably. At the present time, the flow in the lower La Grande for the period of record which is available fluctuates from approximately 12,000 cubic feet per second up to 160,000 cubic feet per second. These fluctuations will be eliminated, the flow will vary from a minimum of 92,800 cfs in July to a maximum of 113,300 cfs in February. . . . It is also interesting to note that this, in fact, almost reverses the seasonality of the maxima and minima. February under natural conditions is a month of low flow, 20,500 cfs, that will become a maximum flow month at 113,300 cfs. July will not be affected greatly, it already has a fairly high flow.

O'REILLY: Now the mean annual flow in the lower La Grande, will this be affected by the implementation of the project?

TAYLOR: Yes, the mean annual flow will be increased by my calculations by approximately 70 percent.

O'REILLY: Could you express any opinion as to the effects which would be wrought on the lower La Grande river system by the stabilization of the flow, and the increase of the flow by 70 percent?

TAYLOR: Yes, the important point about the stabilization of the flow is the fact that it is created by damming the river and forming a forebay behind the dam. When a river is dammed to form a body of standing water, the water flowing into this standing water loses its capacity to transport sediment. Any sediment that is being carried by the river from upstream is deposited in this standing water. The effect downstream is that the water coming up below the dam is clear of sediment. Now, a river has a complex system. A river, to remain in a condition of equilibrium, achieves a balance between a great variety of variables, including both the supply of water to that reach of the river and also the supply of sediment to that reach of the river from upstream. For a river to remain in the same condition, the amount of erosion which takes place should be balanced by the amount of deposition which takes place, not necessarily in the same place in the river, but elsewhere.

O'REILLY: Will this have any effect downriver?

TAYLOR: Yes, the river will try to regain the sediment that it loses behind the dam. It will need to do this in order to establish a new equilibrium form. One of the controlling factors in the river system has been changed, the river must adjust to that change. . . . I feel that the tendency of the river will be to erode. The rate at which it does erode is indeterminate at this stage, as far as I have been able to establish. There is no data on the exact grain size composition. There is no data available to describe the strength of these channel banks or the strength of the material on the bed of the river, and this information is essential if one is to attempt to quantify the magnitude of the changes that would take place.

We asked Job to tell us what he thought of the LG-2 camp. "Okay, I will tell you how I feel about it," he said. He rubbed his hand over his brow, through his close-brushed hair, let his hands fall by his sides, raising them and letting them drop in a gesture of disavowal as he spoke.

"It was never like this before they came. It was a beautiful earth. The people really liked to look at this beautiful earth, but now it has been destroyed." The food needed by animals had been destroyed. The white man had chopped down trees that the Indians guarded carefully because they were important to the animals. "When we flew over here I saw how many trees are being destroyed. And it is all white man's destruction.

"They are robbing the land from us, the owners of the land. I am worrying and wondering about that all the time. There is going to be a lot more destruction, I feel. It is only beginning." He folded his arms in a magisterial gesture; his face creased in a sort of half smile, as if trying to prevent himself from weeping. With Gilbert he walked over to a ditch. "Is this what it's going to look like, all of it? Speaking as an Indian, I don't like this at all." Some people could tell what would happen in the future. "I am always thinking about the future," he said, adding that these were signs of what the future would be like: he was now seeing the future.

"It's just like ripping something apart, it doesn't look good," he said, folding his arms again as he looked along the ditch torn out of the land by a huge machine. "It looks like people have been fighting, everything is shattered. I have seen it before in the mating season for the bear: they fight, and when they do that, they usually tear up a lot of land." He bent down to pick up an exposed root of a small tree, decimated, stripped, raw. "They are killing the roots, and in my opinion nothing will grow here again. This is the way it's going to be. The white man is only thinking of himself. Many people are saying that. The white men are not even thinking about the land they are destroying, they are thinking only of money."

KANAAUPSCOW

Brian Deveney flew us another forty miles inland to the abandoned Kanaaupscow post of the Hudson's Bay Company (more than $1,000 worth of helicopter time). The La Grande River dips southward in a V just inland from the construction camp, and at the bottom of the V it is joined by the Kanaaupscow, which flows in from the northeast. We did not follow the river inland, but in effect flew from the top of one arm of the V to the other, across an immense tract of low-lying, forested lake country. It is estimated that there are 100,000 lakes in the La Grande river basin alone—there must be millions of lakes in all of northern Quebec —and it was only when one flew over this enormous wilderness that one could begin to conceive the scale of the proposed hydro-electric project. All of the land that we covered by helicopter in our one-hour flight would eventually be covered by water, and that, of course, would be only a fraction of the proposed flooding. For 500 miles inland from the coast the countryside would be turned into an almost unbroken string of vast reservoirs. Even to a casual observer the scheme seemed ill-conceived: in this low country the dammed water would simply spread all over the countryside, filling its innumerable valleys and washing out its lakes, ultimately to be stopped only by the proposed eighty miles of retaining dykes.

Deveney set us down on a point jutting out into the broad, flat, swift-flowing waters of the Kanaaupscow, which, though only a tributary of the La Grande, had by this point run some 200 miles from its source farther inland and had become a major river. It was here that Glen Speers had built the Kanaaupscow post to service the Indians who trapped and hunted far inland: they were able to take their supplies from here in the fall and would return to this point in the spring. It meant that they did not have to complete the two-week journey to Fort George to dispose of their furs and get new supplies. As a younger man, Job had been one of those Indians, for his own territory was in the area south of Lake Bienville, the lake at the headwaters of the Great Whale River, which was not far north of the headwaters of the Kanaaupscow. So this was a place that Job had known all his life: he knew every Indian who had ever used it in the last half century, he knew all the pathways around it, he knew the people who had died and been buried here and the children who had been born here. He remembered when the shorelines surrounding the hill on which the post is built were crowded with the canoes of Indians who had come downriver from their hunting grounds, and he knew the huge rock on which all the other Indians would sit in the spring, watching the ice floating downstream and waiting for the first view of the trappers' canoes as they emerged from their winter in the bush. Now that he was an older man, his relationship with Kanaaupscow had become even more intimate: for, as is the custom, he had been invited by a younger man called Clifford Bearskin to join him in his hunting ground in recent years, and that hunting ground was right across the river from the post.

The great machine lumbered in like an ungainly bird, yet surprisingly delicate as Deveney chose a landing pad on a small spot of open land among the three abandoned buildings. We ran out of it crouching: I was wearing an orange waterproof suit, and the moment I stepped out of the helicopter it was covered with those tiny blackflies that are the bane of the Canadian north. Almost before we knew, the machine had lifted and disappeared over the hill. It would be back for us at five o'clock Tuesday, in exactly two days.

The abandoned buildings had a slightly ghostly air: the white paint was chipped and the red facings and roof were faded. The windows had been knocked out, and from a distance the old store, with its empty eyes and gaping mouth of a door, looked like a melancholy and lonely face abandoned in this wilderness. The whole knoll had once been cleared and busy, but nature was claiming it back, and the old paths were just barely discernible, running down to the shore and farther up the hill to the burial ground. We walked down the hill toward the river, and found Job and Gilbert, who had come on an earlier flight, squatted there over a small fire, drinking tea and talking quietly. Gilbert had been away from home in the white man's schools for fifteen years. He had left when he was only six or seven. And these days spent with Job were proving to be like a rebirth for him. The two never stopped chatting away in low voices as the old man told the young one about the past, and explained to him the Indian tradition.

We walked up the hill toward the ghostly face of the old store. We pushed open the creaking door and went in. In the eleven years since it had been abandoned the animals had taken it over. The businessmen had left behind them items that may have seemed worthless when they moved out, but now seemed rich in associations: fur tallies, old invoices, old balances and scales, jars that all seemed to speak across the years to us of the people who had handled them, signed them, handed them back and forth across the counter, investing each of them with some sort of meaning. The house was in better shape than the store: the door still closed and the windows were intact. But someone who had moved through there only a year before—"the Kanaaupscow canoe expedition, July, 1971," as they had written on the wall—had left the door open, and the animals had moved in. Besides, there were now many leaks. The sofas that had been left there were musty and torn. The drawers were full of invoices, signed by the last agent to man the post; he turned out to be our old friend Josie Sam. Job remembered all the agents: the best of them, he said, wrote with his left hand. He had been a good fellow, friendly and fair to the Indians. (As I discovered later, watching Speers write invoices in his Mistassini office, he was left-handed.)

Through the dirty front window one could look down the hill and across the river to the forest on the other side. Near dusk, the trees were sharply etched against the reddening sky. We went for a walk along a trail that led around the hill toward a big rock from which one could see far up the river: this, said Job, was the story rock, the theater rock, the center of all activities during the busy years. According to a local legend, at this rock a monster once stood, a cannibal, and watched the people gathering across the river to defend themselves. Job indicated an island, the place where nowadays he always made his winter lodge, where he caught ptarmigan by the hundred, where he trapped beaver, muskrat, caught porcupines and rabbits. He had, he said, already seen signs of the bear having been around. The Canada goose and ducks laid their eggs on that island, and the fish reproduced in the waters of the river. "That is why we call this whole place our garden," he said. "Around here you do not see any fences preventing people from trespassing. We let all others help themselves. I am happy when other Indians kill something because it helps me. And that is why we care for this place. This is the place where the animals and plants reproduce every year. The whole place is a garden, and they have done well this year."

BIG PLANS FOR A RIVER

ROBERT LITVACK: What may be anticipated in that particular stream [the Kanaaupscow]?

ROLF KELLERHALS [civil engineer]: The flow will be increased tremendously from virtually nothing [near the source] to a very sizable river. . . . [The increase] is probably in the order of well over 100 percent. An increase of that order will bring about tremendous erosion. It might even try to put the river onto a completely new course. The river might take a different path down the valley, like, bends that are now quite stable may be cut off. Besides the increasing widths with increasing discharge, a river also likes to flow at a lower slope. It likes to flow flatter so the increased discharge will cause tremendous erosion or scouring of the bed.

LITVACK: Where will this material go?

KELLERHALS: It will be brought into the LG-2 reservoir.

LITVACK: In the form of added silt?

KELLERHALS: Added silt, trees, everything, and this is 100 miles length of river, as the bird flies, so it's probably 120 or so miles measured along the thalweg of the valley. That is a tremendous length of river.

LITVACK: Would this phenomenon have an effect on the shoreline of the Kanaaupscow river?

KELLERHALS: What is presently the Kanaaupscow river will just sort of disappear. There will be tremendous erosion, particularly in the upper parts, the valley will be unrecognizable after the diversion. In the lower parts, this may just amount to greatly increased bank erosion. Some banks will remain, other banks will erode. This has never been done on this scale, and it is very difficult to say what will happen.

LITVACK: Well, when we say bank erosion, are we talking of, like blueberry
    bushes being washed up, or are we talking of spruce trees being
    washed up, or somewhere in between?
KELLERHALS: Well, just everything. Blueberry bushes and the spruce
    trees. I mean the spruce up there are not terribly resistant. They're
    relatively shallow-rooted.

THE STORY ROCK OF THE CREE

In the legend the people used the big story rock to kill the cannibal: they
killed him with the big rock and buried him under it, using the small
stones around the edge of it to cover him completely. In the fall the
hunters would take off from this rock in three directions for their territories and return to it in the spring. Before they left for the bush, everyone
would gather around the rock and hold dances, and when they returned
in the spring they would meet again to dance. "It was a happy time for
everybody."

Job's father, very old, was the man who would preside at the spring
ceremonies. One of the most important of these, which is still carried out
in each settlement of James Bay, is to initiate the children who have
"come of age," two years, during the winter. Until that age the children
are not allowed to go around the village or the camp by themselves. But
in the spring the child is given a pouch of tobacco, and he gives some to
the oldest man in the teepee. He goes ceremoniously around the teepee
kissing everybody, then walks out of the teepee by himself for the first
time, walks once around the woodpile and then returns to the teepee,
where a big feast is held.

The old man would also preside over the initiation of the young
hunters: when a child has killed its first animal, whether it be a fish, a bird
or a rabbit, pieces of the animal are distributed to the old people around
the teepee for them to eat. Then a bone or the beak of a bird is preserved,
stuck in goose or bear fat, and passed around for everyone to admire, and
again there is a big feast.

These ceremonies would be held when the hunters returned from
their hunting grounds with plenty of food in the spring and everyone
would get together for the first time. "I guess you heard," said Gilbert
to Job, "that all of this will be seventy feet underwater?"

"We were never told that," said Job. "I used to come across white
men in the bush and I would ask them what they were doing, but they
never told us. They would just reply: We are working. It wasn't until they
started to really destroy the land that we were told what was happening.

All the Indians were instantly opposed to the project because they were thinking about the land that would be destroyed. This is our land. We own this land. We think we have been robbed, that they're stealing from us."

We struck down from the rock toward the river and walked around the soggy shorelines back toward the clearing where the buildings stood. Job leaned over the river and pulled up a branch. "The beaver has been this way," he said. "If we look at this branch we can tell how old he is by the size of the tooth marks he has left." The beaver had eaten through the branch and apparently had brought the branch to that point and left it there. Usually, said Job, the beaver were to be found across the river. "We can see some signs of the beaver all along the river. Where will they go when all this is underwater?" When we returned to the buildings, Gilbert said, "Job enjoyed his walk. He told me he has learned such a lot from what he saw along the trail."

O'REILLY: Are beaver able to eat other things than willows and alder?

GARRETT C. CLOUGH [American beaver expert]: The beaver uses other food in other parts of its range. Beaver is primarily known in the United States as an animal which eats aspen. Aspen is available in the Fort George area, but not to the beaver because it's too far away from water. They will eat birch . . . but birch occurs very sporadically, and again usually not near the water. There are some water lilies, roots of sedges, which they will eat sometimes, too. But this only forms a portion of their diet in the summer. Beaver, if they're starving, will try to eat almost any tree . . . but [normally] they don't eat spruce. They eat primarily willow and alder.

O'REILLY: Did you notice whether there was lots of this willow and alder along the La Grande river?

CLOUGH: Yes, this wetland habitat lines the La Grande river and the tributaries that flow into it, but it's not a very wide band of vegetation. It may be as little as five feet, it may be at the most 50 feet.

Like Dr. Spence with his sample of the ovaries of the female whitefish, Clough produced a sample—beaver food, "to bring a little bit of James Bay to the court," as O'Reilly said. He held two twigs about as thick as his small finger. The corporation lawyers again objected, on the grounds of irrelevance, but Clough said the small size of the beaver food illustrated an important point: farther south beaver would eat poplar, alder and birch that might be five or six inches in diameter, but along the La Grande, eating only these small twigs, the beaver could chew out their food resources quickly. They cropped the twigs rather in the way that the people crop the spruce boughs, taking only some off each tree: they would, said Clough, take as many as twenty twigs off a tree three or four feet from the ground. This allowed the alder and willow (which grow about a foot a year) to regenerate quickly, so that within four or five years there would be enough food for the beaver to reinhabit that area. He estimated that beaver stayed in an area only two or three years before moving on to another place, whereas in places like northern Michigan, beaver would live in a single place for ten or fifteen years. He had estimated that the La Grande area had .75 beavers per square mile.

O'REILLY: Then what is the relationship between your figure of .75 beavers per square mile, and the total carrying capacity of the area?

CLOUGH: In my best professional judgment, the beaver are at about the carrying capacity of the land. There's always lots of area suitable for beaver which is not occupied by beaver at any single time. But this doesn't mean that more beaver could live there because as I pointed out there has to be some of this land lying fallow. The beaver actually use a rotational cropping, just as a farmer would let the soil lie fallow and nutrients build up at different cycles in maybe two or three years. The beaver actually crop their food on this rotational basis.

O'REILLY: Then how does that figure of three quarters of a beaver per square mile compare with areas to the south?

CLOUGH: Well, beaver can reach a maximum of eight or nine animals per square mile in very good habitat, such as northern Michigan or certain areas in Maine or southern Ontario or southern Quebec. The beaver is essentially the same animal in Russia and Scandinavia, and a beaver population can go much lower than the estimate I gave. It can get down to maybe one animal every three square miles.

O'REILLY: But in your opinion can that be increased very much?

CLOUGH: I don't think it could be. The beaver are not just limited by food. Beaver are very complex in their social organization. So one of the main determinants of where beaver live and how many can live in an area is the social factor. The characteristics of water bodies and the availability of food are important. A third factor, just as important, is their tolerance for each other. The beaver ecology is based on a family and one colony is comprised of one family. Even further than this, it will have only one mature male, and mature males actively keep away other beaver. They do this both by overt aggression and by signaling the beaver scent and so on. So they defend their territory and in all cases there has to be some living space between beaver colonies. They can't be adjacent to each other. In some cases a beaver colony might be half a mile along a small stream, and then there'll be a space of another half mile which beaver don't use. There has to be a buffer zone between the animals, and this would determine the amount of crowding that could be tolerated. Some animals can tolerate crowding and others can't and the beaver is one that can't.

o'REILLY: So when they migrate, these beavers who have to go out after a certain time, which migratory route do they follow?

CLOUGH: Well, in the life history of this colony, as I said, there can be only one sexually mature male. The young take two years to mature. In the springtime, June or so, when they're two years old they are becoming sexually mature. There would be aggression going on between these males and their father. And at this point they would be driven away from the colony, would have to seek some empty space in another colony. This annual spring or summer migration of the two-year-olds would be a regular feature of their life history. This is the time when many of these wanderers would perish. Any animal away from home is more susceptible to predators, to starvation and all the mortality factors. So there's a heavy mortality at this time. There is of course some natural mortality, the old animals. The beaver's average life would be seven or eight years in other places. I don't know about Fort George. Beaver can live longer than that. They can live up to about 18 or 19, but the average life would be seven or eight years. So there are always male beaver dying away and some of the two-year-olds would find a place. Of course, when there's trapping there's an annual harvest and removal of certain male beaver. So there are some empty spaces where these two-year-olds would move into, but many of them would perish.

JOB RELAXES

Though Job was happy to show us Kanaaupscow, an area of wilderness that has played an important role in the Cree perception of the world, from the point of view of an Indian hunter it is worse than useless to be taken somewhere by helicopter and dropped off, white man's style, for a couple of days. A hunter has to be able to move, and we were immobilized at the post. Job's canoe was stored for the summer on the other side of the river, where he had his winter lodge, and without it he could not begin that methodical process of catching food which seemed so effortless and impressive when we moved inland, on his terms, in the canoe. The fish can be counted on to feed an Indian while he is preparing to catch other game. But if you cannot set your nets—and he could not at Kanaaupscow—then you cannot catch fish.

Still, back at his old home, Job was in great form, and Gilbert, in intimate relationship with the old man, and seeing all this country for the first time, was in his seventh heaven. Job laughed a good deal at the bumbling efforts of Borremans and myself to cook food, chop wood and help erect the tent that he had brought along. Even when wakened by Borremans cursing about the mosquitoes in the middle of the night, he was cheerful. Gilbert had chased five million or so blackflies and mosquitoes out of the tent, leaving only 5,000 or so to bother us overnight. Within a couple of minutes of Borremans' anguished outburst, Job, squatting on his heels and giggling away almost uncontrollably, had whittled four sticks, stuck them into the ground and erected a net under which Borremans was able to sleep at last.

It was raining when we woke in the morning. It wasn't just raining: it looked as if it were going to rain for the next ten years. The entire sky was black, everything was soaked and dripping, and the rain that kept pelting down was cold. We went over to the house, cooked some breakfast and sat around for a long time talking to Job about his family. There had been ten brothers and sisters, of whom four were still alive. Three had died in the bush and one brother was buried here at Kanaaupscow. Job had come here with instructions from people in Fort George to count the number of graves. He believed there were eighteen people buried here, and it was a matter of concern to everyone that these people would eventually be under water.

Job said the rain would stop and in the early afternoon it did. We got out our equipment and went for a walk. Job took as lively and detailed an interest in every rock, tree and plant as we would in every shop if we were walking along the Champs-Elysées. Here, a porcupine had been eating at the topmost branches of a tree. Lower down, see, even the beaver had been around, he had come for food. There a rabbit had left signs of his recent presence. And over here, under this rock, is one of those places he had told us about, where the bear goes looking for ants. You see, the bear has overturned the rock to make a bed for the ants, and sometime later, perhaps even a year later, the bear will return to find the ants. Job took a dried cone from a pine tree. "This is what the squirrel likes best," he said. "The squirrel and the ptarmigan eat from this tree." Job set a rabbit snare for the camera: by now thoroughly happy and absorbed in his role as actor, he laughed and sang snatches of song, chatted and commented on the various qualities of the rabbit, joked over the various ruses he must use to try to outwit the animal. These were among the most cheerful scenes ever filmed of an Indian in Canada, and an intimacy and a directness about everything he told us made it all profoundly moving. I had never been with an older Indian who relaxed in this way. Nothing could affect Job's pleasure in our strange expedition.

THE DRASTIC EFFECTS ON ANIMALS

The blackflies and mosquitoes had miraculously disappeared with the high wind that had blown away the rain. I went down through the dripping grass to the bank of the river and squatted down by the shore. The skies had cleared somewhat, but were still intermittently full of black clouds, which drifted over the setting sun, sparkling and darkening the great, silent river by turns. The trees on the opposite bank were sharp-etched against the sunset. The river was 200 yards wide here, swift, a big river by any standards. Just along from where I squatted, Job had picked out of the water the willow branch that had been left there by the beaver, and had asked what would happen to the beaver when the river was flooded.

O'REILLY: What about habitats which are totally flooded? What happens to the animals in those habitats?

DON GILL [Professor of Geography, University of Alberta, specialist on boreal, or northern, environments]: Well, above a dam, it stands to reason that the preferred alluvial habitat is inundated, flooded out, and therefore, this has drastic effects on animals, this does force them to move, they either have to move or drown, die out, and you might say, well all right, if you raise the water levels, then this may be good because this is creating more wetland habitat. In fact when you raise the water level behind an impoundment, in many cases it's been shown that this actually reduces the number of miles of shore-line which then, of course, reduces the potential for wetland habitat to form. . . . It also takes a number of years for wetland habitat to re-establish . . . itself, and also, of course, behind a reservoir, no one really knows what's going to happen environmentally or ecologically along this greatly fluctuating water line.

O'REILLY: Are you in a position to make a preliminary appreciation of what might happen in a case of reservoirs fluctuating 20 or 30 feet, for instance?

GILL: I don't think I can comment on that, I don't think enough has been done. Certainly I haven't done enough to make a valid judgment on that. I have a personal feeling based upon past experience, I would say that it will have a very hard time establishing good aquatic vegetation but this is only a personal opinion.

O'REILLY: How do you compare the productivity along the large river systems with the smaller lakes and rivers in the north?

GILL: In general in the northern environment, the larger the river the more significant and important are the alluvial habitats along them, because of a greater flood stage, the greater input of nutrients to sedimentation, greater perturbation. . . .

O'REILLY: What is the respective capacity to support mammals?

GILL: . . . The larger the river the larger the quantity of animal life along it, in general terms. Again, I'm referring to aquatic animals especially. I'm not referring to caribou, which is a different animal.

O'REILLY: Have you any comments with regard to other kinds of aquatic animals, other than waterfowl? For instance, ptarmigan and spruce grouse. How would they be affected by controlled river regime?

GILL: Well, virtually all animals in the north are affected to one degree or another by impoundment. Ptarmigan, the so-called white bird or white grouse, which occurs very abundantly and is used by a large number of natives for food, especially during winter time, disperses very widely across the tundra and upland surfaces throughout the boreal forest during the summer time. But during the winter, the same birds . . . make short migrations down the river valleys, where they feed almost exclusively on willow buds. . . . If anything happens to the willow, in terms of its quantity, then obviously this is going to affect the winter habitat requirements of ptarmigan. You can say much less about spruce grouse. They use primarily such things as coniferous trees, which grow very widely throughout the north, and therefore spruce grouse would not be nearly as affected by impoundment as ptarmigan.

JOB LIGHTS A FIRE

I knew there must be some way of lighting a fire when everything in the world is wet, and that Job would know it. But I thought I would try it myself first. I went into the bush and looked for the dry twigs that the Indians usually use to kindle the big logs into flame, but all were hopelessly wet. Gilbert and I put some together under some logs, but we sputtered our way ineffectually through a box of matches. Job watched us for a while and then disappeared into the bush. He emerged in a minute or two with an armful of what looked like dried moss. He took the wet logs and with his razor-sharp ax shaved them, carefully keeping the shavings joined at one end so that they curved round in a nest, almost as thin as a sheaf of torn paper, and somewhat resembling those Scandinavian Christmas decorations that one finds in the expensive import shops. He placed this over the moss, with the heavier logs on top. He took out one match, skillfully lighted it in the high wind and ignited the moss. The shavings blazed, and within a minute we had a warming fire.

O'REILLY: Would you describe what the effects of the flooding will be [on the small mammals]?

MELVILLE BROCK FENTON [Professor of Biology, Carleton University, Ottawa]: The rising water will replace the terrestrial animals. . . . They will move away from the flooded habitats wherever possible. If they have no access to an upland area, one which is not to be flooded, they will drown. Those that can retreat to the upland areas will do so. However animals which are restricted to the flooded habitat will die because they have been displaced from the habitat to which they are adapted, which, in the case of the northern lemming mouse, is the very rich, closed-canopy forest immediately adjacent to the waterways. So these animals will die whether or not they can get to the high lands. Other animals which are not so choosy about their habitat will find themselves under conditions of extreme crowding. The upland areas have a lower carrying capacity, that is to say, they are able to support fewer animals than the rich lowland habitats which have been flooded. Therefore in these upland habitats . . . there will be a concentration of the animals that lived there before plus the animals that could retreat there. In this situation of crowding, there will be competition and death probably from starvation and a population crash. . . .

O'REILLY: What about managing the animals or replacing the animals?

FENTON: The animals that are adapted to living on dry land when the land is covered with water can't live there any more, and you must realize that this area has probably had about 8,000 years to achieve an ecological equilibrium and therefore the adjoining habitats are going to be filled. There are as many animals in these habitats as can be there. Therefore moving the animals that are going to be displaced by flooding will only cause worse problems. There is no such thing as managing an area when it's under water with respect to terrestrial animals. In my opinion that is impossible.

BURIAL TO THE RISING SUN

The burial ground lay on the brow of the hill, looking down over the post toward the river. It was open and covered with light green lichen. There was nothing formal about the cemetery, just some crosses placed in the ground, in a couple of places a little wooden fence to mark off a grave. For years, plants, grasses, lichen, trees had grown at random around the graves, but the open aspect to the river had remained: it was by far the most beautiful burial place I had ever seen.

Job described to Gilbert why the people were buried there. "We place the bodies toward the rising sun because when the Indian was alive, it was a beautiful thing for him to see the sun come up in the morning. When the Indian would get up he would come out of his teepee and watch the sun rise, just to look at the beautiful sight." The old man was speaking softly. As when he was introducing his songs, one could tell that he had modified his speech and delivery to the subject: of such subtleties is the Cree language made.

"The morning is so beautiful, and that's why we lay even our dead in a position to see it. Though they are dead, they are still part of nature, they are still with us, though dead, and that's why we place them toward where the sun comes up in the east." He walked around, pointing at the graves. "Here is the first child who died here after the post was established." They wandered slowly down the hill, looking for a grave a little apart from the rest. "Farther inland there are more graves, at least three of them, closer to the river, and all these will be drowned when the flood happens. We had an idea that this flood might occur, even before they told us about it. But we are worried about it. Who could wish that these people would be under water?"

JOSIE SAM, GUIDED TOURIST

By the time we returned to LG-2—Brian Deveney appeared out of the skies as scheduled, at five o'clock Tuesday—the atmosphere of the camp was already beginning to get to us. Though the reality was that the whole of the James Bay territory, including the land around LG-2, had always been occupied and used as hunting grounds by the Indians, the workers at LG-2 were convinced that Indians no longer hunted and were in no way discommoded by the project.

Surprisingly, we met Josie Sam, the most acculturated Indian in Fort

George, who was on his way back from having been taken on a guided tour of all the camps and the works in the area. As a gasoline salesman, Skidoo dealer, garage proprietor and taxi driver, Josie was no doubt considered reliable by the people running the construction work, who hoped he would give a positive account of the works to his fellow Indians in the village.

o'REILLY: Did you have occasion to visit that [road construction] camp?

SAM: Yes, I did visit the camp, and this is where we used, my older brother used to hunt there.

o'REILLY: Your older brother used to hunt?

SAM: Yes, on the lake.

MALOUF: What's the name of that lake, you say?

SAM: They call it Mile 17 in English.

o'REILLY: It's on the shores of the lake, and this is where your brother used to hunt and fish?

SAM: Well, that's his trapline.

o'REILLY: What's his name?

SAM: Norman Sam.

o'REILLY: Has he gone hunting and fishing on his trapline over recent years?

SAM: Yes, in fact, I met him when I came back from LG-2 and I went up there.

o'REILLY: You met him, where was he?

SAM: At the Mile 17.

o'REILLY: And what was he doing?

SAM: He was trapping.

o'REILLY: Now in that camp did you have occasion to see anything in particular?

SAM: Yes, there was a lot of damage in that area where they got the gravel.

o'REILLY: What kind of damage?

SAM: Well, getting the gravel for the road.

o'REILLY: Yes.

SAM: The reason I want to express myself here is because this is where the women they used to pick blueberries in this area.

o'REILLY: Yes.

SAM: It's all damaged.

o'REILLY: What kind of damage? Is it plowed over, or what?

SAM: Yes, plowed, getting the gravel.

MALOUF: You mean there's a gravel pit there?

SAM: Yes, gravel pit.

O'REILLY: Now did you notice anything else in particular? What about the water in the lake?

SAM: Well, they built their water station, the pump station.

O'REILLY: A pump station?

SAM: On the lake, they've made some sort of a wharf.

O'REILLY: That's built with what kind of material?

SAM: Well, they've put gravel on it, and boards.

O'REILLY: Like a causeway?

SAM: Yes, a causeway. That's where they used to fish, that's where they set their fish nets.

O'REILLY: Who set their fish nets there?

SAM: My brother. And the women that I'm talking about, they used to go and fish in the summertime.

O'REILLY: Now did you notice anything else about the water in that surrounding area?

SAM: Well, the thing I noticed right away is that the sewage discharges there. That is, discharges into the creek.

O'REILLY: Into the creek?

SAM: Into the creek. That comes out from the lake and the creek goes down to Fort George river, and the sewage system is drained out of the creek and down to the Fort George river.

O'REILLY: Now, at the LG-2 camp, did you notice anything particular in connection with water?

SAM: I noticed that the sewage is discharged down the bank towards the Fort George river.

O'REILLY: What is the distance between the camp and the place where this sewage is discharged?

SAM: Well, the cabin they have for the purpose of the flush toilets would be about 150 feet from the bank, it could be 200 feet, around that distance, and at the edge of the bank, that's where the pipe is that discharges the sewage down . . . the bank towards the river.

O'REILLY: Now, have you had occasion, yourself, or anybody whom you may have seen from the Fort George band, to hunt in the areas alongside the roads or in the immediate vicinity of the roads other than your brother, whom you've mentioned?

SAM: Yes, at the LG-2 airport, there's another trapper from Fort George.

O'REILLY: Yes.

SAM: . . . that traps in that area besides my two brothers. There's another brother of mine that traps in that area along the road, but at the LG-2 airport there's another family.

O'REILLY: And over the past two or three years, has he used his trapline?

SAM: My two brothers are using it every year, and also the other trapper, he uses it every year. In fact he came out last month from the bush.

A MEETING OF THE ELDERS

Job was anxious that we should hear the opinions of other hunters besides himself. Back in Fort George he gathered some of his friends to a meeting in his cooking teepee, beside his house. Mary and Job laid fresh spruce boughs, decorated the teepee beautifully with a weathered old root and set out their best china on a colorful tablecloth. At first the old men were nervous before the camera. They made a remarkable spectacle as they sat around the fire, the smoke drifting lazily across their gnarled, copper faces. Three of them were very old indeed: David Cox was small, his leathery face wizened from a lifetime in the open air, the very picture of an ancient Indian sage; Johnny Bearskin, remembered by Glen Speers from years before as the best hunter in Fort George, had a long, melancholy face, also leathery and tough in texture; and Samson Nahacappo was a gentle little man with glasses, with European rather than Indian features, but the greatest talker, the finest singer, one of the superb storytellers of James Bay.

The session lasted for three hours. By the time it ended, the teepee was crowded with people—the women had arrived to have their say, the children were jammed between the old people and were blocking the door. What started out in an atmosphere of diffident shyness ended in a burst of emotion and Cree eloquence. One after the other, the men gave me a rerun of what Job had said up the river.

"I want to know," said Samson Nahacappo, "what these guys are doing here, what they are up to. What the hell are they doing taking pictures and everything? Who are they?"

"They're trying to help us," said Job. "We should be grateful that they're trying to help us."

"If you have any power," said Samson, mollified, "you should try very hard to stop the project."

"As far as we're concerned, we don't care for what the white men have," said a younger hunter, Thomas Pachano. "But they are coming here and messing with what has enabled us to survive."

"The only thing I can suggest is that we take them to court," said Job.

"When the dams are built," said Samson Nahacappo, "where will the animals go? The caribou won't know which way to go."

CARIBOU OF THE UNGAVA

ROBERT LITVACK: Have your studies indicated the existence of caribou in the Ungava?

ALEXANDER BANFIELD [Professor of Biology, Brock University, St. Catherines, Ontario, and acknowledged leading Canadian expert on caribou]: Yes, the previous literature indicated that caribou were once abundant, and then had greatly declined. By 1954 no one really knew how many were around or whether they even existed, and this study indicated that there were perhaps 6,000 caribou in the Ungava. Since then other studies have shown that there are more today. The caribou herds have greatly increased since that time. A low point was probably around 1950 when the caribou may have been down to about 3,500 animals. . . . Later studies, 1958, indicated that there may have been about 9,000 caribou in the area of the Labrador boundary alone. And later studies still suggested around 30,000 caribou in the central area, and I believe this has now risen to about 40,000 to 45,000 caribou in the central part of the peninsula.

LITVACK: You mentioned studies of caribou on the Labrador side and on the Quebec side. Now, these herds of caribou, do they respect the political frontiers or do they move back and forth?

BANFIELD: I'm afraid not. They have dual citizenship, I guess. There's a considerable problem here which has not been solved to anyone's satisfaction. Starting with the original literature it was supposed that there were four distinct herds. One on the Labrador-Quebec boundary in the George river area, another on the north shore of the St. Lawrence, a third in the central interior, and a fourth in the western Ungava peninsula north of the area we're speaking of at the moment. . . . Studying the caribou west of Hudson Bay, we've found that there is a great transfer from one herd to another, year by year. One herd would swell and another decline. At the moment we feel that this has not been clarified in northern Quebec. There may be a shift from east of the Schefferville area to west of Schefferville [where] at the moment there are very large numbers of caribou.

LITVACK: Is there any particular part of northern Quebec which is particularly noted as a caribou habitat, or do they live pretty much the same all over the area?

BANFIELD: No, they're not evenly spread all over, as perhaps mice are that we've been discussing earlier. They definitely occur in herds and are found in groups. And you . . . fly over long reaches of land where you see no track or sign of caribou life. However there is a belt of woodland that stretches from about Lake Bienville on the northwest right over . . . to the central Labrador area . . . a woodland belt that is particularly rich in arboreal lichens, lichens growing on trees, and this belt seems to be the favorite winter area for these caribou. We do our counting in winter when they may be seen against the white snow. But even in that area there are favorite spots which the caribou occur in year after year, and the most favorite spot of all is in the Lake Delorme and Caniapiscau lake area.

LITVACK: When you say favorite spot, obviously this wasn't an expression of opinion of the caribou. What precisely do you mean?

BANFIELD: Well, I guess it was an expression of opinion of caribou and since they can't communicate with us, it's rather difficult to appreciate why they choose certain spots. We can make certain estimates based on scientific judgment that has to do with the richness of the forest, the vegetation, the amount of food available. There may be microclimates that make certain areas more favored than others. The one thing we know for sure is that they do not frequent burned areas, and that particular area is still quite green. The area immediately west towards Lake Bienville and Great Whale river had a large burn in about 1955 and this seems to restrict the caribou from moving westward in that direction.

LITVACK: Is there a particular type of woodland that is a favorite habitat of the caribou?

BANFIELD: Yes, the caribou do have quite a restricted habitat. We say that they are animals of a climax forest. They are the very opposite to the moose, that prefers open country, but the caribou, particularly in winter, must live in a green mature forest, or open woodland type, where they find the lichens that are their mainstay in the winter. These plants are, you've also learned about them, My Lord, very slow growing and very specialized and after a fire it could take from 30, perhaps to 100 years to regenerate. So this restricts the caribou quite a bit to a very definite habitat.

LITVACK: Do they stay there throughout the year, or do they migrate?

BANFIELD: These caribou migrate to some extent, but as a matter of fact, we don't know to what extent, they are there in winter mostly, from about October through to May. Even in summer a few caribou have been seen in this area by pilots, but we believe that the bulk of the herds migrate northward to the Larch river drainage, crossing the Caniapiscau but moving generally north westward to the forest. . . .

LITVACK: You're aware that there will be a north-south road from

Matagami to Fort George, a road parallel to La Grande river east-
ward to the site of LG-2, you're aware of the location of the dams,
the location of the reservoirs, the flooding pattern? Will this, in your
opinion, have any marked effect on the migration patterns of the
caribou?

BANFIELD: These roads will have a marked effect on caribou distribution
and populations, even the main road which runs east of James Bay.
In that muskegee country, there are a fair number of woodland
caribou, and undoubtedly they will wander on to the roads. The risk
is perhaps greater on the roads running up the La Grande river,
because these roads will be running at right angles to the normal
north-south migration pattern. We have found that they are deviated
from their normal migration route, and through this enforced deviation, the Indians and the Inuit may well miss the caribou at their
traditional hunting spots. Caribou are very traditional, and return
year after year to certain river crossings and to narrows in lakes to
cross, to swim across, and the native people lie in wait for them. Any
deviation from this may bring about hardship to the native people.
. . . Considering all these factors, what remains unknown is how
these will add up. For instance, forest fires . . . are known everywhere to be very damaging, but I couldn't tell you how much land
will be burned either through nature, or accidentally set by man. But
considering the obviously good range, the selection of that area by
the caribou and these other factors of interruption of migration
routes and everything . . . if all these other factors come up negative, then . . . there is no doubt in my mind . . . that damage could
go from a measurable significant damage to a most disastrous damage to the caribou range.

A MYTHOLOGICAL FLOOD

Samson Nahacappo responded to an appeal that he might tell us a story
about a flood. It was a long story, and greatly interrupted by the laughter
of the people who crowded into the tent as the word got around that he
was storytelling. But Samson himself never cracked a smile from beginning to end.

"The story goes that this guy, whose name was Mark, destroyed men
and animals, killing two people, and this caused a flood. He tried to
escape, but the water kept catching up with him as he ran. He had made
a raft and was running toward it. The raft was very big. The young people
here have never seen such a raft, but we old men know how they were

made. I made one myself and I used to paddle with it. We dried the wood, tied the logs together and used a long pole to push it. I think he made it the same way.

"Finally after much difficulty he reached the raft. And then he saw all the animals swimming toward him. Those were all the animals that we survive on today. They climbed on the raft and the raft began to float away, as more and more animals came aboard. They came aboard because they had nowhere else to go. All the land was flooded. The raft was the only dry place. And then he saw something really huge swimming toward his raft. I think if we saw such a thing we would be afraid of it. He told the monster, before he tried to climb on the raft, that all the animals on his raft were his brothers and sisters. But the monster climbed on the raft anyway, and tilted it. So Mark told him, 'My brother, turn around and face in the other direction.' "

David Cox muttered: "The monster must have been quite a sight if he was that scared and asked him to turn away."

"The monster grabbed one of the animals and Mark told him: 'My brother, leave this animal alone. You'll be doing him an injustice.' But the monster started to eat the animal and Mark told him: 'My brother, leave the other brothers alone.'

"Then he said: 'My brother, I think you have destroyed the land.' Because by now they could not see land anywhere, only water, and they were floating around. The monster was flooding the land foolishly, just like these white men now plan to do. Mark had everything aboard the raft, trees, grass, everything that we see around us today. He checked to make sure that everything was there, but two things were missing, white moss and sand. Those two things were missing, so he told his brothers: 'We are really in trouble. We are really destroying the land, my brothers.'

"Mark knew that he would have to test his brothers the otter and the beaver, and that he would have to kill another brother, the deer, to do so. He killed the deer, tore the skin off and dried it, and then he made strings which he tied around the waist of the otter and beaver so that he could send them underwater to fetch moss and sand. He instructed the beaver: 'My brother, try to get sand and white moss from the bottom.' The beaver and otter went, but could not stay very long. They had to be pulled up, and when they came up they were all stiff. Mark revived them. But there was only the mink left.

"We all know that the mink cannot stay too long underwater. He tied a string around the mink's waist and told him: 'Try to bring up moss and sand.' The mink jumped in the water and went farther down than the beaver and otter. The line moved at an angle, stopped, began moving again, stopped again. Mark started to pull the mink up, and when the mink came on the raft he was all stiff. His two front paws were closed tight. Mark forced them open. In one he found sand, and in the other moss. Mark revived the mink and said: 'You did it.' He helped the animal up, then he put the sand, the moss and the other plants that he had altogether and blew them out onto the horizon. Then he could see land as he blew and blew, and suddenly his raft was sitting on land again. He finished blowing and said: 'We'll leave it like that for now.'

"All the animals jumped off the raft and started to eat. I guess they must have been hungry. But Mark told them to stop eating, because they might eat all the land away. He told the deer, who had been eating moss, to go and check how much land there was. 'Deer, my brother, if you come to a hill and you're going down, start running. When you're going up the hill, verify the land, stop every now and then to test the ground." The deer did exactly what he was supposed to do, and came back. The deer reported that he had seen land as far as the eye could see, and Mark said: 'This is where the deer will roam and feed.'

"Then he sent out the loon, and all the other birds, in every direction. The birds came back and said, 'We can't get to the end of the world.' So Mark sorted out all the animals, blew once more in every direction and assigned each animal a territory. While he was sorting out the animals he heard a splash. It was the rabbit. Mark said, 'What are you doing?' The rabbit answered: 'I want to be a beaver.'

"Mark said: 'No, you cannot be a beaver because you will multiply too fast.' Mark began twisting the rabbit's kidneys every which way and told him: 'Quit fuckin' around.' Then Mark heard somebody crying. 'Who's cryin'?' he asked, and found out it was a squirrel. The squirrel said: 'I want to be a bear.' Mark told him: 'You can't be a bear because there will be too many of you.' The squirrel cried so much that this is why his eyes are white. 'You'll remain a squirrel,' said Mark. 'Men will have fun chasing after you. You can do anything you want, fart across the trees if you want.' "

Samson paused as the laughter in the teepee died down. "I used to have fun chasing squirrels with a bow and arrow," he said. "I was used to hunting with a bow and arrow." A ripple of interest ran through the people listening to him. Then they shouted with laughter as he said. "I'll finish the story here because what I am saying is not true anyway." He stopped, and finally, for the first time, gave a merry chuckle.

THE DAUGHTERS ARE CONFUSED

Now the women began to talk. They were a strong, handsome lot, show-ing none of the nervousness that their menfolk had shown when we began. They are powerful women, and they do not hesitate to say what they think. I asked them to talk about their children, and anything else they wanted. They described how, in a hunting camp, a girl begins to learn the Indian ways by fetching water and keeping a constant supply of firewood ready for the lodge."Then," said Mrs. Samuel Pash, "she is asked to check the snares and the nearby traps. Finally she is able to bring fish back, though at first she cannot take it off the hooks. The water is too cold for her, but she is allowed by her mother to put the hooks back in the water when the fish have been taken off."

"She learns as she goes," said her husband, another hunter who had joined the discussion. "She learns how to hunt by watching her parents. But for a boy it is different. He sets traps close by. Before he learns to shoot a gun a young boy likes to set the traps, but once he can use a gun, that's what he likes best, to go out and kill some game with the gun. The boy likes to go with his father and learn by watching him."

A remarkably beautiful, strong woman, Juliette Bearskin, spoke out forcefully from among the knot of women standing by the door. "I have two kids in school. I went to school myself and I understand what they are going through. I went through it for seven years, and before I was sixteen I was asked to leave school so that I could learn the Indian ways.

"I had lost most of what I had learned before I went to school. I had to learn it all over again. I had to learn how to check the hooks on the ice and set the rabbit snares. When I came back I felt as if I had lost everything during my short time at school, and it is the same with these kids now.

"One of my two girls is going to school down south. She was thirteen when she left, and the other one is here with us in Fort George. The two are very different. I know they are both learning many other things, and forgetting their Indian ways. We try to teach them when they are both here in the summer. The one who has stayed in Fort George learns our ways much faster than the other one. It is as if she doesn't remember anything she knew before she went down south. She has been slowed down. She seems to have a hard time understanding what we tell her. She's now slow in learning, compared with her sister. I think what caused her to slow down is what she learned at school down south. The sudden change in her environment, the change of cultures, is the cause of it. She must have seen a lot of different things when she was down there. Something must have swayed her. The white man's ways have swayed her. That's all I want to say. I'll end it there."

THE INTRUDERS

The people who crowded into the teepee talked for a while about the white man, and his negligence in the bush. The land is vast, but these Indian people know every inch of it, and little happens in it that they do not notice. They had, over the years, run across a lot of disturbing things on their hunting territories, particularly since the people from Hydro-Quebec had been roaming over the land, making studies for the proposed dams. Not only are the Cree used to asking permission to enter the hunting territory of another. They are used to leaving their equipment in caches, confident that it will never be disturbed. But in recent years they were finding that white men would come across their equipment and help themselves to it. Their ice chisels, traps, snowshoes, even their canoes would be stolen, a terrific shock to them after thousands of years in which the possibility of such an event never entered their heads. Their lands were being invaded, their equipment stolen, their ancestors drowned. The world was collapsing around their ears.

Juliette Bearskin spoke up again. "The first traces we saw of the white man were when we saw the things they had killed lying along the shore of the lake. We saw where they had tents downstream from Kanaaupscow, and they had just pulled a lot of fish onto the shore and left them there —big fish, too. Of course the area did not look very pleasing because of this mess." The tent was quiet now as she spoke, for her subject was serious. "They wasted all that food. And at the same place we found a dead beaver. We know they had killed it and just left it lying there. Because we found an ax handle nearby. They had just hit it over the head and left it there.

"It was the first time I saw traces of the white man, and right away I did not like the idea of having white people up here after having seen what they had done. All of the white men's camps that we have seen abandoned were all the same, left with things lying around all over the place. Things that the Indian has survived on, and still survives on, the white man was just throwing them away. The white man shows disrespect. I am not the only person who has seen this. Many people have seen what they have done: some have even seen moose that they have killed and just left lying there."

There was a rustle of indignation among the people in the teepee, and others, men, joined in with more stories of pitiful dead moose left lying around, of traps stolen and mink stolen from traps.

"We recognize that they have no respect for the animals," said Juliette Bearskin firmly. "There is no way we can agree. When we kill something, we do not leave it lying around like that. We do not know what they are up to now, but we know that they have stolen our equipment." She pointed to Job. "That old man there, they stole his wife's snowshoes when they went through our stuff. The stringing was made out of the best caribou hide, and there was also a young boy's birchbark canoe missing. Everything that is useful to us was missing."

SAMSON TELLS ABOUT THE BEAVER

Samson Nahacappo started again: "I am going to talk about the beaver. He is very smart. At this time of year he is just starting to collect his food for the winter. He starts building his lodge, too. But first he starts to gather all his winter food. He wants to have his lodge warm. He gathers so much food that it extends far in front of his lodge. He collects twigs and bushes and then gnaws through trees and puts them on top of the bushes so the ice won't reach them and he will be able to eat them when he wants them in the winter. Even if there are several beavers living in the same lodge, the food he has collected lasts all winter, and there is usually some left at the end of the winter.

"The beaver stays inside his lodge all winter. He never comes out unless somehow he has run out of food. But he also runs around during the winter in tunnels under the ice, along the shore. When I am looking for beaver dens, I start singing and count them as I go. I walk along tapping the ice with my chisel and, depending on the sound it makes, I can tell whether the beaver has traveled underneath.

"There are places where there are waterholes, no ice, and that's where the beaver comes out sometimes to check if the snow is melting, if he is running out of food. Before the beaver is two years old he starts having young ones. He lives with his young ones for one year and then they take off on their own. Sometimes they go in pairs, especially in February, when they start mating, and sometimes you can catch two of them in a den. The young ones are born in the month of June. Every time a beaver has young ones, the number in the brood is different, though some beavers are sterile."

Samuel Pash then chimed in, as the teepee full of young and old listened agog to the lesson. "My father once caught twelve beaver from a single lodge. That is when the beaver was plentiful. Now you hardly ever find that many in one lodge. The beaver lives with the young ones for only one year. The young ones are called pups. Beavers mate in February and have their young ones in June.

"Sometimes, if the young ones are of different sizes, they don't leave their parents, but stay and mate at home. They split and build lodges. But sometimes there's only one living by himself in a lodge. When the beaver starts having young ones at the age of two, she has two the first time. Three-year-old beavers may have three or four young ones. The following year the beaver is a fully grown adult.

"Nowadays adults may have four young ones, but when my grandparents were trapping beavers they told me that adult beavers might have as many as six at one time. When my parents were trapping and they found twelve in one lodge, four adults, four three-year-olds and four two-year-olds, it was the time when there was no disturbance and the beavers had peace. We began noticing away back that the beavers were decreasing. And we came to the point where there were hardly any beavers left.

"We knew why this was happening. It was because everything was growing old. We were reaching the end of a cycle. Nature had to renew itself somehow. And then we had forest fires, and afterward everything began to grow again. The beaver came back. Then they multiplied very fast, and soon we had lots of beavers again."

After three hours I had discovered that the Cree people, in Cree, are great talkers. Though they seldom talk to outsiders, they have a lot to say when the conditions are right.

OVERWHELMED BY WHITE INFLUENCE

But the young people are not such great talkers. When we tried to record a similar session with a roomful of young people the next day, they proved to have great difficulty expressing themselves, either in English or in Cree, though they are supposed to be the better-educated group. Many of them do not speak Cree well, having been forcibly divorced from it during the long years at residential schools. And they do not speak English with any flair, either. Their values, unlike those of their fathers, are confused. They have lost that sure cultural base which makes of their fathers such impressive, self-assured personalities. We had used up nearly all our remaining film, shooting around Fort George during the day, our last full day in the village. The scenes were rather different from those we had shot up the river. We had seen workers coming back from the housing construction with their tools over their shoulders, and hunters returning from a trip into the country to get game. We had gone inside Billy's house, in whose three bedrooms, we now discovered, eighteen people lived, a not untypical Fort George situation. The band council had taken over management of the houses and was charging Billy $100 a month, one of the techniques by which the authorities assure that Indians are tied into the wage economy and pulled off the land. Sharing the house with Billy, Martha and their four children were Martha's brother, his wife and their four children, two of Martha's other brothers, a mother-in-law, and a sister and her two children. Every bed was jam-packed every night. The other men were not working, and Billy had to find the money somehow every month. He had taken a mechanics course in Montreal and was dreaming of opening up a snowmobile agency. But it never came to anything because the distributors demanded $75,000 as the minimum guarantee for establishment of a well-stocked agency, a sum far out of the reach of any Indian, unless Indian Affairs could be persuaded to contribute. So it was not surprising that, a month or two after we left, Billy took a job helping to build the road to Fort George, working on the project which he, along with everyone else in Fort George, hoped would never materialize. No one could blame him: it was the inevitable consequence of many years of government policies designed to urbanize the Indians and detach them from the land which is so coveted by outsiders for resource extraction.

Only when the camera was turned off did the young people begin to talk more fluently in Cree and with a bitterness which was missing from the more measured discontent of their elders the day before. These youngsters are trapped: though most of them are great drinkers, they know that as drink becomes more easily available, the Indian family will degenerate and the social life of the village will be destroyed; though most of them are locked into the white man's cage of mortgage payments, wage packets, utility bills, they know that as the white man arrives in greater numbers, the quality of Indian life will plummet; though they are but part-time hunters, they know that in full-scale competition on the white man's labor market they cannot compete, that only by living the Indian way could they really command their own lives. And they know that for them this is now impossible.

Billy, who a month or two later would be building the road, made a long, worried attack on the effects the road would have on Fort George: "When the road is finished the white men will be coming up in throngs in their cars. They will be stopping at different lakes to go fishing without anybody knowing about it, and they won't want to have anything to do with the Indians. It's already like that: the Americans fly here in their own planes and they use any lake or river without bothering about the Indians. After they finish fishing they stop over at Fort George only to refuel and go home. When the white men flew here at the beginning, they would stop at Fort George and pick up an Indian guide. The Indians showed them where there was good fishing in different rivers and lakes. Now they know all the spots and they don't need the Indians any more. So they don't even bother to stop at Fort George. If they don't need to refuel, they go straight through. Sometimes they go straight back home without anybody knowing that they even came. The only way we know that they have been here is when we discover the mess they've left behind.

"It's going to be worse after they complete the road. Then drinking is going to be a lot worse. If we think that Fort George is bad now, wait till the white man starts coming in in large numbers. The way it is now, there's hardly any white men even though it seems there is a lot. Our children will be easily swayed by the white man. We can see how the young ones act now. It's going to be a lot worse in the future.

"When they came up with this project they said it would be in the best interest of the people and that they would be creating jobs for us. Sure, it will help the Indians who are going to be working there. But those jobs aren't going to last. After the dams are completed, there will not be any more jobs for the Indians. Once the dams are completed, all the Indians will be laid off and only a handful of white people will be working. I'm sure that the majority of those who will remain employed will be white." The jobs were usually impermanent and low-paid. Even $100 a week did not go far, for the workers had to buy food from the Hudson's Bay Company, the only store in town. And with children, $10 worth of groceries disappeared in the blink of an eye. "He'll be paying out money all the time, and none will ever be left. With the food bill, hydro, mortgage, fuel, he'll spend over $200 a month. Even if one is working all the time, he'll always be in debt."

For the younger Indians, debt was the vision of the future, and they knew that the bigger the project, the more white men around, the worse their relative position in the new society would be. The project would complete their disorientation: their glorious wilderness would be destroyed, the culture to which they still clung tenuously would be demolished, the tranquility of their small communities shattered. These young people may never have been hunters as their fathers were, but they had lived all their lives in a small village into which they had seen flowing the year round a perpetual stream of fresh food—fish, ducks, geese, rabbits, caribou, moose. Not only did this food still provide half of the diet of all the Indian families in Fort George (as the anthropologists discovered with their questionnaire) but it was their link with their heritage. It meant far more to them than just the saving it enabled them to make at the store.

They were embittered by the arrogance with which the white men made all the decisions about their land without bothering to consult them, and they deeply resented the denigration of the superb Indian competence in the bush which was implicit in all of the white man's actions. "The workers who come up from Montreal," said Billy prophetically, "are probably people who have been in jail." (Less than two years later I was told when visiting a jail in Montreal that the long-term prisoners had been offered parole to go work on the James Bay project.) "They do not care about the land. They have the idea that they own the land up here and can do what they like.

"The white man thinks that the Indian doesn't know anything. Yet when he comes up here he is like a little child. He has to be led by the hand or else he gets lost easily in the bush. When the white man comes up here he cannot put into practice what he was taught in his schools. The Indians take care of him. Last winter when the white men were working inland the Indians had to do their work for them because it was so cold the white men couldn't work. The Indians had to build fires for them all the time, and if there hadn't been any Indians with them they wouldn't have been able to do what they had to do." I had checked these stories out, and they were true. There had been some horrendous tales of groups of white men stranded in the bush because their motors would not work in the intense cold, helplessly depending on the Indians they had taken along as laborers to keep them alive until help arrived or their motors could be got going.

A PLAN FOR NO WATER SUPPLY

Besides, these young people were beginning to realize that even within the white man's own technology his work was sloppily done. They knew by now that the project might well wash away half of the island they lived on at the mouth of the La Grande, that it would ruin their water supply and that their needs were the last thing that anybody in authority gave a thought to.

COLIN TAYLOR [hydro-meteorologist and geo-morphologist]: . . . According to my calculations the present plan is to shut off the La Grande River at LG-2 for one year, and during that one year the average flow of the La Grande River will become approximately 1,200 cubic feet per second, which compares with the present mean annual discharge at the mouth that is in the order of 60,000 cfs.

ROBERT LITVACK: 60,000 reduced to 1,200?

TAYLOR: Yes, and the 1,200 being contributed solely by tributaries which enter the La Grande below the dam at LG-2.

LITVACK: I understand, Dr. Taylor, that this is not a permanent condition?

TAYLOR: No, this is not a permament condition. After the completion of LG-2 and the regulation of flows, the flow in the La Grande river will never decrease below approximately 92,300 cfs. . . .

LITVACK: Will the altered regime of the river affect the island itself, the banks of Governor's Island [on which Fort George stands]?

TAYLOR: . . . The fact that the overall average flow has increased significantly will mean a total net increase in the ability of the river to erode during a period of one year. With respect to Governor's Island, I feel that there could be significant erosion on the upstream end of this island. . . . I have located the line of deepest and fastest flow in this section of the river. A river normally flows at its highest velocity where the water is deepest. In any river this line of deep fast flow swings from one side of the river to another, and in the case of this stretch of the La Grande river, this line of deepest flow comes from the south bank, swings against this island here (indicating on the map another small island) which I feel will be eroded, but swings back right up against the northeastern edge of Governor's Island, up very close to the island, then swings to the other side. If the river is adjusting to an increased capacity to erode sediment, to erode the channel banks, it will be concentrating its attack precisely in such a location.

LITVACK: Would this have an effect of reducing the area of Governor's Island, in the point of maximum erosion?

TAYLOR: In fact it probably would because with the increased flow, there will probably be a decrease in deposition at the seaward end of the island as well.

HOW CAN I FORGET ABOUT THE ISLAND?

No one from the corporation or the government had ever appeared in Fort George to explain the plans to the people, and these young people knew that their needs would be—indeed, were being, daily—ignored. They, being Indians, were the insignificant factor in the gigantic equation of the province's schemes for their lands.

O'REILLY: Is there any particular significance to that place [where a road has been built] at the west end of [Governor's] island?

JOSIE SAM: The west end of the island is . . . that's where we used to hunt, spring and fall hunting, and also it's a feeding ground for the geese and ducks.

O'REILLY: So the spring and fall hunting is of, what, the geese?

SAM: Yes, the geese.

O'REILLY: Now you say it's a feeding ground. Would you describe to the court exactly what you mean?

SAM: Well, this is where the geese in the spring and in the fall, this is where the geese feed on the west end of the island. This is where we go goose hunting and duck hunting.

O'REILLY: Now there are many geese which come there or a few geese, or what?

SAM: Well, there's a lot of geese come in the spring and in the fall too.

O'REILLY: And do many members of the Fort George band kill geese?

SAM: Well, I'll say all the members that are able to hunt.

O'REILLY: What effect has the road had on these particular grounds?

SAM: Well, the thing is, they're going through our camping grounds where we used to set teepees there, and also we had in mind to have a . . . to bring out a course for the kids to teach them how to hunt and how to hunt geese and . . .

O'REILLY: Were the children from Fort George?

SAM: From Fort George.

O'REILLY: Why won't the geese come to that place?

SAM: Because of the damage they're doing, like they're pushing the trees away and they are scraping the land . . . it's a mess. The geese will never land any more there.

O'REILLY: What did the geese feed on?

SAM: Well, the feeding grounds are the plants and grass on the shore.

O'REILLY: Has the grass been affected at all?

SAM: Yes, it's affected by bulldozers because this is where, in the fall the geese come there and they feed on wild berries.

O'REILLY: You mentioned that there would be a course for the children. Is there any special significance to children and goose hunting in general? Why teach them goose hunting?

SAM: The reason why we want to teach the younger generation for goose and duck hunting is because they don't, they don't exactly know how to hunt when they're educated down south. They don't really know how to hunt the way they should hunt geese.

O'REILLY: Are there any geese in the surrounding area? You mentioned feeding grounds. Are there other geese feeding grounds right around Fort George area?

SAM: Around the Fort George area there's a lot of feeding grounds, but here at the west end of the island, it's been like that for many years and during the breakup, as you know, this is an island, and during the breakup we can't go across the . . . during the breakup we have to stay here. We have to hunt there until we can go across by canoe.

O'REILLY: So in point of fact, do many members of Fort George hunt geese at that particular spot, the west side of the island?

SAM: Yes, yes, that's what I mean, yes.

JACQUES LE BEL [cross-examining]: You will agree that this part of the island is not the only geese feeding ground, there are other feeding grounds all around?

SAM: Yes.

LE BEL: Right?

SAM: Yes.

LE BEL: And at springtime they still come back to the west portion of the island?

SAM: Yes, they still come back for the . . .

LE BEL: They still come back?

SAM: . . . but in the fall when they come back in the fall there'll be no feeding ground for them, there'll be no berries there.

LE BEL: But in the immediate vicinity of the island, there are also other feeding grounds, right?

SAM: No, no, there's no feeding grounds . . .

LE BEL: That's the only place?

SAM: That's the only part.

LE BEL: All along the coast, the geese don't . . . there is no feeding ground along the coast?

SAM: I'm not talking about the coast, I'm talking about the island.

LE BEL: Yes, but I'm not only talking about the island. I'm also talking about the coast in the vicinity of the island. The coast is not too far from the island?

SAM: Yes, I know, I know, but what I'm interested in is the island.

LE BEL: Well, never mind the island now.

SAM: How can I forget the island because I was born in the island?

WHAT'S WRONG WITH STRESS?

When we finished the young people's discussion on that Friday evening we were within sixteen hours of climbing into the plane for Montreal. We adjourned to Rose Sam's room in the federal school residence, about half a dozen of us, and I listened for an hour or two as they developed their discontent: they were not only embittered against the white man, and particularly distrustful of the French-Canadians and the Quebec government, but they felt helpless, too, before the control exercised over their town by the old people. Religion was strong among the old people, and life was made difficult for anyone who didn't go to church. The old people dominated the band council, did not speak English, could not relate to the white man's world and were easily manipulated by the Indian Affairs Department. Everywhere they turned, these young people faced nothing but frustration. And though they were a charming and gentle group of people, I knew that they were having many personal problems. A few months later, one young man tried to kill himself. Other youngsters from Fort George quickly dropped out of school when they came to Montreal, and hung around town drinking and smoking pot; one of the young men had cheerfully sired five bastards, and one of the girls had two. With the increased flow of transients through the town, already the older girls were having a hard time keeping their younger sisters—fifteen and sixteen years of age, and many of them astonishingly beautiful—out of prostitution. In this group was located the point of breakdown in that solid, coherent and profound familial structure which had meant so much in the Cree hunting culture. The difference between a group of hunting families in their winter lodge, commanding the wilderness and confident and proud in their skills, and the somewhat harassed group of families who jammed into Billy Bearskin's house and scrabbled around every month to find the rent, was profound: yet in the eyes of our governments it was the latter group which had progressed the furthest.

ROBERT LITVACK: Dr. Berry, what's wrong with stress?

JOHN BERRY [cultural psychologist]: In itself stress is very uncomfortable. It's a very unpleasant situation. It manifests itself in a number of minor complaints, irritability and so on. Personally and psychologically that's very important: legally, that's up to the others to consider. However I think that stress is also conducive to inter-personal and societal difficulties. People who suffer stress are irritable in their interpersonal relations. Numerous disagreements, bickering and so on are frequently associated with high levels of psychological stress. And at the societal level, stress and marginality are associated with apathy, disinterest in working. It has also been implicated in the use of alcohol, and I think that, although I don't have any particular data of my own on this, one set of data which I'm familiar with does tie the level of stress and particularly cultural stress, with rates of incarceration, criminality, delinquency, and so on. And so, there are psychological problems associated with stress which are in themselves unpleasant. As a psychologist I'm happy to say that I'd be satisfied with that as something wrong with stress. But in addition it has interpersonal and societal consequences, or societal associations which I think are also awkward and can lead to disintegrated communities and disintegrated individuals.

LITVACK: Is stress an indication of absence of mental health or a less healthy person mentally?

BERRY: Stress measures are very often considered to be the inverse of mental health scales. In the literature they're referred to in this way: community mental health scales or psycho-social stress measures. Insofar as you are stressed, you are less mentally healthy, insofar as you have mental health you have less stress. They are reciprocal and inverse relations. And so the word stress is very clearly related to a state of community mental health.

Eventually, when the effects of the liquor took hold, we all stopped talking and began to dance. For hours half a dozen of us whirled around the little room in frenetic and riotous competition. I collapsed, exhausted, half a dozen times, only to drag myself back onto my feet and begin another contest. I was still on my feet when we broke up at 5:30 a.m. When I boasted to Gilbert the next day about beating them, he laughed in that mocking way of his and said, "You were the only one competing."

THE DEATH OF THE CULTURE

A young man came knocking on our door early on our last morning in the village. His father, he said, had shot two bears for the white men at the LG-2 camp. He had brought them back to the village to distribute the 400 pounds of meat to the people, but the meat was poisoned. Would we go to his house, where the meat was, and look at it?

We walked over to the house of George Shem, who had skinned the bears and hacked them to pieces only to find that the meat had turned green because of the bad food the bears had been eating at the garbage dump. These were the bears Job had taken a shot at, and whose health he had judged at a glance to be poor. The kitchen floor was covered by this meat, which filled the house with an awful stench. The family stood around gazing down at the poisoned green meat. I noticed that one of the girls in the family was a pretty young girl who worked as a cashier at the Bay store. Every one of them, however, was subdued. They did not feel like talking as they helplessly participated in what should have been the most joyful experience a group of Cree people can have, the arrival of fresh meat and its distribution to the people. The culture is entirely built around the killing and eating of wild animals, and the arrival of a bear—in this case, of two bears—is as important an event as ever occurs in the life of a Cree family. What was lying on the floor was more than just a bunch of green meat. It was almost as if the destruction of the culture itself had taken material form in the poison that had infected the most powerful and respected of all the animals in the Cree world. It was a really terrible and heart-breaking scene.

The son took me over to the father, George Shem. "I am sorry for what I have done," he said. "They asked for two men to kill the bears that were worrying them at the garbage dump, but if they ask again, no hunter will go to help them. It is as if they have poisoned the meat. It is not good what they are doing there. They say the bears are worrying them, but they are worrying the bears, and the bears are one of the crops which grow in our garden."

The gall bladder had been promised to a friend so that it could be used to cure the illness of his wife. But it, too, was poisoned, full of green stuff, and could no longer be useful as medicine. "I am very sorry about that. We were so happy when we brought the meat in, we thought we would be able to feed our children as usual with this good meat. We have witnessed the damage these people are doing, and they have only started. We know what it will be like in the future. But we are all angry that they are destroying our food."

Job, Billy, Gilbert, Rose and others were at the airstrip to see us off. We shook hands gravely with Job and thanked him for the help he had offered us. We kissed Rose and assured her of our love. We laughed with Billy and Gilbert. I went to sleep as soon as the plane was in the air, exhausted after fourteen hectic days, and when I awoke we were far to the south, far from Job's garden and all the living things—the spruce, alder, willow, lichen, beaver, otter, geese, ducks, moose, caribou, bear and Cree—who share it.

10

1972: LAC TREFART

Philip Awashish failed in his effort to escape from the hassling of the white man by going fishing with his father that summer. It was no longer enough to disappear far into the wilderness. The tentacles of the bureaucracy followed him even there, and by the end of the summer he was up to his ears in legal consultations, claims and squabbles about the fishing.

I flew in to see him during one of several summer visits I made to Mistassini. Their camp was on his father's territory among the headwaters of the Rupert River. They had—as the Indians always do—chosen a beautiful spot on a sandy spit at the end of a small island in a lake. Just up from the beach in a shaded and sheltered spot under the trees they had built four or five of their comfortable tents. As our plane flew in they hurriedly hid some caribou meat they had shot—you never know if you can trust the pilot—and the evidence of other good food, of fish nets hanging out to dry, ducks hanging from trees, showed that they were living well. Philip, wearied by the duplicity he met as he tried to confront the James Bay project on behalf of his people, and disgusted by the inaction in the Indian defense, had found a quiet place where he could get his head together again.

At the suggestion of a bureaucrat from the Indian Affairs Department—always on the *qui vive* for an economic-development project that will look good in an annual report and prove their unceasing work on behalf of the Indians—Philip's father had decided to embark on a commercial sturgeon fishing operation for the summer. He had recruited four fishermen—Philip was not one of them—and together they had produced a lot of sturgeon for the Montreal market. It was hard work. The sturgeon varied between nine and sixty pounds, and when they were caught in the nets they had to be handled carefully so that they could be kept alive and brought back from the fishing places to the camp. A rope was then tied around the tail of each of the fish, which were in effect tethered in locations where there was plenty of food, tied on a rope long enough to allow them to wander around eating. When the fishermen had caught enough for a plane load, they would radio Chibougamau for a plane. And in the few hours it took for the plane to arrive, the men had to kill the sturgeon and the women gut and clean them so that they could be delivered quickly and fresh to the Montreal fish shop Waldman's the next day.

Their only problem was that they didn't know what price the sturgeon were bringing. Though they asked every time a plane arrived, and sent radio messages requesting the information, they never received a reply from Robertson, the Indian Affairs bureaucrat in charge of the project. They were worried and restive because they had no way of knowing whether they were putting in a summer's profitable work or merely providing jobs and money for other people. Philip asked if I could find out how much Waldman's was selling sturgeon for: the next week I sent him a note that the price was sixty-one cents a pound.

When I next saw Philip a few weeks later, he was drinking with his father in the Waconichi and I heard the end of the story. They received their answer from Indian Affairs only when they finished fishing. Then Robertson told them that, unfortunately, after the plane trips and the handling of the fish had been paid for, no money was left for the fishermen. They had worked all summer and had provided fish for Waldman's, jobs for the men who handled the fish at Chibougamau, orders for the plane company, a full-time job for an accountant in Chibougamau, and had occupied the time of several bureaucrats, only to be told there was nothing left over for them.

"That no-good Lobetson!" cried Isaiah (there is no *r* in Cree), breaking into English as he would occasionally do when really exasperated. "No-good goddam Lobetson!"

"My dad went storming into Robertson's office," said Philip, "and really blew his stack. It was the first time I'd ever heard him raise his voice like that to anyone."

They were with Robertson in the Waconichi when I came across them, arguing bitterly, and Philip was about to leave for Montreal to see what could be done. He was going to enlist the help of James O'Reilly and was ready, if necessary, to sue Indian Affairs. They were bitter: it was the kind of treatment that Indians had to endure almost as a matter of course. And the next morning Philip and his father were still in the Waconichi, still drinking. "I'm saying goodbye to my father," said Philip thickly when I tried, at his wife's request, to get him home. "My father is going into the bush tomorrow. He'll be gone all winter, and I won't be seeing him until the spring, and I am having a little drink with him before he leaves."

The old man, clad in his customary dirty old maroon jacket, was drunk and cheerful by now. Usually extremely reserved, he now patted my shoulder (the last thing I ever expected him to do to me) and said in English: "My friend."

Isaiah, though an intensely private man, often had bizarre adventures in his relationships with outsiders. "One time he was spending some time in town with me," said Philip, "and he happened to get up early and went to eat some breakfast, and as he was leaving the restaurant he was stopped by these two white hunters, who asked if he could find them a moose. It was the moose-hunting season for white people, and they hadn't been able to find one. So my father had a bit of time, and he decided he would go with them and find them a moose. He went with them in their car and they went off to the Chibougamau park area.

"These fellows had liquor with them, and they were drinking away at it all the way along, and they kept asking him to tell them when to stop the car. My father said to them, 'You know, it's not good practice to be drinking when you're hunting. We don't do that. Hunting is a serious business.' But they just kept on drinking. Finally they came to an area that had been burned years ago, open land, that was good moose country. My father told them to stop the car. It was an area that was hilly, with green little trees and other vegetation growing back, but not so forested that if you got up on the hill you couldn't look around for miles. These guys are still drinking and my father is still telling them they shouldn't be doing that. But they just carried on.

"So he takes them up on this country and finally he finds a moose for them. They are so pissed by this time, they are blazing away at the moose, but they can't hit it. And my father gets really uptight with these guys. He says, 'Give me the damned gun, I'll get it for you guys.' He took the gun and was goin' to shoot the moose for them, but there were no shells left. They'd shot off all their shells, so my father threw the gun in the bush and told them they were crazy people, and demanded they take him back to town."

ANOTHER FILM PROJECT

My visits to Mistassini that summer were preliminary to the making of a
film built around a Cree hunting camp—an idea now coming to fruition
after having worked its way laboriously through the Canadian National
Film Board bureaucracy in the previous several months. For this, I was
working with a cameraman-director on the NFB staff, Tony Ianzelo, one
of the finest cameramen in the country, and a cheerful, equable person
to work with. Every hunter we spoke to in Mistassini was willing to take
us on the trapline with him, provided we paid him for the interruption
we would inevitably cause in his hunting and trapping schedule, and we
finally selected a group centered on a man called Sam Blacksmith. His
territory, which he had occupied for thirty years, lay fairly far north, about
185 miles as the crow flies from Chibougamau, halfway between the
Eastmain and Opinaca rivers. He had invited two other families onto his
territory this season, so there would be not only the traditional multi-
family hunting camp, but a good collection of older and younger hunters,
and a full range of teenagers and children. The hunters took off for the
bush early in September as usual. We arranged that we would fly to their
camp at the beginning of October. That would give us two weeks to shoot
their fall activities, and allow us plenty of time to leave before the freeze-
up, which in that area, they assured us, never occurred before about
November 4. We had to be out before freeze-up because in fall the little
planes land on floats in the lake, and in winter on skis on the ice. When
the water begins to freeze, neither of these landing methods works during
the six weeks until the ice is thick enough to carry the weight of the plane.

On the last day of September we went again to Chibougamau, where
we met Philip, who had agreed to go with us as interpreter and translator
(and who, indeed, proved to be the indispensable link in the making of
the film). We went to the Waconichi for a last drink before plunging north
into the wilderness. Business was booming, had been all day, and already
some Indians were asleep over their tables. One youngster from Mistassini was circulating around the bar in a strange way, following Philip
around, talking earnestly to him. Philip said, "He's on mescalin." He
came to our table, pathetic and cringing. "Philip! Philip! I need help. I
want you to help me. I want to get off the goddam stuff. Can I go to the
center, Philip?" Philip was still director of the Indian Friendship Center.
"I can't take him there," he said. "You can't tell what he's going to do.
If he starts to break up the place and we have to call the police . . . well,
we can't take the risk, the center gets a bad name, we'll be hassled by the

police all the time. Already we have a bad name. There's quite a few chicks from Doré Lake who hang around town and specialize in shoplifting, and they hang out at the center, and the police come looking for them there."

An old man with a long, dark, strong face under a worn fur cap came over to our table. He had been in the bar all day and was heavy-lidded from the effects of the beer. He talked for a long time to Philip. "He says he's sorry he can't talk to you people," Philip said. "He can't talk any English."

"Tell him not to worry. We're sorry we can't talk any Cree."

"He says that when he is sitting close to people, he likes to talk, even if he can't understand the language they speak."

The young boy came back, jumped to his feet unexpectedly, began shouting and kicking out at a chair. The old man said something to him, and the boy leaned over, hit the old man on the shoulder and knocked over some chairs.

"These old men think he shouldn't drink so much when he can't handle his liquor," said Philip. "They don't know anything about drugs, they've never heard of any of these things." Though plenty of the Mistassini youngsters smoked pot, this was the only one who ever used hard stuff, at least locally. He was also the only Mistassini youth who spoke French well, and was therefore able to mix with the white youngsters around Chibougamau. He wandered off eventually.

"I was named after this old man," said Philip. "This is Mr. Philip Schecapio, senior. His wife was there when I was born in the bush, and she delivered me. So I was named after her husband. My father was out on one of his trips around his traps at the time. He realized that I was being born, and he interrupted his trip to return to the camp, but he didn't arrive in time.

"He has just been telling me that he's making me a pair of snowshoes," said Philip. "The old man has been talking to me about the bush. He has just told me that the most beautiful sight he knows is the sight of his son carrying a rifle and a packsack as he walks through the bush."

DEEP INTO THE CREE HOMELAND

The pilots of the little bush planes fly by instruments and readings, but also by landmarks. They get to know the look of this extraordinary land, can recognize many of the bigger lakes, become familiar with the Rupert and Eastmain rivers as they ramble through a maze of indeterminate

lakes. We needed two planes, for we were five men, plus the 738 pounds
of camera equipment we had brought up from Montreal, plus box after
box of food, gasoline, plates, pails, lamps and so on that we had bought
in Chibougamau the day before. The pilots, of course, fly the hunters into
their camps, and know the locations. But they were looking for Sam
Blacksmith's camp on an island in a smallish lake called Lac Trefart, and
the island was deserted. The hunters had changed their camp to the north
shore of the lake: they came onto the shore and waved as we flew around
looking for them.

We ran in on the water and pulled the planes up to the shore: the
men and women shook hands shyly. We had met them only a couple of
times before, and this decision to take a group of white men onto their
hunting territory was for them an adventure full of possible pitfalls. The
men got busy straight away unloading the planes. As package after package, case after case, emerged from the two planes, we began to feel more
and more foolish. There for only two weeks, we seemed to be traveling
like Jackie Onassis, with enough stuff to last any reasonable person a year.
We carried the stuff up from the shore, past the three tents that were
spotted among the trees, onto a piece of higher ground above the camp.
We had brought our own tent, which Philip had had a woman in Mistassini make for us, and as soon as it was unloaded the men set about
erecting it. I have been camping all my life, but very much *à la française*,
with an orange-and-blue tent erected on a metal frame, and this, of
course, was quite different. The free availability of wood is the key to the
comfortable camps that the Indians make. The men quickly hacked down
some tall, fairly thin trees, erected them into triangles at either end of the
tent, slung a supporting pole between the triangles, anchored them with
corner posts, slung horizontal side poles from each corner, and before
we knew it we had a tall tent with high walls held in place by sixteen logs.
The canvas walls were anchored with more heavy logs, staked to the
ground, and within an hour shelves and platforms were built around two
of the walls and hanging rails were slung from the center pole. Sam
Blacksmith cut a hole in the canvas roof, attached the lid of an old lard
pail to it, assembled the chimney of the stove and shoved it up through
the lard-pail top. Some girls arrived, two of whom were very dainty and
beautiful though shy, and quickly laid a magnificent floor of spruce
boughs, which they had gathered in anticipation of our arrival. Sam took
our stove, enlarged the draft hole in the front of the door, threw some
earth on the bottom, jammed it full of logs, threw a match into it, and
by some magic it began to blaze immediately.

I wandered down to the lakeshore and sat on a rock for a few minutes. The trees around me were nearly all black spruce, but they were not very tall or thick, many were in poor health, rather overgrown with heavy lichens, and a good number were either dead or dying. The thin muskeg does not provide a good anchor, and many had been keeled over by the wind. As at Kanaaupscow, I had been dropped down in the middle of a vast wilderness, more than 160 miles from the nearest settlement (East-main, on the James Bay coast). But this had been home to the Cree for thousands of years. To us it seemed empty. To them it was pulsating with life, with trees, fish, birds and animals, all of which were willing to help them in their journey through life.

Sam brought us some bear meat and fish, and after supper we asked if they would like to visit us in our tent. They all came in, sixteen people from the three families. We opened a bottle of Scotch, just enough in forty ounces for a couple of drinks for all the adults. As we drank I ran over again our purpose in being with them. I told them we wanted to show outsiders what their lives were like, so that people outside could come to understand what was involved for them when the white man's world interfered with their way of life. I told them again that before we finished the film we would return to show it to them, so that they could make any changes they wished, delete anything they thought should not be there or ask for anything to be added. Sam Blacksmith and Ronnie Jolly, the second hunter, were both quite talkative. They did not think there would be too much for us to see. They had been around checking the animals, and had not seen anything except the signs of the beaver. They had caught two bears, but nothing else. They had not seen any moose, since the moose did not stay in this area but had to be caught while passing through and it was a matter of luck to come across them while they were passing. It was the same with the caribou. Before they left they said they would have a bear feast the next night. They would be in camp all day, they said apologetically, cooking the bear. "Only the men are allowed to cook the bear," said Philip, "since it is the most powerful and important of all the animals."

The men spent all the next day cooking the bear. Sam fried the bear's fat in a big pot in his tent. This bear's fat, a valuable commodity, is boiled down and kept for use during the season. The harder fat, on being fried, comes out like pork gristle. Ronnie Jolly cooked the bear's head in a large pot in his tent, and Abraham Voyageur, the third hunter, cooked some of the meat, while, on a stove outside, the bear's paws were roasted in an oven.

Philip and I wandered off into the woods looking for firewood. I had assumed that we would just pick up any old wood lying around, but Philip said we must look for a dead, standing tree. We tramped a considerable distance. The bush was fairly open and we had no difficulty making our way among the trees, but the muskeg was soft and springy and hard to walk on. Finally we found a suitable tree, cut it down, hauled it up onto our shoulders and headed back. We walked a long time without coming to the tents, and then heard some noises off to our right. We had missed our way, and had to lug the tree all the way back: we had gone no more than a mile, and had already been almost lost. With a two-handed saw we sawed the tree into foot lengths and then split them: but neither of us had mastered the delicate touch that splits a log effortlessly.

SAM IS THE BOSS

As we discovered at the feast that night, there is a marked hierarchy in a hunting camp. The hunter who owns the territory is the leader in all things. He makes the decisions and the others follow unquestioningly. Sam Blacksmith, as the leader, was the head of the feast. Everyone squatted around the walls of the tent, the floor of which was covered by a printed pink tablecloth, on which the food lay in large enamel dishes. Sam took some bear grease in his hands, rubbed them together and passed his hands through his hair. He took a piece of food from each plate and a cigarette and dropped it into the fire in the stove. The bear meat was passed around. It had a strong, tangy taste. The bear fat had a sharp taste, almost like a condiment. It was used, explained Sam, instead of butter, which they didn't have. The tea was served lukewarm in enamel cups, and the meal was finished off by handing around a large enamel plate full of cream-filled biscuits, bought at the store but by now injected into the Cree feast as an adopted tradition. Tony shot the feast under a brilliant light. I asked if the light was bothering them, worried that somehow they might feel it an offense against their traditional occasion. "No," said Philip. "Sam was just remarking that the women would like to have that light when they were working over the skins in the winter."

When the eating ended we sat around the tent for a while, smoking and chatting, and then Sam gathered all the remaining food around him and began to share it out: tradition demands that what remains must be shared equally among all the participants. In this case, into four parts, one for each family and the film crew.

We asked them to come over to our tent again when they were ready. When they arrived—all sixteen of them again—we opened our second and last bottle of Scotch. "Sam says he likes to visit you people," said Philip, "and have a shot like this."

Sam told us about the bear grease in the hair: "Long ago it was established by the old people that by rubbing bear grease through your hair at a feast, you are enabled to think good thoughts. We put the food into the fire so that we can give it back to the outside, where it came from. We also send the tobacco back to the outside through the fire, and we rub our guns with the animal fat, too."

Had we noticed, asked Philip, that certain parts of the bear had been eaten only by the men? The women were not permitted to eat meat from the bear's head and front paws. The bear was the most powerful animal, the one which had to be most respected by the hunters, the one with whom a hunter who had attained powers had the most intimate relationship. "The bear likes to think of himself as the most respected animal, he wants to be well respected," said Sam. "The bear knows everything, even when you talk about him, and if he feels he is not well respected, he will not allow himself to be caught." Two or three days later, when the bear meat was finished, all the bear bones were placed on a bed of soft spruce boughs high on a specially constructed platform, so that the dogs could not violate them. The same day Sam wrapped the bones of the bear's front paws in birchbark and tied them to a tree on the shore at the entrance to the camp. "I do not know why they are tied along the shore," he said, "but it was established long ago that this must be done."

Sam and Ronnie were again talkative. We took out the maps we had brought, and they showed us their territories. Sam had a vast area of thirty miles by forty, at least 1,200 square miles, covered with lakes. Ronnie's territory, almost immediately to the north, about forty miles as the crow flies, was a similar size. His territory was on Goose Lake (known in Cree as Pubunshoon, or The Lake That Withstands The Winter). The men, of course, never use maps, and though I found the maps of the area fascinating with their numberless lakes (later I counted at least 850 lakes on Sam's territory alone), the maps in the men's heads were even more interesting.

The Eastmain River, one of the major streams flowing into James Bay (and one which is slated for eventual hydro development), lies about twenty-five miles south of Lac Trefart. The southern boundary of Sam's territory is about three miles north of that river. The northern boundary of Sam's territory is the Opinaca, a tributary of the Eastmain which is to be dammed and diverted northward into the La Grande as part of the first phase of the James Bay scheme. And north of the Opinaca is Ronnie Jolly's territory, between the Opinaca and the Sakami (a tributary of the La Grande).

I was able to tell them that both the Opinaca and the Sakami were involved in the first phase of the project, though I couldn't say whether their own lands would be affected by the Opinaca reservoir that was to be created. But whether their own lands were touched or not, they knew that they would be personally affected, for the effect on the animals would extend to their land, whether or not the water reached that far.

Every lake has a Cree name, associated with some event in the past or with some physical characteristic. Lac Trefart is called Mikwash-auhebanan, The Lake Where The Red Fish Is Caught. Opinaca is called Wootskibeck, Where The Muskrat Is. And so on. Only a very small number of lakes have been given names by the white surveyors and mapmakers.

A GROUP OF SHY TEENAGERS

They began to talk about their children. Sam had taken his boy Malick, fourteen, out of school to try to teach him something of the Indian ways, and the daughter, Rosie, nineteen, had been out of school since she was sixteen, as she was needed by her mother. Malick was a small, cheerful lad and Rosie was sitting on the other side of the tent with her back to us. She was quite tall and thin, with black hair falling down the back that she always turned to us. She always wore a shapeless blue sweater and bell-bottom trousers and when we first arrived she wore glasses which tended to obscure the shape of her round, olive-skinned face. She had a tendency to squint when without her glasses, as if short-sighted, and had the most delicately shaped lips I had ever seen on a woman. Physically she was a considerable beauty, but so painfully shy that she could never answer a question. She would giggle with the other girls when asked something, and, if we were lucky, would turn and say brusquely, "Yes" or "No," as if it were more answer than we deserved. "Twice the principal has asked for her back," said Sam of Rosie. But he would not allow her to go, and he told us she was happy in the bush.

Sam was also looking after a five-year-old boy, Abraham Mianscum, one of the children of François, his son-in-law, the trapper who testified in court about the road running through his territory. Abraham was a charming boy, full of fun, as spontaneous and cheerful in his behavior as Rosie was involuted and taut.

Neither hunter seemed convinced that education through the school system was doing their children any good. Ronnie and Mrs. Jolly had also taken their son Eddie out of school the year before, and he was becoming quite efficient as a hunter and trapper. Eddie was a taciturn, plain, though friendly enough lad with a twisted jaw. After so long away at school he had had some difficulty adjusting to the bush in his first year, but now he had settled down and apparently liked the life. Ronnie also had taken his daughter Philomen, fourteen, out of school. Philomen, like Rosie, was shy. But, unlike Rosie, she was not shy to the point of apparently wishing never to be looked upon. She had delicate Oriental-type features, high cheekbones and the suggestion of a slant to her eyes. Her face was not round and soft like Rosie's, but sharp, and from time to time it was irradiated by merry laughter. She had a touching little figure just about to grow into womanhood, and though she seemed totally unaware of the fact, was a wonderfully beautiful female. The third member of this trio of slim young women was Ronnie's elder daughter, Mary, perhaps the most interesting one of all. She had never been to school, since the grandmother who brought her up did not believe in school, and she spoke no English. She was about the same size as the other girls, delicate and womanly, but seemed to be supremely competent, better adjusted to the bush life than her companions, and without that indefinable air which they had of not quite belonging to what they were doing. The three girls moved in convoy. They would often come to our tent, but usually together, and in the first days they would sit down on the spruce boughs with their backs to us, pick up a book or magazine and pretend to read it for fifteen or twenty minutes without looking at any of us. Often they would just get up and leave without ever having uttered a word.

Ronnie and his wife were looking after—that is to say, were bringing up—a two-year-old child called Abel, one of the fourteen children of George Coonishish. This is quite a common thing in Mistassini: a family with too many children will "give" a child to another family, who will "adopt" it as their own. Thus it is not unusual to find a Mistassini child referring to a couple of women as "my mother," and it sometimes takes a while to get them sorted out. Abel, unfortunately, appeared not to be used to white men, and I especially alarmed him. He would burst into tears and howl piteously every time I appeared.

Ronnie had already "lost" two children to the outside world, and a third was going the same way. His oldest boy, Johnny, unlike Eddie, had continued with his education, and was now working for the Indian Affairs Department in Ottawa. His daughter Kathleen was living in Vancouver, and Winnie, two years older than Philomen, was still going to school in La Tuque, 200 miles south of Mistassini.

The Voyageurs had three children with them. Two, Jane Clara and Helen, were younger than school age. The third, Louise, also fourteen, had left school early because, said her mother, the teachers used to beat her and said she was a bad girl. Louise, like the others, was painfully shy:

indeed, from beginning to end of our experience with these families, she scarcely uttered a word to us. Shirley and Abraham Voyageur, the parents, bore the marks of considerable infusions of white blood: they both had European features, though neither spoke any English, and Louise was a fresh-complexioned girl with curly brown hair. Though she had a large, somnolent body, much more completely developed than the other girls, she was very much a child, and usually stuck by her mother. There was enough variety among the sixteen people in this camp to keep any novelist busy, and certainly they offered a rich subject for the documentary film maker.

EXPLORING THE LAKE

On our third day there, the men decided that they would go to set a bear trap, and that we should accompany them only the following week when they went to check it. They said they were not going far, only a couple of lakes over from the camp, but they would be quicker without us. The morning was completely still, the water of the lake glassy, reflecting the still trees and light clouds, as they set out in two canoes at around eight o'clock. We had assumed that they would use outboard motors and had brought twenty gallons of gasoline for their use. We were wrong, however: they had no motors and paddled everywhere in light canoes. They could not go far without having to portage and they didn't need heavy motors. Our two red ten-gallon drums of gasoline, therefore, were quite useless.

In the absence of the men we decided to explore Lac Trefart. With Tony in front, Philip steering and me squatting inelegantly in the middle, we paddled a canoe across to the island and around it, passing an abandoned winter lodge from a previous season, now collapsing, overgrown with moss and vegetation and, like most ruins, looking unusually beautiful. We had seen on the map that rapids were marked at either end of Lac Trefart, which was a lake about five miles in length. "If we find these rapids," said Philip, "we'll get fish."

Toward the east end of the lake we stopped a couple of times to throw out our lines and then headed up a small river, one of the countless streams connecting the lake system together. As we went deeper we could hear rapids roaring in the distance and eventually we could see them. We pulled the canoe onto a little island right below the rapids and were able to stand in the middle of the rapids and fling our lines into the foaming water. We saw a little animal sitting on a high branch of a tamarack tree, but none of us—three brilliant outdoorsmen stranded in a gigantic wil-

derness—could identify it. Was it a small bear? A porcupine? To try to
get it to turn around we shouted: "Get out of that tree, you god-damned
bear!" But it turned itself into a small ball and never moved again. We
filled the air with Tarzan cries, but it ignored us. We ran even closer to
the rapids, halfway around them along the north shore of the stream, but
we couldn't get a good look at it, so we headed downstream toward the
lake again. It was a glorious day, cloudless and still, and as we headed into
what now seemed like the amazingly broad waters of Lac Trefart, we
decided to try to raise the tent on the walkie-talkie sets we had brought
with us. We could not raise anyone in the tent, but a babel of voices
invaded our wilderness: Spanish voices, pilots' voices, people ordering
taxis in Los Angeles and someone saying, "Okay, come back from Camp
Four and we'll talk about it then." Occasionally we would allow the canoe
to drift through the wilderness, the only sound being the gentle lapping
of the water against its sides: then at the flick of a switch the voices would
come roaring in from all over the continent, as if we had picked out of
the silence of the wilderness an extra-sensory convention of madmen.

ROBERT LITVACK: Now, Doctor, these northern waters seem to be replete
   with fish, there seems to be boundless, unlimited quantities of fish:
   it's a fishermen's paradise. Your predictions, do they take into account this super-abundance that apparently exists in the north?
GEOFFREY POWER [professor of biology]: This is an unfortunate impression about the north, that even some of the biologists going into the
   north come back with the impression of great abundance, but you
   have to understand what has happened there, these fish populations
   have been practically unexploited, and they consist of a large number
   of large, old fish. These fish have habits which are known—the con-
   gregating below rapids and waterfalls, that they're hungry, they are
   voracious, and one of the reasons that the native peoples have ever
   been able to survive in the north is that they know of these congre-
   gating places, and these fish are, in the angler's terms, stupid.
   They're very easy to catch. I've been in the north in places where just
   putting your finger in the water and wriggling it around, you can get
   trout to come and nibble it, big trout. This is the situation you see
   and that leads to the impression of abundance, but if you investigate
   this thing scientifically and try and count the numbers of fish there
   are, you find this impression is completely wrong. There are very few
   fish produced in these northern areas. The sort of production rate
   that one would expect for lakes and in the La Grande system would
   be no more than three pounds of trout produced per acre of water.

LITVACK: How does that compare with more temperate climates?

POWER: Well, that's right at the low end of the scale, because in richer lakes further south, you can produce anything up to 40 pounds per acre. You can never produce a great deal of trout because trout like clear, pure waters, and if the waters get too rich, you substitute another species of fish for trout. I think we have to be careful here, because the behavior and habits of these fish, which make them very valuable to native peoples, also make them very susceptible to exploitation by more sophisticated anglers, and sportsmen going into these areas can make tremendous catches for a few years. Then they've caught all the big, old and stupid fish and they've reduced the populations and because of this they've cut down the number of eggs that population can produce, because the big fish produce the most eggs, and the population crashes or is reduced drastically and there are few fish left. A few fish grow a little more rapidly because there's fewer of them to share the food that's available, but they never recover. We've all witnessed this kind of thing: wherever a highway is built and access is made to good fishing waters, we quickly see that the trout populations in the province of Quebec are reduced. The further north you go, the faster this will happen.

A SHORTAGE OF MOOSE

As they had promised, the men began to build their winter lodge earlier than usual so that we could photograph its construction. Usually they would leave this task until the water started to freeze, occupying their time before that with the census of the beaver. The hunters have two vital tasks when they first enter the forest: one is to gather the food on which they must live day by day until the beaver trapping begins in November, when the fur is in its best condition (and at which time, of course, beaver meat becomes plentiful). Fish is perhaps the staple food at this time of year, but every member of the family, especially the women and girls, sets out his or her own rabbit snares. Each family places its nets in a strategic spot in the lake, and we noticed early that Abraham Voyageur seemed to be particualry skillful at trapping a large number of fish. Every few days he would bring into camp a bucket full of big pike, trout, sturgeon or whitefish, and his wife, Shirley, a gentle and most dignified woman with a luminous smile and a quick intelligence, was always busy preparing this fish and smoking it for later use. The fish was smoked over racks about four feet from the ground over a slow wood fire that would smoke away all day. The Indians could, if necessary, feed themselves on this small game until the beaver began to come into the lodge: but the job was much simplified if they caught some big game on arrival in the forest, as they had caught two bears this season. There was no doubt that farther south in the Mistassini hunting territory there were more animals, particularly more moose and woodland caribou.

ROBERT LITVACK: Doctor, have you studied the moose? Would the effect of the works be significantly different on the moose from the caribou?

ALEXANDER BANFIELD: Yes, I have studied the moose several times in my career. The effect on moose would be quite different from the effect on caribou. Moose live in a different habitat, and however the river valleys are changed it would have an undesirable impact on moose populations. The moose is a lowland creature and very much involved in water and the edge of the forest. The alders along river banks are favorite feeding grounds.

LITVACK: So then the destruction of this wetland habitat would have an adverse effect?

BANFIELD: . . . The destruction of wetland habitat in river valleys would have an adverse effect on moose populations, but the factors would be entirely different from the factors affecting caribou.

LITVACK: To your knowledge, has there been a discernible trend of movement among the moose northwards in the Ungava peninsula?

BANFIELD: Yes, there was some mention today about 8,000 years since glaciation, and the moose has not occupied this area until very recently and it is, in fact, now migrating into the area. The densest population is in the south, from about the Otish mountains southwards, Lake Mistassini, that area, southward. There are rare moose, rare animals, right up to the treeline, to the Leaf river, but they're extremely rare, and the moose is just working along the north shore [of the St. Lawrence] beyond the Schefferville railway. The Schefferville railway seems to be an eastern boundary, and at Lake Plitipi is sort of the northern boundary of fair moose populations.

A SUPER-LUXURY LODGE

Sam's territory lay about thirty-five miles north of this northern limit of the good moose range, and the lack of moose and caribou was to make life difficult for the hunters throughout the season. As it turned out, they did not mind having to build their lodge earlier than usual, because the signs were that the winter was early. One of the reasons we had chosen to accompany Ronnie Jolly into the bush was that we had heard that he still built a dome-shaped, teepee tent lodge, a traditional Mistassini structure which was mistakenly being spoken of in the past tense by the anthropologists twenty years earlier. Many of the decisions we made about the filming revealed only our ignorance of the hunting life, and this was one of them. We did not realize the extent to which decisions in a hunting camp rest with the *auchimau*—the boss man, or owner of the territory. Given that this was Sam Blacksmith's territory, there was no way in which Ronnie Jolly, an invitee for the year on Sam's land, could make the decision about what sort of lodge would be built. That decision would be made by Sam. He might, of course, consult the others, but his decision would be accepted as soon as it was made.

Sam now decided to build an unusually large lodge. Abraham Voyageur, a big, strong man of thirty-nine, was responsible for choosing the logs, while Sam, Ronnie and Eddie cleared the muskeg, removed rocks and leveled ground. Abraham was using a small chain saw, the first time any of them had ever taken such an instrument into the bush (it was, incidentally, the only machine they used), and the whine of the saw rent the wilderness and echoed across the lake. He chose four especially big trees, trimmed them quickly and expertly, and together they carried them to the clearing, which was about fifty yards from the shore of the lake and 200 or 300 yards from the tents. These were placed around the clearing as anchor logs—a foundation, as it were—and the rest of the logs were cut to size and placed in position, one on top of the other. This choosing and trimming of logs was a considerable work which occupied three or four days during which it rained and snowed several times, the changes in the weather being totally ignored by the hunters. Some logs were split and driven into the ground to retain the walls. And as the walls mounted, the bark was trimmed from the interior of the logs, the purpose being to have the white walls of the interior reflect the maximum amount of light from the opening that would be left in the roof. Interlocking grooves were cut in the ends of the logs so that they could sit comfortably on top of each other, and they were trimmed to give a neat fit. The men were remarkably skillful with their axes, which were used as accurately in trimming and cutting as if they were knives. While this work was going on, the women, in between their usual tasks of preparing food, smoking meat and getting the men's hunting gear in order, went into the forest and began to collect huge piles of moss, which grows in great clumps over the muskeg in very many parts of the forest. They slung it into bundles to carry back to the lodge, very much as they carried bundles of spruce boughs. Sometimes these bundles were piled far up over their heads, and it was now that we discovered that the three willowy young girls, apparently so thin and fragile, had almost the strength of men. Having observed their carrying feats during the day, we challenged them to some Indian wrist-wrestling at night, and found that even Rosie had wrists like steel.

From the very first interlocking groove on the wall, the moss was used plentifully as insulation, and as the walls rose, the women, using a shaped, blunt-ended piece of wood, stuffed the moss between the logs. The walls therefore were solid and windproof. The moss did not dry out enough to become brittle and fall out before the snow arrived to be piled up around the walls as a second line of insulation.

Sam had asked us to bring forty pounds of nails for use in construction of the lodge. Along with a few sheets of plastic that they threw over the rafters to cut down the snow falling into the interior during construction, these nails were the only materials that did not come from the forest. (The plastic, which we had also donated to the hunters on their suggestion, was later pinned around the inside of the walls and in places covered with old cardboard, providing third and fourth layers of insulation against the cold winter.)

The rafters themselves were a cunning arrangement of single, double and triple rafters hammered together to make a kind of elongated teepee shape. The roof was completed by split logs, run diagonally, and in teepee fashion, from the top of the walls to the peak of the rafter assembly. The logs, however, did not meet in the middle: a long hole about a yard and a half wide was left at the top just as a hole is left at the top of the traditional teepee. The greatest utility of the chain saw for the men was that it reduced the work of splitting logs for the roof: in all previous years these had been split by hand, being laboriously and carefully opened up, wedges driven into them and split longways, a tremendously painstaking work.

The roof was covered by about a foot of moss, which proved to be excellent insulation once the snow had fallen and lay permanently on top of the moss. This was, however, another reason why normally the men would not build their lodge before freeze-up: the moss could not keep out the rain, and a lodge prematurely occupied could be miserably flooded by a heavy rain.

It is doubtful if James Bay ever saw such a splendid lodge before. It was three or four times as big as the lodge Sam had built fifteen years before half a mile around the lake, which was still standing but was declining into that charming state of reversion to nature that makes these lodges seem a natural part of the landscape. The earlier lodge had been occupied for a season by sixteen people, too. Sam's new lodge was begun on October 5, and though the men did not work on it by any means every day, it was ready to be moved into nine days later.

"The men said it would last for fifteen years," remarked Rosie as we stood looking at the finished structure (it was one of the few remarks she ever volunteered out of the blue), "but they said what's the use of building something for fifteen years if they're going to flood the land?"

WE CHECK THE BEAR TRAP

A week after the men had set the bear trap, they went to check it, taking us with them this time. We paddled along Lac Trefart and into a small, rocky river running into the next lake to the north. Within about three minutes I felt as if my arms would drop off, but after half an hour I began to get my second wind and, though far from comfortable, was able to keep paddling. The little river ran into rapids, and we pulled in to the shore for the first portage. I had been expecting that we would all lift the canoes and carry them over the hill. But each of the three hunters slung a rope round his forehead, attached the rope to the thwarts of a canoe, whirled the canoe onto his head and headed off into the trees. The snow on this day was about two feet deep, and the walking on the snow and muskeg was extremely difficult, for the snow obscured all sorts of fallen branches and traps for the unwary walker. I had been given the tripod to carry. I pounded along behind the fifty-eight-year-old Sam Blacksmith, who was carrying his canoe, and only with difficulty was I able to keep up with him. Sam had not waited for us to get our cameras ready to photograph this event—one so commonplace for him as surely to be of no interest to anyone?—and Tony, hurrying along behind me carrying his camera, tried in vain to catch up with the hunter and get in front of him so that he could get a shot. We walked for about half a mile and when we got to the other end, huffing and puffing from our exertions and finally catching up with Sam, who had already slipped the canoe back into the water, the only thing I could think of to say to Tony was: "If these Indians don't learn how to work, you know, there's no way they'll ever be able to make it on equal terms with the white man."

We made three of these portages, none of them being considered by the hunters to be worthy of the name. We probably traveled about twenty-five miles to get to the bear trap, a trip these men regarded as going just around the corner. They had not been expecting there to be any bear in the trap: the weather had turned unseasonably cold, and the bears would have too much sense to be walking around at this time of year in this sort of weather. And sure enough when we arrived we found that the meat which had been wrapped up and tied to a tree over a leg-spring trap was intact. We did not pause long before heading back. This time, to cut out one of the portages and give us a look at some more country, we made a little detour—a detour that seemed to me to occupy some 500 miles, though in reality it was only a few miles, perhaps four or five, longer than on the outward trip. We stopped halfway through for lunch. They pulled the canoes up on a sandy shore, and the men immediately fanned out into the bush. Within a couple of minutes they returned, hauling wood, dried moss and brush. They had a fire going in a minute or so, they whittled a tripod of sticks on which to hang their kettle over the fire, and the brush was arranged on top of the snow along the side of the fire as a seat for all of us. They had not brought much to eat, just a few mouthfuls of bannock, but they sat around a little while, sipping tea, looking at the fire and gazing out over the peaceful lake. Once again I had that sense I had experienced on the La Grande with Job Bearskin of a strong perceptual satisfaction in the moment of the daily task.

On the way back they had agreed to run the rapids for us. The members of the film crew tramped across a hill and downstream, and one by one the men came in their canoes down the rapids. They ran down a fast, gurgling little stream as though it were the simplest thing anyone could do. Though none of them had ever learned to swim, these northern waters being too cold for pleasure swimming, they had never tipped out of their canoes, and the idea that there was anything dangerous in the feat just didn't enter their heads.

Three strong, competent men

Though the autumn is a quiet and unspectacular time in a hunting camp, we had seen enough of the men to appreciate their supreme competence, the high intelligence of everything they did. Whether in setting and checking their nets, manipulating their canoes, using their axes, designing and building their lodge, whether in catching game for their families to eat or carefully preparing for the winter that lay ahead, these men were all masters of several crafts who had adapted their talents and needs beautifully to the environment they knew so well.

We understood that we would never be able to see the forest as they see it. We were blind, and would remain blind, to the many signs of life that lay around them as they walked among the trees. The irony and tragedy of their situation was that the outside world remained ignorant of their enormous capacities: however masterful the men might be in this environment, it was obvious to us that if they were to end up in a small Canadian town or village as government policy would have them do, they would be qualified for nothing except perhaps to collect garbage.

The three men were quite different: Sam Blacksmith, though serious about his momentous task of ensuring the survival of his family, was by nature a clown, and very often when we set up our cameras he would go into some clownish routine, such as pushing over a tree instead of cutting it down. A remarkably handsome man with white hair, he affected a taxi driver's cap, which gave him a certain jaunty and even comic air. Perhaps to protect himself against the ridicule of the white man, he adopted a facetious joking air with us, and in response I would always elaborately pretend that he was exceptionally favored to have attracted to his hunting camp five such brilliant providers and hunters as the members of our crew. I would make a lot of jokes, but Philip would very seldom translate them, saying, "It's not funny in Cree," or "I can't translate that."

Ronnie Jolly also adopted a cheerful, laughing tone with us. At least on the surface a very merry person, capable of great fun, he also liked to talk much more than did Sam, seemed more interested in both the political and religious aspects of their life and was never too tired—as Sam sometimes was—to engage in a conversation. None of these men was ever idle, but Ronnie was particularly busy and creative. He was especially capable with his ax and his whittling knife, and later on in the winter was almost always occupied in mending a hand-made sled, making snowshoes or fashioning snow shovels.

On the other hand, Ronnie seemed to be shrewder and more argumentative than the other men, a bit more conscious of his rights and willing to stick up for them in argument. We heard about this vaguely, but did not get to see any of it close up. The Cree world is a private world and even in the hunting camp we could not expect to penetrate very deeply into it.

Abraham Voyageur seemed in many ways the most admirable of all. Younger, around thirty-nine, he was quiet, undemonstrative in the most dignified way, but friendly, always willing to talk if he felt there was anything useful to say. He had a large injection of white blood—he was one of the Voyageur family from which, on another side, Philip Awashish was descended—and this was reflected in a more pragmatic attitude both to the Cree culture and to wage employment, education and the future. He was a tremendous worker, as anyone who watched the building of the lodge could see, and a skillful provider. But to boot he was a devoted father—he had three more daughters at school in the white man's world —and was clearly the steady anchor of a warm family life. Apart from the young people, who spent quite a bit of time in our tent playing a ludicrous card game called Monkey most evenings, we did not get to know the people in the camp particularly well during our first visit. They were feeling us out, and by the time we left we had the feeling we had passed some sort of test with them and they were beginning to warm to us. Our efforts to get to know them, however, remained more theoretical than real: we admired their behavior and their account of how they felt about life, but from a distance, since we could not penetrate any closer.

No one can predict anything

One night Philip and I recorded a long conversation with the three hunters. I asked how they felt when they heard that white men regarded the land on which they live as worthless and empty. "They are saying," said Philip after a good deal of conversation, "that the white man's argument is bullshit. It may appear to white people there is nothing here. But if you will leave that land alone for a year or two, the animals will come back. That is why we are close to the land. We love the land because we know it will bring back animals, even if to white people it looks as though there is nothing there."

QUESTION: Can you ask Sam if he regards himself as the owner of the land?

SAM: I feel I cannot really say I own the land in the full sense of owning it. I am old, and I do not know how long I will live. After I am gone, I do not know who will follow. I cannot commit myself to say I own this land here.

QUESTION: If you do not own it, what is your relationship with it?

SAM: What I expect from the land is to be able to support myself as a trapper. I expect a lot from the land. And when I come down to discuss it, I always refer to it as my land, as long as I am depending on it.

RONNIE: It is quite ridiculous, this idea of the white man that a person can own all of the earth, and everything that's under it, and everything that moves on it.

SAM: We cannot know what will happen to man. That is the way things are while we are living. Even myself, today, I cannot know what will happen to me. If I was to fly back to Mistassini, I could be gone for days, for a long while. I cannot say, "Well, I am going back because I have territory to look after." For no one can tell what might happen to a man while he is in Mistassini. One cannot predict life or death, so how can one say, "This is the way things will happen to me on my land" or, "This is what the land will be doing to me"?

QUESTION: Well, what are your basic thoughts about the land?

SAM: For one thing, a man who is trapping and hunting always hopes that things will continue to grow on the land.

QUESTION: If this is your trapping ground, do you own it?

SAM: We are told that we own or possess it, but really nobody can own or possess it, the land. Even if one says that he does own or possess it, he cannot do so. Nobody can, because eventually everyone dies. In this way, no one can really predict anything.

QUESTION: Is it the land you relate to, or the animals on the land? If the animals all left it, would you be interested in this land at all?

SAM: My relationship with the land depends on the animals. The animals support my family, they establish my pattern of life. If there were no animals here, I would not be here.

A UNIQUE CONCEPT OF POSSESSION

For many years anthropologists have been trying to discover the fundamental concepts of land and of ownership among the Mistassini people. The longer the investigations have gone on, the more complex the ideas have seemed to be. For the one thing that is sure is that the Cree do not have the white man's concept of "ownership," though they do have territories which they think of as theirs, territories which are handed on to others, young men, when the older men die. The geologist A. P. Low in 1894 discovered that the Cree hunters divided their territories into segments which they exploited in rotation so that the animals would have a chance to renew regularly. But as late as 1965 Rolf Knight reported: "Over a period of some years, the same region is exploited by different hunting groups. Some families, since the beginning of their trapping life, have trapped in almost all of the band's territory, and even outside of it. It does happen that hunting and trapping groups leave the settlement without having decided where they will spend the winter. Other groups which cannot reach their destination begin to trap where they are caught by the freeze-up. And others still trap in a particular place at the beginning of the autumn, and afterwards decide to move to another place, and sometimes join in with another hunting group."

Adrian Tanner, the anthropologist who gave such valuable evidence for the Indians in court, tried to establish the lines of inheritance for each trapping territory by going back to maps prepared as long ago as 1915 by Frank G. Speck. His investigations show that there is a tendency for the succession of hunting territories to pass to sons, but this is by no means a rule. It often happens that a young man from a neighboring territory will look after an older man whose sons have died or moved away, or that a son-in-law or perhaps even a nephew or young cousin will move in with an older man and become the person with the closest relationship with the hunter. Usually such a relationship will be rewarded by the handing on of the territory to the young man, whatever the blood relationship may be, or even if there is no blood relationship. Tanner's conclusions are quite startling: the anthropologists of the past who spoke of fixed territories owned and occupied by one hunter or family have been mistaken. It is the special relationship that an experienced hunter attains with the animals which controls the movements of his group. The hunter defines the place of the group in space, and only for a single winter. The hunting space is what makes the link between certain animals and certain hunters. In 1947, Tanner recalls, an anthropologist called Julius E. Lips was told by the Indians that it is the animals who possess the land, and his own investigations tended to confirm this idea. When an old man dies, his "territory" is transmitted to the young man who has established the closest relation with the animals of the region. This "territory" is defined by the movements of the animals.

The way in which Sam and Ronnie came by their territories illustrates the flexibility of the ownership patterns. Sam's land had been occupied by an old man from Eastmain called Noah Shanush. When he died many years ago, the land was vacant, and Sam moved in to take it over. He had begun to hunt in that land thirty years before, was already there in 1948, when the provincial government, as part of the establishment of the beaver preserves, gave each territory a number and registered it under a certain person's name. This government system has perhaps strengthened the idea that certain lands are owned by certain people.

Ronnie Jolly got his land from his father, but that also came into the Mistassini territory by interchange with Eastmain hunters. This territory is directly inland from Eastmain, about 165 miles. Perhaps because of the greater concentration of the coastal hunters on the fall and spring goose hunts, the coastal people have not penetrated to the sources of the major rivers far inland, leaving this territory, a huge one, to be occupied by the hunters from Mistassini. The territory Ronnie now owns and controls was in the hands of a man from Eastmain when his father, who had married an Eastmain girl, came to hunt with this old man. In that way the territory was handed on to Ronnie's father, who handed it on to Ronnie. A similar process can be seen at work in Job Bearskin's territory: now that he is old and has moved in with a younger man, his vacant hunting territory near Lake Bienville is being used each year by hunters from Great Whale River, the next village to the north.

The matter of succession of territories, of course, has been very much complicated by the tremendous increase in the Mistassini population. In 1915 Speck reported there were 175 persons at Mistassini. By 1930 this had grown to 288 in fifty-four families, by 1952 to 450, and to some 1,600 today. Much of this increase has been caused through the urbanization of the subgroups of the Mistassini Cree, the people from Doré Lake, Nemaska and Nichicun having been encouraged by the government to settle in the village of Mistassini. The increase in population has meant that many hunters have remained landless. Both Sam and Ronnie had brothers who would often come to share their territory with them for a season: Abraham Voyageur did not have a territory registered in his name. He would often hunt with his brother Reuben, who had inherited a territory from their father. But essentially he was landless, though an expert hunter, and depended upon other people taking him onto their land.

QUESTION: How are the arrangements made for hunters to join other hunters on their land?

RONNIE: When we are in Mistassini for the summer and we discuss the trapping resources of our territories, we hear about a man whose animal resources have diminished, and that man will be invited, either directly or through gossip.

ABRAHAM: We arrange with Sam . . . we were asked by him. . . . He has to give his consent.

SAM: I mention it to them first, more or less hint to them that I would like to have them live and trap and hunt with me. They didn't mention it to me, I mentioned it to them first. It works out well for me to take someone along. I have taken a lot of people into my ground, and also young men.

RONNIE: You see, a man cannot easily trap by himself. He likes to have another man along. There is too much work for one man. Things will not go well for a man who is alone.

ABRAHAM: The well-being of the whole group depends on the leader. If one man is not going too well, there are always others he can depend on. Everyone appreciates the others' efforts. Everyone has to be in good health so that everyone can work together.

RONNIE: Sometimes we will leave our ground alone for a year or two. I have left my territory alone now for two years, and later in the season

I will probably go there to check out the animals. There is always something there when you come back to it.

SAM: It's not that a man will throw away his ground. Even when we are not there it is still ours.

NICHICUN HUNTERS

The extent of this sharing was investigated by Adrian Tanner in the Nichicun area. Of seven proprietors of land in that area, over the decade of the sixties none worked their own land every year. Four of the seven men spent more than half of the decade on someone else's land: of seventy man-years of trapping by these people in the ten years, just forty-two were spent by the owners on their own land. The vitality of the sharing ethic, the collective view of the use and occupancy of land, can be gauged from this figure, for there is no sense at all among the hunters that a man who spends his time on another's land is "getting away with" animals that really do not belong to him, or taking an unfair advantage in some way.

When Sam was young he would hunt with his father in the Nichicun area. One day his father went out to check the nets and returned with two suckers: that was all the food he had to feed his family. He had none of the white man's goods as insurance, no flour or lard, tea or even sugar, and it was then that Sam decided he should pitch in and help with this business of ensuring the survival of the family. That gave him the will to become a hunter. Sam, too, had his stories about the hard times, when animals were short and the Cree had little or nothing to eat. During those days they ran out of food and he and his brother decided to walk to the Nichicun post for supplies. They had one and a half pieces of fish between them, and they had to travel for four days. His brother collapsed, bleeding, at the beginning of starvation, and Sam put him on the sled and pulled him the rest of the way to the post. During that same time Sam remembered going on a hunting trip with half a partridge leg to sustain him until he could catch some food. In those days they would travel north from Mistassini by canoe. The trip would take them a month and require 100 portages, on some of which they would have to make four trips to transport all their goods to the next water. Often they could not leave their equipment overnight for fear that animals would get into it, so they would sometimes work throughout the night and paddle throughout the day, passing two and three nights without sleep. It is highly ironical that these are the very people who, in the Canadian public mind, are regarded as lazy, shiftless and unwilling to work.

o'reilly: In Nichicun, you mentioned there is presently a Hudson's Bay Company store?

adrian tanner [anthropologist]: Yes.

o'reilly: Is there a man at that store on a fulltime basis?

tanner: No.

o'reilly: Are there store supplies there?

tanner: Yes.

o'reilly: Then how do these supplies get to the Indians?

tanner: The Indians go into the storehouse, take what they need, and leave a record of their name and what they have taken, and when the Hudson's Bay manager arrives he takes note of these and charges the accounts of the Indians for these.

o'reilly: Hudson's Bay hasn't become bankrupt lately to your knowledge?

tanner: No.

Men who live close to animals

All the major occasions of life are built around the animals, the killing, preparing and eating of them. "When we go out for big game," said Sam, "the women will make something special for us, like a new cartridge pouch, a gun case or even new mittens or moccasins, and we wear what they have made when we start hunting big game. Then, of course, when we take a beaver, we have a special rope, decorated with ribbons, with which we pull the beaver back to the lodge."

question: Is it part of your respect for the animals that you must not kill too many?

sam: Yes.

question: And how do you know when you have killed too many?

sam and ronnie [after a long consultation]: As far as the bear and otter are concerned, we cannot overkill them, there is something in their characteristics which prevent it. But with the beaver we can tell by the marks on those that we kill. We keep track of them. We can have an idea from the marks on a grown beaver how many young ones she has borne and how many of them have lived. So we always know how many beaver there are, even if we have not seen the litter.

question: But if for some reason the animals are shown disrespect, how exactly is it that they become scarce? Do the animals decide just to disappear for a while?

sam and ronnie [after consulting]: The tradition that has been passed on from long ago is that the animals are aware that a certain individual does not respect them, and they avoid that man.

QUESTION: Do they go away, or do they just become scarce?

RONNIE: They simply cannot be caught. It is not that they leave the territory. But we cannot really thoroughly explain the religious aspects because we have lost some of that already, and it was established long ago by our forefathers.

QUESTION: Does that apply to all the animals? Or are there some more than others that react to disrespect? For instance, if you did not respect the bear's bones this year, would you expect not to catch any more bear?

SAM: Yes, if we do not show any respect, we would not expect to kill any more bears, and that applies to all the animals. There are priorities: the bear is the most respected. There are small animals like rabbits and partridges of which we can feed the bones to the dogs, but of the rabbit we keep the skull, and certain parts of the beaver we hang up.

LAND THAT IS FULLY OCCUPIED

Within an area 100 by 130 miles from our camp at Lac Trefart eight other camps had been set up by hunters from Mistassini and Eastmain. Thus, from the Cree point of view, an area that I calculated to be roughly 13,000 square miles in the immediate vicinity was being fully occupied and used in the traditional manner. From Mistassini alone that year Glen Speers was serving seventy-four hunting camps like these, the largest number for many years. When the planes had first arrived, a few years before, he would fly to a central camp and the people from surrounding camps would walk there with their furs. Farther south, where the traplines are not so big, this was practical, though the method was abandoned within a year or two when he decided to try to visit every camp. Certainly such a method would never have been practical in the north, for the camps are too far from each other: the hunters have such a crowded life during the trapping season that they cannot afford to take time off to walk thirty, forty and up to seventy miles to the neighbors' just for a visit.

Sam's camp was on the western edge of the Mistassini territory, and west of him the land was occupied by Eastmain people. To the south on the Eastmain River and just beyond the headwaters of the Eastmain, Charlie Coonishish and Charlie Jolly (who was cousin to both Sam and Ronnie) had set up camps. Immediately east, Sam's brother, Billy Blacksmith, had decided to leave his land for the year and had moved many miles south to share a camp with Reuben Voyageur, Abraham's brother. Immediately east of that land, John Loon was trapping around Fire Lake, and north of him George Coonishish (Ronnie's brother-in-law, the father of little Abel, the crying baby in our camp) and Samuel Mianscum were working the territory stretching as far north as the La Grande River. Ronnie's territory immediately north of us was vacant this year, but north of that, between the Sakami and La Grande rivers, John Neeposh, Edna's uncle, had set up his camp.

These camps had no means of communicating with each other or with the outside world, and it was rare for the hunters to visit each other: they might drop in if they happened to be hunting close by, but normally that would mean travelling over the other man's territory,and the hunters do not usually have time for that sort of social travelling. A camp which was thought to be in trouble, was manned by inexperienced men or was being operated on land already known to be low in animals would be more likely to have a visit from the neighbors, who would be extending beyond the limits of their own territory their duty to look out for other people. None of the men in this area, however, would need any such visit.

If the other camps had about the same number of people as we had, the 13,000 square miles would be supporting some 150 people. This is one of the toughest points to explain to white people in the south. Since eighty-eight square miles of the island of Montreal support 2 million, most people cannot see how such huge territories can in equity be left to so few people. If these lands are needed in the public interest, it is argued, then unfortunately the Indians will have to adjust. The question is, however, how are the lands to be handled and how will the Indians be allowed to adjust? The manner of doing the deed reveals our quality as a civilization. For the Indians have always been there: it is not their fault that other people have arrived, making claim to the lands on which their whole culture and way of life has been built. In decency and justice, they cannot simply be trampled over: our own society would be infinitely richer if we could find the means to give Indians a genuine choice, some real control over the pace of change, as they face the inevitable arrival in force of this alien civilization.

The way to do it was certainly not the way being pursued by the authorities, who were busy denying, in public-relations handouts and in court, that the Indians still occupied the land and still lived on the land, at the very moment that the people in all of these camps were preparing to feed themselves throughout the winter from the animals they could catch.

A DEEPLY DIVIDED FAMILY

The culture shock that has been visited on Indians by Canadian government policies was evident in Ronnie Jolly's family, which embraced the
cultures of several centuries within the experience of half a dozen people.
There are probably no two sisters on the continent whose experiences
have been more dissimilar than those of Mary, twenty-four, and Kathleen,
twenty-two. Mary, brought up by her grandmother, a woman from the
very heart of the hunting culture, and never permitted to go to school,
now spoke only Cree, was capable and strong in the work around the
camp, expert in handling the animals, constantly busy making moccasins
or mittens or threading snowshoes, and was a petite, womanly and cheerful girl destined no doubt to make an excellent wife for a trapper and
hunter. One could not feel that she had been deprived of anything essential by having been kept out of school—so long as her culture was allowed
to continue, a matter on which she, along with others, seemed to have
grave doubts. Sometimes she wished she had been to school, she said. It
would be better for someone to go to school long enough to learn how
to communicate in another language, provided they returned before they
had lost their ability to survive in their own world.

Perhaps through not having been educated in any philosophy of
doubt, she now seemed to have few if any uncertainties about the validity
of her life, and when, later, we came to interview the young people about
their attitudes, she was the one who had the most to say and who seemed
the least tortured by embarrassment in talking before the white man. "I
am taught everything that a woman should know," she said in answer to
a question. "Like preparing the beaver furs and lacing snowshoes. Also
how to sew and make things like mittens and moccasins. These are the
things I am taught."

A stranger to fashion, cosmetics or consumer obsessions, Mary
showed none of Rosie's interest in pop music and exhibited none of her
restless desire to get away. Nor was she constantly brushing her
hair and striking poses, as was Rosie, who was always very conscious of
her beauty, though trying to hide the fact. Mary was still a Cree woman,
and found it unnecessary to reflect unduly on her life; her sister Kathleen,
living in Vancouver and working in a clothing factory run by Koret of
California, was a Cree girl only by origin and no longer by culture, and
was bewildered by the vast gap that had opened between herself and her
family.

Kathleen, larger and rounder in the face than Mary, but with the

same shapely figure and gentle manners, had retained from her upbring-
ing that lack of assertiveness and that tolerance of others which are
charming elements of the Cree character. But she was now a city girl,
quite at home anywhere. Taken away from her parents at the age of seven,
she had never really been home since. She spent four years as a small
child in Moose Factory, where the children were not allowed to speak the
only language they knew, Cree, and then in Brantford, Ontario, for a year
before being moved to La Tuque, the little French-Canadian mill town
a couple of hundred miles south of Chibougamau, where the Indian
children live in a residence and go to the local schools. After six years
there Kathleen went to high school in Montreal for two years. During all
those thirteen years she was never home for more than two months in the
summer: and each year the gap between herself and her parents seemed
more and more unbridgeable.

"When you first start going to school, you miss your parents, you cry
a lot, it's very hard on you," she told me, "but when you go home you
notice that things are not as you are taught in school. My parents would
ask us what we were being taught in school, because when they would ask
us to help, it was usually things we were not capable of doing. In these
residences, you know, all through school, you are always told what to do
and when to do it, you have to ask permission before you are allowed to
do anything, and when the time comes to leave, you've had no basic
training about how to look after yourself. I lost contact with my family
because I began to lose my Cree language and I couldn't communicate
properly with my parents. I remember one time in school I got a letter
from my parents, who had had to write it in Cree, because they didn't
have anybody to write it out in English for them; and when I received it,
I had to have someone translate it for me."

Kathleen, however, was lucky to find a family which helped her over
the years of her disorientation. Mr. and Mrs. Jacques Aubé—he an amiable French-Canadian, she a determined, protective, talkative and kindly
lady from England—were house mother and father at the residence, and
they took to Kathleen as if she were a daughter. They received permission
to have her live with them in their home in La Tuque, and when they later
quit the residential system in disgust and moved to Vancouver, Kathleen
followed them there, living to all intents and purposes as their daughter.
Mr. Aubé worked as supervisor for a housing development. Kathleen's
home in Port Moody, a suburb of Vancouver, within sight of the snow-
capped Rockies, was the essence of middle-class Canadiana—culturally
speaking, about as far as she could go from the bush camp of her parents.
But she was lucky: for she was deprived, as all the Cree children

in school are, of her culture, language and background, wrenched away
from her forcibly by the system, but through the kindness of her adoptive
foster parents she was given a stable base from which to build a meaningful life in the dominant society. She was not left alone to swing halfway,
purposelessly, between the two. Yet she was still racked with self-doubt.

"At one point you are not quite sure if you want to go home or not,
especially when the time comes and you don't have the experience to
work in the city. I mean, your Indian culture has been taken away from
you, but you're not white, you're sort of in-between, and you don't know
which way to go. But it's too hard to just come out of school and say you'll
stay at home and live the way they do. They're happy the way they live,
I admire them a lot, they don't have all the hang-ups that people have in
Montreal, they enjoy what they're doing, and I think they're better off
than white people living in cities.

"And sometimes I'd like to go back. I'd like to go to school again and
perhaps I could learn to be a teacher or nurse and then I would be able
to be useful back home. . . . You see, I've been through it myself, I don't
really agree with what's happened to me, and I don't see why the younger
generation should have to go through it. You know, at school we were
always made to feel like the bad guys. Sometimes I feel bitter. . . . The
history books didn't tell the truth about the Indians and how we lost our
country. The white guy was always the good guy, and the Indians, when
they won a war it was always a massacre. . . . I used to resent it really, but
once you know the truth of it you learn to accept that we've lost our
country. . . .

"I've often wondered what I would think about my sister if she had
been the person who went to school and I would be the one at home,
helping my parents, doing all those things that she likes to do."

Kathleen, said Jacques Aubé gently, should really go back home. She
didn't know whether she was Indian or not, and she should give herself
the chance to find out.

"I want to go," said Kathleen, "but I'm a little afraid. I've always
thought of going back and actually going in the bush with my parents. It's
been a long time since I was in the bush, when I was six, and I don't
remember all that much about it. I often wonder what it would be like.
But I would have to learn everything from the beginning, all the work the
women do, the trapping, skinning, things like that. I wouldn't be any
good at it."

But Kathleen never went home. That is, she did go, but for one night,
the following summer. By that time she was married to a young man

from Vancouver, and she didn't stay. "She came on July the sixth," said her mother, "and stayed one night."

HISTORY OF THE PROUD AGES

In the residence at La Tuque, sixteen-year-old Winnie Jolly had very little to say for herself. In a little room shared with some other girls from Mistassini, the walls covered with pictures of David Cassidy and Donny Osmond—"David says, Love Me Tonight"—Winnie flipped through a movie magazine, reading the letters to the advice column ("Dear Paul, I am a senior in high school, and I am in love with my teacher, and I do not know what to do about it . . ."). She knew that she would never again see a trapline, never go in the bush, and was destined for a future, some kind of future, in the white man's world. Painfully I tried to get Winnie and her friends to talk, but before strangers they withdrew into nearly total incoherence. At least three of the five girls were beautiful, dark and slim like Rosie and with a suggestion of Asiatic features in their high cheekbones and narrow eyes and thin eyelashes.

QUESTION: Do you feel that this school is trying to kill off your Indianness? [No answer].

QUESTION: Do you think they are trying to help you appreciate your Indian background at school, or are they trying to make you think like white men?

REPLY [after looking at each other and giggling]: They're trying to help.

QUESTION: Trying to help you do what? [No answer.]

QUESTION: Have you ever lived in the bush on a trapline?

REPLY: No.

QUESTION: Do you think you will ever live on a trapline when you're out of school?

REPLY [after looking at each other and muttering together]: No. I don't think so.

QUESTION: What do you think you'll do when you leave school?

REPLY: I want to be a nurse.

QUESTION: How about you?

REPLY: I want to be a secretary.

QUESTION: How about . . . ?

REPLY: Same as Winnie, a nurse.

QUESTION: And . . . ?

REPLY: Secretary.

QUESTION: Would you all like to work in Mistassini or elsewhere, in the
    city?

REPLY: In the city.

QUESTION: You'd like to live in the city? Do you have any problems with
    your families when you go back to the reserve?

REPLY: No.

QUESTION: Do you think the trapping life will die out, or would you like
    it to carry on?

REPLY: I don't think so.

QUESTION: Do you have any respect for it? Do you admire the way of life
    of your parents? [No answer.]

QUESTION: What do you think of your parents, the way they live?

REPLY: I admire them.

It took ten minutes to get this much out of them, and we were no
more successful when we tried to find out what they were learning in
school.

QUESTION: What kind of geography do they teach you? Is it about this
    area, or some other place?

REPLY: Ours is on Ontario. . . . And ours is on Europe. . . . Ours is called
    general science.

QUESTION: What kind of history do they teach you?

REPLY: Proud ages.

QUESTION: Eh? Proud ages? Where? In Canada, Canadian history, you
    mean? [No answer.]

QUESTION: Do they teach you any history about Indians in Quebec?

REPLY: No, not in ours.

ARROW TO THE MOON

Philip told us a story about a play he had appeared in when he was in
school at Sault Ste. Marie, only five years before. The play, *Arrow to the
Moon,* written by a minister, had been staged to celebrate Canada's cente-
nary in 1967, and received such acclaim that CBC television had broadcast and recorded it. "It was designed to show how Indians have been
fitted into Canadian society," said Philip. "Act One was all about the
plight of the Indians long before the white man came. It consisted in
Buckley Petawabano [Philip Petawabano's son], Peter Coonishish and
myself, all boys from Mistassini, dancing around all over the stage in our
costumes and playing the scene of shooting down the caribou and carrying it back home to the camp. The second part is about when the white

man comes. There's all sorts of people sitting on stage in a circle, and Petawabano is describing what he is seeing, men wearing black robes and bearded. And then he picks up two sticks and makes a crucifix, and he holds it high above his head while the light fades out slowly, except for one big spotlight which is finally fixed on the cross."

In the next scene Father Jean Brebeuf was living among the Hurons. A character called Iron Brave approached him as he wrote a letter to his superiors in France. "Father . . . often have you told us of the mighty Gitchi Manitou whose Son's birthday comes now in the winter moon." Brebeuf: "Yes, my son, we sing soon the songs of His birth." Iron Brave: "But, father, these songs are in your tongue, and we do not feel them ours. Can you not give us children of the forest a song in our tongue to sing for the Boy of Manitou?" Brebeuf: "A Christmas carol in the Huron tongue? Why not? My friend, I will think about it."

"Brebeuf is sitting there writing this carol," said Philip.

> 'Twas in the moon of wintertime
> When all the birds are fled
> That mighty Gitchi Manitou
> Sent angel choirs instead. . . .

"It goes on and on like that. Brebeuf is listening to the cattle and behind the curtains Petawabano and I are making special sound effects for this production, like going 'Moo—moo.'. . . The first few rows were reserved for Indian students who came to see us perform, and there were some Cree people there. When the narrator said that Brebeuf could hear the chit-chattering of the Hurons, there's Buckley and I chit-chattering away in Cree, saying things like, 'Motherfuckers, go fuck your mother tonight,' and all the Cree members of the audience are laughing and laughing, and all the white members of the audience can't figure out what's so funny about Brebeuf sitting there writing this carol."

As the carol swells majestically, three Indian chiefs enter reverently and kneel at the feet of Maggie Bearskin, a girl from Fort George, who is sitting before a manger holding a baby.

"Peter Coonishish is the first chief, he walks in with this gift of fur and pelts, delivers it at the feet of the mother and kneels before the baby. The second chief, which is myself, I come in with my horns, I represent some chief on the prairies, and the third chief is Petawabano, who is a feathered chief, he is wearing this bonnet thing, and now we are all kneeling before Maggie Bearskin, and the choir is singing higher and higher until all the lights fade out."

The picture-book man with the stick cross has by this time arrived and changed everything, and there is a long scene in which Billy Diamond tells an old chief that the "new ones" have brought many good things. "The new ones showed us how to use the wheel. . . . The guns and the horses brought by the new ones have made it possible for us to run faster than the buffalo and to kill him faster. . . . If we do not change our ways and even our paths, our villages will be empty as well as our stomachs. . . . The books of the new ones show the paths of change." The old chief says, "I am too old to change. I die as a hunter, even as I lived as one. . . . My son, if you be wrong, our people will be no more in the time to come," but Billy Diamond concludes that "the new ways show a way to work and live, but the old ways have shown us how to die."

Two years later in Montreal we managed to borrow a copy of the videotape of the play. We had to rush it back to Sault Ste. Marie, because the videotape was still in great demand for educational use with Indian students. It was even more grotesque than the script, and when Buckley Petawabano screened it for a meeting of Cree chiefs from James Bay, they were shocked to see the extent of the brainwashing to which their sons had been subject in school. Most of these young men were already eighteen or nineteen when they took part in the play. Within a few years they were called upon to confront the power of provincial and federal governments in defense of their society's cultural traditions.

WHERE THE GREAT SPIRIT COMES IN

I was interested, I said to the hunters, in the idea I had heard from many Cree hunters that man was just one of the creatures of nature, equal in status with the other creatures. What was their sense of the relative importance of men in the scheme of things?

After some discussion Philip said the reply was that in their society women were the most respected persons. I said I had something different in mind. "I know that," said Philip, "but in the Cree language there is no idea to express the attitude of Chinese, black, Indian. There is only man, in the sense of mankind."

I said I had been told by Job Bearskin that the land was like a garden, which they loved because everything multiplied there, and the crops which grew included men, bears, plants and rocks.

SAM: We agree with the idea that everything grows from the earth.
QUESTION: So where does man get his power to kill the other animals?

Man seems to be the most powerful being in the complex of nature.
RONNIE: That's where the idea of the Great Spirit comes in. He has given

man certain characteristics which have made him stronger. He has given man the power to kill what is on the earth, so that he may survive.

QUESTION: Do the older men get additional powers from the relationships they establish with the animals?

RONNIE: When everything comes from the earth, it takes a great power for the earth itself to produce what it has. And anybody with some power has to keep this whole concept in play in order to remain powerful.

QUESTION: Do they still have this kind of power in Mistassini?

SAM AND RONNIE [after consulting and laughing]: Yes, there are some people in Mistassini who still have powers.

QUESTION: What does it mean? What can you do with it?

SAM AND RONNIE: Long ago the people with power, some abused it, and some put it to work to be able to control the animal resources. But now the people with these powers are just simply abusing them, because they curse other people and so forth.

QUESTION: Maybe you should curse the James Bay project? [They laugh.]

BEAUTIFUL AND TRAPPED

We all of us began to develop a special feeling for Rosie, so beautiful was she, so strange, so much a mixture of coquetry and tongue-tied timidity, so clearly had she been damaged by the experiences she had undergone. I tried several times to interview her. If anyone else was present, it was impossible, for she would turn her back, ponder each question for two or three minutes while shaking her knees agitatedly back and forth, and finally make no reply. But eventually I caught her when no one else was around. And what she told me appalled me. Like the others, she had been taken away from home at the age of seven and sent far away to Brantford to be educated. There she was punished for speaking the only language she knew, Cree, and was schooled into accepting white man's food, china on the tables, knives and forks, and a shower every day. When she returned home in the summer to live with her parents in their little Indian Affairs shack in Mistassini, she found their life "poor and hard," and wanted no part of it. She was sixteen and in grade six, five years behind her age group, when her father and mother decided she should leave school. She had been coming into the bush with them ever since, though she hated the life and was not capable in it. Everything she did, working with the skins of the beaver, mink and otter, still had to be checked by her mother: she did not learn easily the things a girl should learn. And

now, if it were left to her, she would like to go away. Where would she
like to go? Rosie looked steadily ahead of her, away from me, her knees
clasped between her arms, swinging them nervously back and forth. But
she could not answer.

"Do you want to carry on your education?"

She could not answer.

"What do you want to study?"

She could say nothing.

"I suppose," I said, "the worst part is that you spend about eight or
nine months of the year here, then two or three months working in tourist
camps in the bush, so you don't have much opportunity to meet men."

"Yes," she said, staring as far away from where I sat as she could,
"and then I'll go back to Mistassini and there'll be a forced marriage."

Such a possibility had never occurred to me.

"They tried to do it with my sister Emily," she said, "but she told
them she would have to be carried screaming into the church."

I asked if it was really possible that her father would try such a thing
on her.

"I have told them the same thing," said Rosie. "I knew a girl of
fifteen in Mistassini who was forced to marry a man of thirty. But I am
not going to agree."

I asked if she cared about the language, the culture, the fight now
to defend the life. Her knees began to move sideways, back and forth.
Would you care, I asked, if the Cree language disappeared?

"No," she said.

"But where do you want to go?" I persisted.

"Anywhere," she said.

I DID ALL THIS FOR MYSELF

When Johnny Jolly was old enough, they told him, "You're going to
school," and he went. At the same time the grandmother needed his
sister Mary around the house, so she didn't go. "In the long run it's good,
you know," he said. "I've got myself a job as a draftsman with the Department of Indian Affairs. There is nothing in the reserve, you know, to fulfill
my ambitions."

An extremely quiet boy living in Ottawa and always in contact with
the many Mistassini students who turn up there every year to go to high
schools, he has great difficulty expressing himself clearly in either English
or Cree, but seems content that he has made it out of the one culture and
into the other. He really feels that Eddie should have carried on at school.

"I talked to him. When I last saw him he'd been in the bush a whole year with my parents, and I told him, 'Why don't you go back to school, get an education, and after that you can do whatever you want.' But he made his decision, to stay with my parents. That's the thing he wants to do, and I can't blame him for that.

"But myself, I want to be able to get out in the world, to get a job and do what I'm capable of doing. Between us, the whole thing is reversed. He just follows the path he knows best, and I go along the path I know best. I like to think: 'I did this for myself, why can't anybody else on the reserve?' We seem to divide right down the middle: half of the people feel like going back to nature, and the other half want to make something out of themselves."

Like many a small-town boy who has gone to the city, Johnny has lost touch with his parents, but with the very good additional reason that after a few years he could not communicate with them easily when he returned home in the summers. Now, he says, he respects their way of life. "They've been living like that from generations back," he adds laconically. "Neither my father nor my mother had any sort of education, so they haven't got the skill to get along anywhere else."

Eddie, for his part—as plain as Johnny is handsome, as rough-hewn as Johnny is smooth—having begun to master the skills needed for the bush, says that he doesn't know one of his contemporaries at school who, having gone on with education, is now doing well.

Tanner makes some suggestions

"To help the Cree to develop the present way of life to a more satisfactory level," said Adrian Tanner, "we would have to start by ridding ourselves of this notion that trapping, or the bush, or the wilderness, or a place away from roads, that any life in such a place is archaic and inevitably doomed. That is what children are being taught in school. The children come out of school believing that their parents are living a meaningless way of life leading nowhere. The education system has never been thought out in terms of Indian life. The presumption has always been of eliminating Indian life and creating a system geared to the turning out of red white men.

"The parents try to keep their options open by having at least some children who are highly skilled in the bush way of life. And the only way they can do this is by taking them out of school. They don't want to, but it's a decision forced on them by the present set-up, where a kid either gets a white man's education or an Indian education.

"The Indians, I think quite correctly, see the bush as a system of education, and it is one that is more important to the culture, to the value system, than the white man's education. We have to understand that just appreciating, getting the feeling of the bush is a value in itself—whether one is trapping or cutting lines for a mining company, being in the bush is itself a value. The children who have gone through our schools come back not only totally unequipped to go into the bush, but with a negative feeling toward the bush which has been imposed on them by our culture.

"They learn that the wilderness is something to be conquered, and not something to be accepted on its own terms. For many of them this creates a conflict. I don't think they actually reject their parents' values, or the values of the bush. But there is a psychic conflict set up over this problem.

"The school system could be arranged so that children could go to school in the summer and perhaps the fall, and come into the bush with their parents at Christmas and stay till spring. Or education through correspondence, through the use of the Cree syllabic writing system, which all the parents use, could be arranged, so that the parents could take part in the education process. At the moment the parents are completely excluded from the education system. They don't understand what goes on at school, though they are fluent in the use of syllabics. We should accommodate the culture as it is now, rather than cramp the Indians into a system which is designed for an entirely foreign culture, urban-based and with totally different traditions.

"In other isolated parts of the world, the notion that health services can be taken to the people, even those living far away, is well established, and we could have such a flying doctor service, with nurses, and radios for calling in emergencies. At the moment poor health prevents many fine old men who are at the height of their spiritual powers from going into the bush, and they are the people to whom bush life means the most. Certainly the camps could have radios so that they could keep in touch with each other and with the outside world.

"You know, the Indians themselves have always tried to take advantage of elements from the white society. They have been doing it for hundreds of years. They have introduced guns, steel traps, all sorts of things, which they use in an Indian fashion, and which are now part of the Indian ceremonial, not alien things considered part of a foreign culture."

For the Cree hunter, things will not always go well. Some things will turn out well, others badly. He waits to see which way things will turn out.

The fathers and mothers took that attitude toward the education of their children, a passive, almost resigned attitude. They had been told that things would work out well if they allowed their children to go away to school. They were waiting to see if it worked out well: since nothing can be predicted in this life, they could not be certain of the result.

Ronnie Jolly felt he had no further say in the lives of his children who had stayed away. They would live in the world of the white man, since they had been taught to act and think like white men. Once he had tried to get Johnny to come back to the bush life, but Johnny had been in school for too long: finally, Ronnie had agreed to let him go. The men all remembered when the first government agent came around, twenty years before, and told the parents that if they would allow their children to go to school the children would help their parents later. "He wasn't telling the whole truth when he said that," remarked Sam. "The children who have been to school for a long time do not help their parents at all. The parents still wait."

THE LODGE CHRISTENED

The men were justifiably proud of the lodge when it was finished, just in time so that they could move into it and hold a celebratory feast the day before we were scheduled to fly out of their camp, October 15. They had built this type of lodge, said Sam, because it was the warmest that could be built. They could have built a dome-shaped or teepee-type lodge, but those had logs only around the side, and their canvas roofs tended to drip: since the winter was early, they had decided to make the warmest type they could.

At the beginning of the feast Sam took some food from each plate, some tobacco and a little of the rum we had given him as a parting gift, and put them in the fire. He took bear fat and rubbed it on the four sides of the lodge, east, west, north and south, in a sort of christening gesture. I thanked them all for the honor they had done us in having us on their hunting territory. It was an experience we would never forget, and we had learned a great deal. We had admired the way they lived, and hoped we might be able to transmit to others what we had seen. We said we would be back in the winter, and promised again that they would have the right to edit the film and withdraw anything they might find offensive. Tony had paid the men the money we had agreed on in advance—$25 a day for each head of family—and Sam now said that they had enjoyed having us there, that we had been no problem and that he appreciated the help (money) we had given them. When we returned they would be in the middle of their trapping season and there would be more for us to photograph.

When we finished eating we sat against the walls on the spruce boughs drinking rum. For the first time Sam and Ronnie seemed completely relaxed with us. They told me how, when we suggested the idea of the film to them, they had discussed it among themselves and with other hunters, and had finally gone to Speers to ask what he thought of it. Speers had given us, as it were, a good report and advised them to go ahead: we had not realized it before, but he could have killed our film stone dead right there. Such was still the influence of the Hudson's Bay factor.

After an hour or so we had to stop talking, for the drum had been brought into the lodge and must now be played. The drum, which had been made by Sam, was hung from the rafters by a string. Sam squatted on his knees, facing the wall, and sang two or three songs in that haunting Oriental chant that is Cree music. He then suggested that everyone should dance, and we all got up and shuffled round the lodge laughing and shouting. They said the drum must be handed once around the lodge and must be played by every hunter before being returned to where it started. The drum was handed to Abraham, who hit his hand against it and passed it on. "They say," said Philip, after a lot of talk and laughter, "that the Voyageurs are not really Indians. They have too much white blood in them, and they have not learned the songs." Ronnie was a better singer than Sam, but had only one tune, which, apparently, he adapted to every subject. Later Ronnie rose and led the dancing: the rum had got to him a little and he kicked up his small feet all around him delicately, laughing uproariously. The best of the dancers among the youngsters was Mary. Rosie was too inhibited to do more than shuffle around a few times. Occasionally the young people would retreat to their own corners of the lodge, lie down, apparently go to sleep and then get up after a few minutes. The children, asleep right after the meal, also awoke and took to running round and round the central stoves yelling. Rosie, who had a transistor radio, at one point turned it on, and WWV in Wheeling, West Virginia—by some freak of nature the radio station most easily picked up in these parts—blared into the lodge, a disc jockey's raucous hyperbole flooding the lodge in the middle of the feast.

The lodge was a picture: the walls, having been trimmed during construction, were gleaming white, and the women had arranged flowered plastic covers on the shelves. The spruce boughs had been laid only that day and the four stoves in the middle of the lodge were matched by a long pile of firewood ready for use. It was almost three o'clock when we drank the last of the rum. We drank to the success of the lodge, and of the hunting season, and now we drank: Death to the dams!

BAD WEATHER

Our radio next morning had difficulty picking up any signal at all, but finally Tony, who usually worked it, got on to Dolbeau Air Services in Chibougamau long enough to hear that it was "zero-zero" weather and they would not be flying in to pick us up. A blanket of bad weather had descended over all of northern Quebec, and indeed far to the south, and there was no sign of relief. We sat in our tent all morning: we had run out of reading material and were now reduced either to going to sleep or to thumbing through *Playboy* for the twentieth or even thirtieth time.

In the afternoon Philip and I decided to look for the rapids that we had not yet seen at the western end of the lake. The winds were heavy and the lake choppy for a canoe, but it wasn't too bad in the lee of the island. Quite a few ducks were flying around, presumably on their way south for the winter, but Philip wasn't able to get a shot at any of them. The rapids ran along almost the entire length of a superb little river running into a lake to the south, one almost as big as Lac Trefart, but unnamed on any map. We clambered out of the canoe onto the rocks and scrambled over the snow and muskeg along the riverbank. Snow began to fall heavily as Philip stood on the rocks casting his line into the rapids, which were, in that empty wilderness, making a tremendous roar, though they were quite small. On that rough day we seemed to be far from anyone, even from the people in the camp about three miles away. For perhaps the first time in my life I was alone (or almost alone) with the elements. That is the normal and desired condition of life for hunters and trappers, those men who, as Willie Awashish told me, live in their totally spiritual world, and though I was slightly frightened by the immensity of the wilderness, the power of the winds whistling ceaselessly through the trees, the surge of the waters rushing over the rocks from lake to lake, I was nevertheless awed that such a place still existed in this overcrowded world, and that people were alive who knew how to survive in it.

We could not make any radio contact with Chibougamau the next day, either, but that evening it seemed as though our troubles were over. The skies cleared suddenly and we had a magnificent sunset. Later the night sky was cloudless and full of stars. But when we woke up next morning it was snowing again and we couldn't see the other side of the lake. Sam came into our tent and pushed the snow off the roof. "He says if this wind stops blowing, the lake's going to freeze over," said Philip. We didn't know whether to believe him or not. We knew him for a great joker. But already the ice was collecting along the shoreline, blown there by the winds, and the temperature was cold.

Five days after we were supposed to be picked up Sam came into the tent in the morning and told Philip that the lake had frozen over. We went running down through the trees to the lakeshore: the familiar choppy blue waters of the lake had been replaced by a sheet of smooth glass stretching to the far hills. The freeze-up, which we had been told never happened before November 4, had occurred on October 19, the earliest that anybody could remember.

We ran back into the tent and started to call Chibougamau agitatedly; though if the lake was frozen, there seemed little enough point to it. The conditions had not improved, and we could not make any contact. We were in quite a predicament. If the freeze-up was permanent, as seemed likely, no plane would be able to land until the ice was firm enough to bear its weight. Greatly though we had enjoyed our experience, we did not welcome the prospect of being locked there for another six weeks with no food and no work to do, just throwing a tremendous burden on the hunters, who would have to feed us if they could. The men must have known the possibilities, but Sam and Ronnie laughed whenever they saw us during that morning, and seemed quite cheerful at the awful prospect; probably to them it was just one of those acts of nature that everyone must accept.

We had porridge for breakfast—a huge bag of oats was one thing we did have left—and one after the other tried to raise Chibougamau on the radio. Finally we realized there was nothing we could do but wait. We lay down in our sleeping bags again. Occasionally a voice would surge into the tent faintly over the radio before being whisked away again by the mysterious forces of nature. One of us distinctly heard a man say, in French, that he was leaving for Lac Trefart. It seemed highly unlikely, but we decided to pack up just in case. Then nothing. Not even a stray voice for the next hour and a half.

At lunchtime we heard the roar of a plane overhead. Marcel Hourdis, the dapper, elegantly mustachioed little bush pilot (a man whose appearance suggested exquisite suppers by candlelight with beautiful women, rather than the rigors of the Canadian bush), had come for us, braving weather that really was not meant for flying. He circled once and then disappeared behind the island. At the far end of the lake there was a stretch of open water we could not see. Eventually he taxied around the far end of the island two or three miles away. As he began to plow through the ice, we rushed back to the tent to finish rolling up our things and haul them down to the lakeside. The men were waiting for the plane: with that feeling for the practical thing to do which seemed to come instinctively to them, they were ready with ropes, planks, axes and armfuls of the inevitable and indispensable evergreen brush to make a working platform, paraphernalia needed for hauling the plane to the shore through the thickening ice, turning it around and then loading the heavy baggage into it. Marcel said we would have to leave as much of the baggage as we could: he could not be sure that he would have enough open water to take off. And we should take our sleeping bags, because he couldn't guarantee that we would get to Chibougamau by nightfall. The controls on the back of the floats had frozen as soon as they hit the ice, and the men had to work around them, knocking off the ice for a long time before we could even move the plane from the shore. We hurriedly said goodbye to everybody and climbed into the plane. It was an odd way to take leave of them. Our relationships had developed so slowly and in such a measured way that there was something wrong with the precipitate nature of this leavetaking. Still, they must have been relieved to see us go. For when we returned in March, Sam told us that the freeze-up of that day was permanent: if the plane had not arrived that morning, we would have been stranded there.

We crashed through the ice to the other end of the lake. There was just enough open water to allow us to get into the air. All around us the lakes were frozen. But the world that day was a severe, unfriendly one, especially from the sky. A huge weather system that stretched over the entire east coast lay close upon the land. Marcel held the plane at 500 feet as he flew a pattern around and through the worst of it. We made it back to Chibougamau, checked into a hotel, luxuriated in a hot bath. We were as relieved and happy to get back into the familiar surroundings of our world as the hunters are relieved and happy each year to get back into the bush.

11

COURTROOM

When Dr. John Spence and his team of scientists returned from their quick survey of the La Grande River area in September 1972, they reported that the James Bay project as presently conceived would destroy all the renewable resources of the river on which the Indians had always depended. They said that Indian life could not coexist with a project on such a massive scale, and recommended that the Indians try to persuade the government to modify or scale down the project to take account of their objections. Particularly, they suggested that the proposed LG-1 dam should not be built on the site of the first rapids (if at all) because that was a place of particular economic and emotional significance to the Cree of Fort George. They said that the Indians might be able to live with a project scaled down from four dams to two, or designed on a less monolithic scale.

The government was not prepared to entertain any suggestion that the project might be changed in any way. Finally the premier of Quebec, Robert Bourassa, agreed to meet with a delegation of Cree Indians and Inuit in October. They were ushered into his office, but he showed little regard for them. As Malcolm Diamond, Billy's father, performed the traditional function of a respected elder by making a formal little speech in Cree, the politician talked loudly with his aides, and showed impatience at the fact that the speech had to be translated. No sooner had the old men gone through the preliminaries than the premier announced he had another meeting in five minutes. The native people emerged from his office outraged at his lack of manners, and insulted by the treatment they had received: they had no option now, they said, but to take every means in their power to try to stop the James Bay project. The court action arising from that decision began in December before Mr. Justice Malouf.

Meantime, Jean-Pierre Fournier and I raised some money privately and finished the film about Job Bearskin which we called *Job's Garden.* Gilbert and Billy came down from Fort George to help with the translation, and with the help of Maggie Bearskin (who had been the Virgin Mother in *Arrow to the Moon*) and her sister Daisy, we finally got the work done, in spite of a disastrous shortage of money and time. The film had many shortcomings, but was later judged one of the dozen best Canadian documentaries of the year, and played a positive role in explaining the attitude of the Cree Indians in their moment of crisis.

JUST LIKE WHITE MEN

Sam, Ronnie and Abraham were tramping their territory on snowshoes, and Job Bearskin, too, was deep in the wilderness when their colleagues traveled to Montreal to give evidence in the courtroom about the hunting life. They were still methodically gathering in the animals on which they were feeding their families when the lawyers for the government and the James Bay Development Corporation began to lead their witnesses through the court. The witnesses were mostly priests, doctors or nurses who had worked in one or other of the northern settlements, and their purpose was to prove that the Indians and Inuit were, in the modern age, living exactly as white men live.

Therese Pageau, a nurse who had spent two years in Rupert House, and Paul-Aimé de Bellefeuille, who had been hospital director in Fort George, gave evidence to that effect. Miss Pageau said she went regularly to the houses of Indians in Rupert House, was very close to them and knew them intimately. In their houses they had beds, a table, some chairs, sometimes a cupboard and sometimes linoleum on the floor. They had dishes in their houses. They went to the cinema regularly, had transistors and record players and amused themselves by playing the guitar.

LE BEL: How do they dress in Rupert House?
PAGEAU: Like everybody.
LE BEL: Like us?
PAGEAU: Yes, they dress against the cold, but there's nothing special for them.

LE BEL: I see. These clothes come from the store, if I understand your reply?

PAGEAU: Yes, yes, all bought at Hudson's Bay, or ordered C.O.D. from Eaton's.

LE BEL: They make a lot of orders C.O.D., eh?

PAGEAU: Plenty.

LE BEL: They are many who make orders C.O.D.?

PAGEAU: They order by catalogue.

LE BEL: You're talking about clothing?

PAGEAU: Clothing and other things. Sometimes even furniture would arrive C.O.D.

Dr. de Bellefeuille testified how in his hospital the Indian patients ate white man's food and liked it, and how in Fort George the young people loved to listen to music on their portable radios, how they jammed into the hall to watch western films three or four times a week and how they dressed in the same way as we, in clothes bought from the Hudson's Bay store or the Eaton's catalogue. They ate not more than 20 percent country food, he estimated.

The flow of witnesses continued in the same vein. A former assistant to Glen Speers at the Hudson's Bay Company store in Mistassini testified that 75 percent of the Indian food there was bought at the store. (Ironically, Speers could not come to the court to testify, as the Indians wanted him to do, because he was at that time engaged on his second round of trips around the seventy-four bush camps, in which two thirds of the Indian families from the post were living on country food!) Residents of Chibougamau, policemen, restaurant owners, doctors, shopkeepers were called to tell how they saw the Indians around town and observed them eating spaghetti in local restaurants. Indians traveled to the town by automobile, and they shopped in the Hudson's Bay store there. Three times LeBel groped for a devastating fact:

LE BEL: You are aware of the Kentucky Fried Chicken restaurant in Chibougamau?

ALFRED FILLION: Yes.

LE BEL: And have you observed Indians using it?

FILLION: Well, I couldn't say.

Finally he found his man. Gaston Boulanger, who sold oil to heat the houses in Mistassini, had seen Indians at the Kentucky Fried Chicken place. Joseph Savage, a municipal councillor, had seen Indians shopping in the Hudson's Bay store. They seemed to buy much the same kind of food as he did.

A difficulty for this line of attack by the corporation was that most of their witnesses had never been in an Indian house while the family was eating, and even those who testified to having gone hunting with Indian guides had to admit that when the day's work was done the guides retired to their own tents, where they had some food, though the witnesses could not say what sort of food. Their estimations of the Indians' dependency on country food ranged from 10 percent to 25 percent, but appeared to be based on guesses as a result of the witnesses having seen so much white man's food in the stores in Indian communities. In cross-examination all of the witnesses admitted that the Indians in all of the settlements still hunted and trapped regularly. The corporation did have one expert witness who could have been most damaging to the Indian case in Dr. Richard Salisbury, an anthropologist who headed the McGill Cree Developmental Change project, and who had been hired both by the Indian task force and by the James Bay corporation to study the likely effects of the project on Indian life. Salisbury, an accurate and painstaking inquirer, had written two reports, based on the same facts, but on different assumptions. For the Indians he assumed that the Indian cultural life, based on hunting, would continue. For the corporation he assumed that the project would be built. Salisbury had agreed to write his report for the corporation on condition it should be made public, but the corporation had refused to allow him to publish his recommendations for three years, possibly because Salisbury had been active in urging them to take cognizance of the immense Indian knowledge of the biology of the area in drawing up their scheme.

For the corporation, Salisbury had assumed that the hunting life would continue to employ 600 people, and had shown that with the quickly increasing Cree population, some other stimulus would be needed to absorb the growing labor force. Salisbury might have proved a damaging witness before a judge who appeared to be open to persuasion either way. But there was a snag for the corporation: Salisbury's researchers had found that the Indian dependency on country food, even in Fort George, was 50 percent, while it went up toward 90 percent in some of the other settlements. If Salisbury gave evidence, he would flatly contradict all the evidence given by the parade of retired priests, doctors and nurses who claimed to know the Indians intimately. Though Salisbury had been asked and was waiting to give evidence, he was never called, and the corporation lawyers fought successfully against the introduction into court of the report he had prepared for them.

On the scientific side, the corporation produced a great number of witnesses, almost all of them French-speaking Quebecers working for the Quebec or federal government. They were well qualified and sound, but while they did not share the predictions of gloom given by the witnesses for the Indians about the impact of the project on the birds and animals, they could not themselves say what the effects would be. They were unable to produce any convincing new facts which would indicate good effects rather than bad, for no such facts had ever been assembled. Indeed, it was an important part of the native case that though the project was already being built, the necessary preliminary studies of the biology of the region had never been made.

$10,000 A YEAR FOR A TRAPPER?

The corporation produced Rita Marsolais, a member of the economic research staff at Hydro-Quebec, with a mass of tables and statistics, to prove how well off the Indians in James Bay villages were, and how much money was being spent on them. By adding up personal income from salaries and all money spent in each village by the Quebec and Canadian governments, including even such items as garbage collection and road maintenance, and then dividing the global sum by the number of households and families, she was able to assert that the individual income per household in Fort George (151 households) was $18,046 in 1972 and per family (268 families) $10,167. A number of hunters were brought in at the stage of counterproof to refute this testimony. All of them said they had never had such a sum as $10,000 in a year. When asked by O'Reilly how much the project would cost him, William Rat, forty, of Fort George, said: "It will be like losing my life. When you talk about money I do not really know the value of it. I do not use it very often. . . . It is the white man who has the money, and on the other hand, the Indian has the land. The white man will always have the money, and will always want to have the land. We have lived off the land, the white man has lived off money." These hunters had been out on the land, living in a subsistence way since the case had started (they included Job Bearskin, who left his trapline to come directly to Montreal to give evidence), and they repeated over and over that they could not put a value on what they were losing, they did not have any idea how much it was worth in money, for "I do not worry about money."

A lean, wiry man (unlike most of the hunters, who tend to be thick-set), Joseph Pepabano of Fort George said that since there were not many

Indians where the white men lived, there should be a place where there were not many whites where the Indians lived. He was worried about the place they called Obidjun, the first rapids on the La Grande. "When I talk about the fishing there, I think about the children. Will they have what I had when I was growing up?" As for the money, "I cannot put a value on what I will lose. . . . Since the start of life my forefathers have lived off the land the way we are doing now. . . . If the project goes through . . . I am very worried about the children, it's they who will suffer most." David Sandy, forty-five, a small, quiet man from Fort George, said: "When you talk about the land, you talk about me and my family. . . . What part you destroy of the land, you also destroy of me."

CLASH OF ANTHROPOLOGISTS

The anthropologists, of course, disagreed among themselves, and in a most interesting way. The James Bay Development Corporation produced their staff anthropologist, Paul Bertrand, who had been hired to handle all matters dealing with the Indians. A thin, strangely tortured person who seemed to be genuinely concerned to help the Indians react to the project, Bertrand had not been particularly successful in his job. He had behaved in an extremely gauche manner in his early contacts with the Indians, had failed to enlist any to his staff and was not trusted by them. But he was an experienced anthropologist who had worked among Indians in Mexico for ten years, and he now gave the court a sophisticated and intelligent interpretation of the history of governmental, religious and commercial contact with the Crees. In his view, this contact had sedentarized a nomadic people:

LE BEL: As an anthropologist, Mr. Bertrand, what is your judgment as to the future of the Cree civilization? You have spoken of seven years. . . .

BERTRAND: I would say seven years. . . .

LE BEL: For what?

BERTRAND: Not for the disappearance of the Cree culture, but for a crisis to occur in the Cree culture. I cannot see how the culture can overcome it with the governmental umbrella that it has now, unless something new happens. I would say seven years, because now in most of the villages, the younger generation, up to 19, is studying. Their theoretical and practical background will certainly not be the same as that of their parents, and that could produce dissension, which I believe most anthropologists have visualized.

LE BEL: Could you qualify the different impacts—institutional, educational, governmental and so on—and say what has been their consequence on the Cree culture?

BERTRAND: I can give you a professional opinion based on my experiences in other groups and also in the north. Naturally, a culture is a dynamic thing. You can't study a culture in a static fashion. Each culture evolves in a given direction, or in another direction, and then there are cultural cul-de-sacs, cultures which, at a given moment, block, for one reason or another. In my opinion, the Cree culture is heading towards a blockage . . . because . . . to occupy the different posts in government, the structures already in place in their villages, would take them 20 or 30 years, and I do not think that the Cree culture is capable of accelerating the process. The impacts they have received to date have not produced what a bigger impact would have produced. That is, a sort of collective renewal, a sensation of becoming, a wish to do something to improve, but an active wish, something felt. What produces this interior disintegration and prevents collective action, I believe, is a slow invasion, and not something brusque. A slow invasion doesn't give enough strength. . . . An immense shock can help a culture to regroup itself in a really developmental way, in the sense in which they can think of supporting their actions with a concordant psychology. . . .

LE BEL: Do you consider that the hydro-electric and other projects on the territory of James Bay will constitute an impact on the culture and way of life of the Cree Indians?

BERTRAND: It's obvious, an immense impact.

LE BEL: Compared to other impacts you say it will be immense. On what do you base that?

BERTRAND: You have 6,000 Crees in a region, then, suddenly, brusquely, immediately perhaps 16,000 or 20,000 newcomers alongside them. As I said before, for a white man you almost have to build a hospital. How many will be needed for 16,000 I don't know. But in bringing up 16,000 whites you bring up in a global fashion the whole society, the shock is going to be brutal. Perhaps it is the only way to make a culture react, and then really begin to participate, to take its own development in hand. I can't see how you could interpret it in any other way.

Harvey feit in refutation

O'Reilly brought back Harvey Feit, the anthropologist who had studied the Waswanipi Cree, to comment on Bertrand's interpretation of Indian life. For Bertrand the Cree culture had withered away and little of value remained; only by an enforced "transformation" could anything valuable be salvaged. For Feit the culture was proving tenacious, was still of tremendous value and could be aided in its current crisis by a more sympathetic approach from the dominant society. Feit challenged Bertrand's view that slow impacts inevitably wither a culture. People are able to change and alter the things they actually do—for example, use a gun instead of a bow and arrow—and still maintain the same view of the world. The Cree people had done this, modified their behavior, adopted new things, accepted changes of their own free will, and had maintained "the coherence and integrity of their beliefs and understandings." They were one of the few groups in North America who had been able to to this. Even recently they had found a way to integrate summer wage employment into their life-style by taking jobs that allowed them to work in their family groups, that kept them in the bush and enabled them to get country food, and that allowed them to use the Cree language. That was not, argued Feit, a disintegration, but an effort to integrate and synthesize new ways of doing things. He acknowledged Bertrand's "important insight" that change does not have to be slow; but more important than the speed of change was the question of whether people were themselves in control of the changes and wished them to happen. Of the idea that the increasing population would lead to a crisis in the culture, Feit said that the Cree had responded by using a much more diverse range of animal resources than ever before "and I think we can foresee a potentiality for a significant number of new people to enter into the subsistence hunting way of life." The integration of part-time wage employment into this life would help. "I can find no grounds to agree [that a crisis could take place in seven years]," said Feit. "My figures show the potentiality for continued use of the resources by a growing population over at least the next decade or two."

Asked to describe the likely effects of the project in the next three years, he predicted, first, that there would be (as in other places in the Canadian north) family disruption, use of alcohol and growing racial tensions following the arrival of Euro-Canadian workers; second, that disruption of the intensive use of the animal resources made by the Indians around Fort George would affect their health and economic well-being; and third, that there might well be a drastic cultural effect. The Indians were watching carefully to see if there were any white institutions ready to protect their interests and uphold them. If they found that there were not, "They're going to suffer a very severe blow . . . that could break the core of their culture. . . . Culturally I see them at a critical stage." He added: "For me, each human group that is able to meet the problems of living and come up with a unique way of life is a very precious thing for the whole of mankind. . . . [Its] destruction . . . is tragic and sad, and I can say no more."

THE REAL HERO: NATURE

The real hero of the courtroom during the six months of the hearing was nature herself. Her wonderful processes had never been described in such detail in a Canadian courtroom before: and the description of her complexities and beauties given by both the native people and their supporting scientists showed that nature has a wisdom that man is very far from emulating. In northern Quebec, particularly, where the ecosystem has never really been interfered with, the balance, economy and lack of waste, the total integration of every living element were revealed in the courtroom to anyone with ears to hear as a thing of the most extraordinary beauty. The contrast between this lovely system and the arrogant, wasteful and foolish plans that man was now proposing for the area was perhaps the saddest part of the whole story.

The system has no real beginning, since it never ceases to change, with life being created and destroyed during every minute of every day. But the climactic event of the year in these northern climates is the immense spring flood that occurs when the warming sun rises higher into the sky and begins in May and June to melt the vast blanket of snow that covers the entire country. The first function of these rising waters is to flush out the ice from the river. The ice has already begun to break during the winter as the river level has dropped, creating a sub-ice space that has been of essential value for many species of little animals during the winter. The ice toward the center of the river breaks from the shorefast ice, under which tunnels form sometimes big enough for a man to crawl through. These tunnels expose sedges and other grasses which are used by muskrat and their predators, such as mink, during the wintertime.

The water now flushes this ice loose from the shore, breaks it into small pieces and carries it down to the sea. The rushing water at this time has great power and energy, so it scours out deposits that already exist all the way along the river, drops new silt in different places and carries much silt with it all the way to the sea.

When the water reaches the sea, its first function is to aid in the break-up of the ice cover and so bring into action at the essential time the regenerative life processes in the sea. The water from the river is warmer than the sea ice—in the Koksoak River, flowing north into Ungava Bay, it is sixteen degrees Centigrade (60°F.) compared with four degrees Centigrade (40°F.) in the sea—so it quickly helps the sea ice to melt. The fresh water flows over the sea ice and, by dropping on top of it some of the silt it is carrying, cuts down the reflectivity of the ice, which then absorbs more solar radiation and thus melts more quickly.

The productivity of the sea depends on fertilizer salts—phosphates, nitrates and silicates—which are formed bacterially by oxidation in the deeper waters. Before the biological cycle can begin, these salts must be brought to the surface where the light is strong enough to enable plants to grow and to synthesize in the euphotic glare. The fresh-water inflow is a kinetic-energy source which helps start the turn of this cycle: it forms a current along the surface which entrains with it water from below by frictional force, thus bringing up from the bottom the water laden with nutrient salts. It is when these salts arrive at the surface that the plant cells begin to grow—cells on which feed the animal plankton, the crustacea or shrimplike creatures that are the basic food for many larger forms of fish life. These crustacea may be as small as one tenth of a millimeter or up to two or three centimeters, but they are the essential diet of a wide variety of riverine fish, such as salmon, Arctic char and whitefish, that spend a large part of their lives in the oceans, during which time they put on most of their weight. The seals, too, depend on the availability of the crustacea. And the balanced and measured breaking up of the ice is necessary for more than just the crustacea: the ring-seal pups are suckled on the fast ice until they are weaned. If for some reason the ice should disappear too early, they would die.

Meantime back in the river the spring flood is changing everything. The animals that have lived along the littoral of the river during the winter have got out of the way, or have made other arrangements (such as the beaver, who has defended his waterside home against fluctuating water levels by creating his own high water, out of season, by building a dam downstream of his lodge). Just as the salts are necessary for growth in the sea, so the nutrients carried by the river are necessary for growth in the river bed and on its shores. The river erodes one side, deposits on the other: according to the contours of the country, this scouring capacity

meanders from side to side, destroying older vegetation and laying down
new vegetation, which, as it grows, becomes the most highly productive
habitat for animals. The newly dropped silt also helps warm the river bed,
because wherever it is deposited it reduces the amount of moss, which
keeps the land moist and cold, and therefore the river valleys are warmer
than the uplands, another encouragement to growth.

During the flood stage the high waters wander into small lakes which
are not attached to the river at any other time of year. The sediment-laden
flood waters maintain the productivity of these lakes, and make them the
best places for waterfowl habitat, both for food and for nesting. The flow
is reversed on many small channels linking the lakes to the river, and as
the flood waters surge in, they make little points of land jutting out into
the lake which the birds, as they arrive from the south to breed in the
Arctic or sub-Arctic, seek out, for they are well protected all around and
are safe from predators there. A reduction in the flow of the flood waters
would lead to the growth of willow and alder trees on these points and
reduce their usefulness to the birds. Spring-spawning fish, such as the
northern pike, also use this spring flood to get into otherwise inaccessible
lakes.

The season for the growth of the new vegetation along the river is
brief, but perfectly attuned to the cycle of life as it now exists. For instance, the sedges and grasses in a full summer can make seed that will
be used the following year for the reproduction of the plant. If the season
was reduced by two weeks through the failure of the river to remove the
shorefast ice on schedule, these plants might not be able to complete this
cycle. In that case there would not be as much food available for the birds
and animals, whose reproductive capacity would immediately be affected.

Amazingly, an animal, fish or bird seems to have been developed by
nature to take avantage of every type of habitat that is offered by the
natural conditions. It is said by Canadian biologists that "as the river
goes, so goes the snowshoe hare, so goes the lynx." And this inter-
dependency of all species is the mark of the natural system. The conver-
sion of solar to chemical energy by green plants is the basis for animal
life on the planet. From very small insects up to bear and caribou (and
even men), many animals eat plants. Shrews feed on insects, hares live
on plants, but animals such as marten, weasel, fishes and foxes eat mice,
rabbits, squirrels and other small mammals, which are also very vulnerable to hawks. Wolves feed on rabbits and mice, but also on larger ungu-
lates (hoofed animals) such as caribou and moose. Birds and aquatic
animals are highly specialized fish-eaters, an activity in which terrestrial
animals cannot compete. Every animal depends on the maintenance of

the natural balance between all the species if they are to survive. The biggest predator of the small animals is man: a mere 150-foot-wide road can trap a population of small mammals, which, unable to roam around in their accustomed territory, will probably have a population explosion until they eat themselves out of habitat, whereupon their population will crash. Such fluctuations also affect the animals which live on these small mammals.

All of the land is divided into animal territories. All land is already filled with the animals it can support. Any changes which might produce animal migrations will create competition for the food available, until in the end the populations will stabilize at whatever is the changed carrying capacity of the land.

Even natural fires, started by lightning, play their part in this system of growth, death and regeneration. Fires are bad for caribou, but good for moose and other animals, for after a fire the first things to grow back are berry-bearing shrubs, aspens, willows and alders, the preferred browse for the moose. If man reduces natural burns too much through excessive viligance, nature will get back at him. By so doing he will have increased the amount of dry underbrush which will become vulnerable to forest fires again. One of the most important decisions man has made in the last fifty years is to have largely abandoned these biological control mechanisms to be found in nature and—confident in his arrogant assumption of superior knowledge—to have substituted chemical and artificial controls against which nature has already begun to react.

All of this is true in the waters as on land. Just as there is an animal for every habitat—the mice and rabbits nibbling away at the low levels, the caribou and bear at higher levels—so in the waters a different fish has been developed to take avantage of each type of habitat to be found in a river or lake system.

The anadromous fish are a wonder: some, like the trout, will return, even after several years, to the very spot at which they were spawned. But each fish has different habits. The salmon spend most of their lives in the sea: they will hang around at the mouth of a river waiting for a sufficient freshet of spring water to enable them to ascend the river. Then they can swim as far as 350 miles upstream to the headwaters of the Larch (on the Caniapiscau system) to spawn. These are far-northern, cold waters, and the salmon are not ready to run up the rivers until August: then it takes three weeks for the vast migration to make its way past Fort Chimo, at the head of Ungava Bay. The one-year salmon stay in the river. They do not go out to sea until as smolts they are up to twenty centimeters (seven to eight inches) in length. The bigger they are when they leave the river,

the better chance they have of surviving in the rough-and-tumble existence of the oceans. It has been calculated that the survival rate for a
twelve-centimeter smolt is 5 percent, while one that has put on another
eight centimeters has a 20-percent chance of survival. They return as
grilse in two or three years to spawn. The young growing in the river are
aggressive: they defend their territory on the river bottom by any means.
A shorter growing season caused by a longer ice cover would reduce the
size of these smolts, and therefore reduce their chances of survival in the
ocean. But scientists cannot be precise about these effects. For nature has
amazing feedback mechanisms. The fewer fish which survived would
produce fewer eggs and fewer young: but these might, because each
would have more food available, do better.

The Arctic char live a totally different life cycle: they stay in the rivers
all winter, sitting on the bottom or in a deep hole. They would freeze
solid in the much colder winter waters of the ocean. They go to the sea
in April or May, spend only two or three months there and return in the
fall, and they do that each year.

The anadromous whitefish, like the char, do not travel far from the
mouths of their home river. They come back every year to spawn. But the
incubation period depends on the regularity of the water temperatures,
and if the river was, for some reason, silted, the eggs laid by these fish
would smother.

Farther inland, a great variety of fish (though not so many this far
north as in the south) make use of the many types of habitats that
develop along a river. The trout require a place where water is flowing
through gravel: this provides oxygen for the developing eggs. They
spawn in autumn and the eggs incubate during the winter. The pike, the
predators of the north, need aquatic vegetation, and are likely to do well
in newly flooded reservoirs. The explosion of nutrients which follows the
flooding of such an area will inevitably lead to an explosion in pike
populations. Deprived of other food, they may then start eating each
other.

In such wondrous ways does nature perform: the caribou munching
away on the lichen which has attached itself to the branches of dead and
dying trees on the uplands; the moose eating the young vegetation in the
burnt areas or along the rivers; the lynx preying on the little snowshoe
hare (from which, when the fur around its hind legs begins to turn brown,
the Indians can tell that the winter is gone and colder weather will not
return); the beaver creating his own permanent high water; the muskrat
burrowing around in tunnels under the shore ice; the birds eating sedges
and grasses, nesting under tender little spruce trees, building their nests

on well-protected points that nobody else needs; the wolf following the caribou herds, culling the old and the feeble, maintaining in this way the strength and vitality of the herd. Even the falling leaves, blown into the lakes during the high winds, play their part in maintaining the infinite variety of life that has adjusted itself to the wondrous rhythm of nature's year.

With this system the people who govern us showed no concern of any kind. They knew only one fact, and cared about no other: we needed electricity for our factories, homes, offices, streets, refrigerators, televisions and hair dryers, and by fulfilling this one need, they claimed, they would be improving the variety and quality of life on this earth.

A DEVASTATING CRITIQUE

But do we in fact need that electricity? This book is not about the James Bay project but about the Indians who live in the land on which the project is to be built. Yet when the fate of the Indians is explained to southerners, the commonest reaction is to weigh their needs against the numerically greater need of the whole Euro-Canadian society for constantly increasing supplies of energy. This is the argument used by politicians and engineers to justify anything they might do in the James Bay area. They have not been punctilious or rigorous in their use of the facts, but whenever the validity of the project is called into question, have tended to rave on about how hospitals, factories and schools will have to close down if the project is not built. Unfortunately, their assumptions have never been subjected to serious analysis by the so-called watchdogs of the public interest, the newspapers, radio and television.

It is certain that this question will be in the minds of many people who read this book, and to deal with it I propose to transfer my focus for a few pages from the Indians to the project itself. As I have shown, the social and environmental aspects of the work of the James Bay Development Corporation, the James Bay Energy Corporation and Hydro-Quebec were sloppy to the point of being disgraceful. And now in the courtroom the Indians' lawyers produced evidence to indicate that even in their predictions of energy needs, these agencies acted in a cursory and self-interested way. Indeed, it is a depressing comment on our public institutions that the James Bay project had never been subjected to public analysis even at this technical level until the Indian side produced its evidence.

Dr. John Spence had dug up from among his colleagues at McGill University a man who proved to have considerable influence on the case,

Dr. Daniel J. Khazzoom, Professor of Econometrics, formerly chief econometrist for the Federal Power Commission of the United States, and the only man in Canada to have developed a sophisticated mathematical procedure for predicting future energy demands.

The specialists from Hydro-Quebec, with masses of tables and charts that were totally incomprehensible to the layman, attempted to prove that the power from the La Grande project would be needed by Quebec in 1980 if the province was not to run short of energy, and therefore that in any discussion of the balance of inconvenience that would be caused by a stoppage of the work, the scales must weigh heavily in their favor because of the magnitude of the public interest involved.

Dr. Khazzoom subjected their projections to computer analysis and produced a version of the facts which seriously questioned their credibility. He showed that their projections dealt not with the entire energy needs of the province, or even with the entire electricity needs, but only with the probable growth of the integrated system of Hydro-Quebec, the province's publicly owned generating and sales agency. Because of the nationalization of private companies in the early sixties, Hydro-Quebec during those years was greatly increasing its proportion of the total energy consumed in the province. Though this integrated system now provided only two thirds of the province's energy demand, and was therefore not an accurate index of the province's whole demand, its unnaturally high growth rates—three times those of the province as a whole—were used to suggest that provincial demand for energy in the future would also be unnaturally high. This was done by a Hydro-Quebec witness, Joseph Bourbeau, who made a simple mechanical projection up to the year 1986 of the maximal peak demand on the integrated system between 1953 and 1970. That is to say, future needs were estimated by projecting forward the highest demand registered on its measuring apparatus in any one-hour period during those years. This of course was far, far in excess of the average demand. Bourbeau analyzed this maximal peak demand and declared that it had been increasing between 7 and 8 percent a year. His method was simply to take the highest one-hour demand for 1970 and project it forward on the basis of a growth rate of 7.96 percent a year. Hydro-Quebec expresses demand in kilowatts of installed capacity, and the difference between a 7.96-percent-a-year projected increase and a 7-percent projected increase would by 1980 be no less than 1,309 megawatts of installed capacity, which is getting on for as much as the LG-2 powerhouse would be producing in that year.

Hydro-Quebec's own figures indicated that the actual rate of increase between 1963 and 1971 varied between 4.82 percent in 1968 and

10.3 percent in 1963, with an average of 7.71 percent. Thus, though the projections were supposed to be based on historical rates of growth of between 7 and 8 percent, in fact they were based on a figure only 4/100ths of 1 percent less than the *upper limit* of historical growth. None of these projections took into account such factors as population growth, changing prices of electricity and of substitute forms of energy, formation of new households, incomes, industrial production or economic cycles. Khazzoom said such simplistic mechanical extrapolations of past figures were "nonsensical." It was the Hydro-Quebec lawyers who had insisted on introducing evidence about electricity demand, and Khazzoom's rigorous criticisms, expressed in a manner of almost overweening self-confidence and arrogance, threw them into a tizzy.

He showed that at no time since 1951 had the rate of growth in electricity demand in the province reached the 7.96-percent figure used by Hydro-Quebec in their projections, and pointed out that the National Energy Board of Canada projected a continuing rate of growth of only 4.3 percent. The National Energy Board prediction for the whole province for 1980 was for an energy consumption of 114,000 gigawatt hours (one gigawatt hour equals 1 million kwh), while Hydro-Quebec predicted 113,344 gwh for its integrated system alone. Translated into installed generating capacity, said Khazzoom, this represented an overestimate of need by Hydro-Quebec of 5,000 megawatts by 1980, almost as much as the total production of the whole La Grande project for the years 1980, 1981 and 1982.

Khazzoom discovered huge gaps in the figures, big enough to drive a coach and horses through. Included in the projections of maximum demand for the province were block sales made to Ontario and New Brunswick, and sales contracted to Consolidated Edison in New York, though the Hydro-Quebec figures gave no indication of how much these sales added to annual demand. Einar Skinnarland estimated that they could have contributed 950 megawatts to peak demand in 1971.

Even if the base figure of 10,356 megawatts shown by Hydro-Quebec as peak maximal demand in 1972 was accepted, the Hydro-Quebec overestimates in comparison with the National Energy Board projections were gross: 24 percent for 1980 rising to 36 percent for 1985, or an over-estimate of 4,606 megawatts in 1980 up to 10,121 megawatts in 1985. The whole La Grande complex would contain installed capacity of 8,330 megawatts, to be working by 1983. Khazzoom's own predictions, based on an econometric simulation study, and giving, in his basic assumptions, as much benefit of doubt as possible to Hydro-Quebec, suggested that they had overestimated demand for 1980 by 14 percent, or

24.5 percent if block sales outside the province were excluded. He concluded, with the National Energy Board, that a more realistic rate of
growth would be around 4 percent. Hydro-Quebec's projections, he said,
were exaggerated and out of line with anything that could reasonably be
expected, and O'Reilly submitted that the figures were erroneous and
misleading.

The native side also argued that a large proportion of the supposedly
needed power was already contracted for, and was available from other
sources. Hydro-Quebec had contracted for Atomic Energy of Canada to
build a 600-megawatt nuclear generating station whose production was
not included in the projected figures, and a 628-megawatt thermal station
also existed which was to be partly used in 1980 and not at all in the
following three years. From these and other sources which would admit-
tedly become available by 1980—from various aluminum and power
companies—1,546 megawatts would already be available of the 1,960
megawatts which Hydro-Quebec claimed would be necessary by 1980.

Khazzoom also revealed another amazing fact about electricity production—amazing, at least, to a layman. This is that Hydro-Quebec maintains 10 percent of its capacity as a reserve: that is, it always has 10 percent
more generating facilities than it uses. Leaving that 10-percent capacity
aside, the top 10 percent of the capacity under that furnishes only .1
percent of the demand on the integrated system, and works only 2.04
percent of the time. In other words, by 1980 these 1,750 megawatts of
capacity would be expected to be idle for 358 days in the year (with
another 10 percent, or 1,750 megawatts, of capacity idle on top of that).
This means that by 1980, even if Hydro-Quebec's generating capacity
were reduced by 3,822 megawatts (20 percent), the integrated system
could still meet 99 percent of the demand on it. Is there a better way,
asked O'Reilly, to avoid this 1 percent shortage than by spending $6
billion (later $12 billion and still later $14 billion) on the building of new
facilities?

Indeed, the natives' lawyers went on to suggest one: industrial consumers readily accept interruptible power if they are given price concessions. Hydro-Quebec produced a list of 194 big industrial consumers
who had special contracts for 3,000 kilowatts or more. By 1972 these 194
customers consumed 99.2 percent of all the energy sold by Hydro-Quebec to industrial clients, and nearly 50 percent of all the energy sold in
the entire integrated system. Tables filed by Hydro-Quebec indicated
that if interruptible clauses were made with all of these clients, capacity
needs would be reduced by 30 percent, or 5,700 megawatts in 1980—so
that the entire La Grande project would be made unnecessary just by the

negotiation of such contracts. To reduce peak demand in the province by about the capacity of the La Grande project in 1980, Hydro-Quebec would only need to interrupt power to a fifth of these clients, or reduce peak power (which covers only 2 percent of the entire year) by one fifth to all of them. Such a course, given price reductions in compensation, would make a great deal of economic sense, said O'Reilly, since what was at stake, "if numbers are considered to be important," was the cultural and economic survival of several thousand natives against the negotiation of interruptible clauses for 194 industrial consumers. Yet all of this assumes that 10 percent of capacity remains idle in reserve.

So where was the overwhelmingly demonstrated need for this vast, expensive and hastily improvised project, a supreme example of the extravagance and wasteful use of energy which perhaps more than anything else characterizes our throwaway consumer society? At least no clear case for the need was made out in court. What was revealed there —to the gaze, it must be admitted, of a totally indifferent public—was the spectacle of a huge public agency, the largest in the province, spending more public money than any other single organization, with a vested interest in, and determining the rate of, its own growth.

The Hydro-Quebec witnesses attempted to prove that hydraulic generating facilities were more economical than proposed nuclear plants (the purpose here being to show that a delay in building LG-2 would have to be compensated by the construction of other, more expensive facilities). Skinnarland and Khazzoom showed that here, too, the agency's figures were shot through with loaded assumptions discernible only to the expert eye.

But perhaps at this point we can return to our Indian theme on a note which indicates the underlying, perhaps subconscious, racism involved in the whole scheme: excluding land required for roads, camps, airports, dams, dykes and control structures, the La Grande scheme would flood 2,179,000 acres more than the 1,280 acres that a nuclear plant built in the south would occupy. But in making cost comparisons between the two types of plant, Hydro-Quebec failed to ascribe any value to the land needed for the La Grande scheme. After all, it was only Indian land, and therefore worthless.

12

1973: LAC TREFART

We spent a week filming in Mistassini before we returned to Sam's bush camp in March, and were able to see something of the creative non-relationship that the Indians had developed with the appallingly insensitive bureaucracy of the Indian Affairs Department that was constantly busy with their affairs. It has been estimated that the 250,000 registered or status Indians in Canada have nearly $400,000,000 spent on them every year, of which, it is said (by the National Indian Brotherhood), some 15 percent makes its way to the Indian people. For every six Indian families there is one Indian Affairs civil servant. The Indians have developed a sort of love-hate relationship with these people, who are perpetually pouring through their reserves with schemes of one kind or another, armed with forms, resolutions and regulations in duplicate and triplicate. Ostensibly, power rests with the band councils, but they are so befogged by the daily onslaught of paper that they seldom offer any direct resistance to the underlying thrust of this bureaucracy, which is to get the Indian people off the land and to keep them busy in villages filling the infra-structural jobs of any small town. Indian Affairs will willingly provide three jobs where one would be sufficient if the result is to transform three hunters into three janitors (indeed, I have seen one school in Old Crow, Yukon, which has five janitors, all former hunters).

Now a team of white men flew into Mistassini while we were there, Indian Affairs bureaucrats mysteriously accompanied by an employment officer from Canadian International Paper Company, to engage the band council in solemn discussion about the future schooling of the children —whether it should continue in La Tuque or be transferred to Chibougamau. Philip Awashish, as a member of the band council, was engaged in these discussions for a day. In the end, it was not clear that Indian Affairs' real motives were unmasked, but the band-council members did not seem to be worried about it, for they had motives of their own. They insisted that the people must be consulted about the future of their own children. And since Indian Affairs was in a hurry for the answer, the council asked that money be made available for a plane to fly around the bush camps, in a sort of ludicrous white-man-style consultation procedure.

"Each councillor who is close to a particular group of families in the bush will go to that place," said Philip with satisfaction, after the day's discussion. "We won't be asking them for a decision right there on the spot. The people can't be expected to make up their minds like that. We will be telling them the sort of decision they will have to make in the summer.

"But the real purpose of the visit by plane around the camps is to bring back meat. We will be putting the Indian Affairs money to good use."

PAVEMENT DRIVERS

It is not usual for rain to fall during the Canadian winter, but if it does, the effect is lethal, for it freezes on hitting the ground. The ice imposes a tremendous weight on trees and plants, which look extremely beautiful but crack and break under the strain, and of course movement becomes difficult. It rained every day during our visit to Mistassini (this was really a most peculiar winter). The roads in the village were covered by hard-packed snow, the sort that echoes as you crunch across it in your boots, and this now took on a protective layer of sheet ice. While driving our equipment from place to place around the village we lurched into snow banks, slewed crazily back to front and got hopelessly stuck at least ten times. Once stuck, we could not push the car out because the sheet ice on the road gave no purchase against which our feet could push. The road to Chibougamau, narrow and high-peaked in the middle, required greater driving skills than we commanded. Our truck slid down a steep bank, and even the Department of Highways grader had difficulty in pulling it back again. As we were being rescued from this disaster, the remarkable taxi driver from Chibougamau, Chabat, a man who had been

driving Indians around ever since Chibougamau was founded, drove
nonchalantly by, stopping long enough only to remark: "Pavement drivers, eh?"

SPEERS TALKS BEFORE THE CAMERA

Our long involvement with Mistassini had by now won the entire cooperation of Glen Speers, who after twenty-five years in and around
James Bay finally agreed to be interviewed before microphone and camera. He had been to Sam's camp since we left and had found them well.
The seventy-four camps set up at the beginning of the winter had now
been consolidated to sixty-five, some of the camps having joined
together. Speers had picked up and shipped out 4,300 beaver skins on
his January rounds, which indicated that the beaver catch would fall
within the usual Mistassini 7,000 to 8,000. The only message he had for
us was from Ronnie Jolly, to the effect that Eddie might be given the job
of translating so that he could earn some money, a hint that Eddie, like
the other hunters, would like to be paid for his contribution to the film.

Speers now told us that the use of the airplanes enabled the hunters
not only to spend a longer time at the post, and therefore to supplement
their incomes with a longer period of summer wage employment, but also
to reach their traplines earlier than when they had to go all the way by
land and water. This gave them a better chance to survey the beaver
resources of their land, and he believed the increased catch which resulted from this covered the extra cost of the aircraft. Though businesslike and unemotional on the surface, Speers clearly had a deep feeling for
the Indian life and greatly admired the capacities of the people. But when
I asked him to describe the camps, he replied laconically: "Well, I like the
camps. I would say nine out of ten have very nice camps. There's quite
a variety of camps, quite large and well made, some are smaller, they're
all warm and they are clean." When reporting the extraordinary fact that
in twenty-five years in James Bay he could not recall a serious accident
having occurred in any of the camps, he said: "They're quite cautious. I
mean, they know they are quite a way out, and they watch it."

Speers had arranged that his visit to Sam's camp would be on March
16 and we had told Sam before we left that we would arrive a week before
Speers on his second visit.

GETTING SATISFACTION FROM JESUS

We decided that we would make a visit to the fearsome Pentecostal minister at Chapais about whom we had heard so much, since his hellfire activities among the Indians were winning converts from the gentler and more ineffectual Anglicans. We drove the twenty-six miles along the paved highway through the bush from Chibougamau to Chapais, and in that small, square mining town found the Rev. J. Arthur Lemmert to be a big, ruddy-faced, plain fellow, a truck driver by profession and a missionary by vocation. He presided with great self-confidence over a plain wooden church. He had come north from Pennsylvania a few years ago to plunge into the battle for the souls of Indians who had been converted to Christianity for more than 100 years. When he arrived at Chapais there were 200 Indians living in shacks around the town. Now there were about 125, and he had personally converted three quarters of them to the Pentecostal faith. All but one Indian family in Chapais attended his church regularly. The Pentecostals were also making deep inroads among the Anglican Indians in Rupert House, and they had some seventy or so members in Mistassini itself.

Lemmert had applied to the band council a year before for ground on which to build a church in Mistassini, but the council was reluctant to alienate Indian land. Lemmert had parked his truck outside the meeting. "Whose is that truck outside?" asked one of the band-council members.

"That is the Lord's truck," said Lemmert.

"Well, would the Lord mind if we borrowed it for a while?"

Now, standing beside a primitive painting of Jesus as a shepherd, he told us that he was justified in his work among these Anglicans because, according to the Bible, it was "faith in Jesus Christ" that saved, not church membership.

"We believe if a person accepts Christ, his life will change, and this is what we have seen among the Indians. All of the Indian people were habitual drunkards when we came here. A big percentage gave up smoking and swearing and things like that, and now seem to live very useful lives.

"In this town we have good compliments from all the storekeepers and the chief of police. They are proud of the work we have done among the Indian people. We save them a lot of trouble. It is not just us, we believe that God has done this work."

Larry Linton, the Baptist missionary who lived in a little house in Mistassini, had told us that there was not really need for both himself and the Pentecostals in Mistassini, "and I wouldn't be surprised that if they got a purchase, we would pull out.

"I didn't come to change the world," said Linton. "I came to change individuals. As long as I can look back on the ministry and see two or three individuals that have completely committed their life to Jesus Christ, and have followed straight through on the Christian life, I feel that my ministry is worthwhile.

"It is possible," he added, in an oblique reference to the Rev. Lemmert, "to get up and preach a hellfire sermon and get maybe fifteen or twenty down to the altar. . . . Indians are stoic on the outside, but they're an emotional people. . . . There is no problem in doing that, and you can do it the following week, and then the week after that. But there's no real commitment. It doesn't last."

Now Rev. Lemmert, his round, plain face beaming as he rattled off his spiel for us, told us that according to the Apostle Paul nothing avails but a new creature in Christ Jesus, and what he was providing to the Indians was "satisfaction in the Lord Jesus Christ. They are finding peace. They testify. We have testimony meetings where they can get up and speak freely of Christ. They can pray. They almost witness as the preacher himself." At least four men in the congregation could preach "almost as good as myself."

Lemmert could speak no Cree, but he thought that the traditional Indian life was okay: in any case, so far as he could see, the Indians were accepting things well. They were adapting to the changes that were being made by the government. "Sometimes we have a little bit of stubbornness in all of us that might be contrary, but you don't have to worry, it will work itself out via each other.

"But as I see it, I think the Indian people, outside of some things that are going on, with their environmental rights and so on, their aboriginal rights, they seem to be accepting things."

BACK TO LAC TREFART

Chibougamau was really jumping, as they say, on the day we flew into the bush, for they were having their annual winter carnival, the highlight of which is a seventy-five-mile race by 2,500 snowmobiles, which gather from all over the continent for the event. This time we had bought a veritable small grocery store to take with us—$311 worth at the Hudson's Bay. Indeed, we had so much stuff that when we arrived at the plane base Marcel said we would need two Otter trips. He wanted to take off early because the March sun tended to melt the top of the snow and cause sticky conditions for landing and taking off.

Overnight all cars were towed off the main street and bulldozers shoveled a foot-deep carpet of snow across the road, pavement to pavement. The bars were open all night and I spent a good deal of the night in one of them with Philip.

Already by eight o'clock in the morning snowmobiles were roaring up and down the main street, making a hell of a racket. We found Philip standing outside the Indian Friendship Center. He'd stayed all night in the bar and had decided to wait in Chibougamau until after the snowmobile race, because three Indians were entered and he wanted to know how they performed. After a bit of discussion he agreed to come along, but he said we'd have to wait until his wife returned to the Friendship Center because she had locked his briefcase in the office and he couldn't go without it. I bundled him into the car and then onto the plane, where he kept threatening to get off so he could get his briefcase. He would come on the second trip, he said. "Who ever heard of a Cree Indian needing his briefcase to go on the trapline?" I said. "Sit still, the plane's about to take off."

He complained bitterly, but fortunately didn't try to get out, and as soon as we were aloft he went to sleep, slumped down in the utterly abandoned posture of someone who is totally drunk and has been feeling good about it.

This northern world in winter is a dramatic black-and-white—snow and ice and an endless pattern of straight black trees. The tiny camps are easier to spot from the air than in summer, because although everything is hidden under snow, the straight pattern of the trees is broken by the things that men have made and left lying around—the long toboggans, the circular frames for stretching beaver skins, the bedclothes airing on a line, the snowshoes, things that demonstrate to the overflying traveler that human influence has been felt here.

Though they seemed pleased to see us again, the Indians greeted us in rather a detached way: it takes a long time to get used to the undemonstrative behavior of the Cree. Sam had decided, rather to our dismay, that we should sleep in the tent. "The lodge is too hot," said Philip in explanation, "they're having problems with it." But the explanation was certainly not true: not unreasonably, they had decided after months of dividing the lodge space among themselves that to inject six strangers and half of their piles of equipment into one corner of the lodge would create too great a disruption in their lives. We carried our bags into the tent which they had erected for us about fifty yards from the lodge, and then Sam ushered me along a little path behind the tent and revealed to my astonished gaze a splendid toilet with a specially carved round seat nailed to supporting logs and covered by a brushwork roof of trees. Sam regarded it as a tremendous joke: but actually it was an act of great thoughtfulness, for details of this sort are what flummox a white man in the bush, and they had spent a lot of time solving this problem (a tough one when all of the ground is frozen) for us. In fifty years anthropologists will probably find that seat and announce the discovery of a subgroup of the Mistassini culture—now extinct, of course—of toilet-seat users.

Rosie was the only one who did not come to meet us: she was pulling a sled full of spruce boughs across the lake as we flew in, and when she returned, too shy to come up to our tent, she went into the lodge and I had to go looking for her.

The tent was, of course, delightful—sweet-smelling with its floor of fresh spruce boughs, and warm as toast with its wood stove. The snow had been piled up as insulation all around the walls, and outside the door a charming carpet of spruce boughs was picked up and shaken off every day by one or other of the people before we got up. This time they provided us with firewood, since the collecting of it on snowshoes was likely to be beyond our skill. They had laid in a huge pile of wood, at least a month's supply, and we helped ourselves to it. We were better prepared this time: we laid space blankets, those paper-thin aluminum products that are a spin-off from the space race, on the boughs, and thick foam rubber on top of that. Some of the crew put cardboard and more space blankets under their Arctic sleeping bags, supposedly guaranteed to maintain warmth in temperatures down to forty degrees below zero. I never thought I would sleep comfortably in a tent at thirty below, but during our whole stay I never felt cold at night, even though when the stove went out at about one o'clock the temperature in the tent quickly dropped to ambient level. Thus we learned that the Indians could keep perfectly warm in an ordinary tent, no matter how cold the weather might get. Indeed, the men spent most of their time in tents, for they carried one with them on their long hunting trips and set it up each night.

THE WORK NEVER STOPS

As they had told us it would be, the camp was alive with activity, in contrast to the slower pace of the autumn. Within an hour or so of our arrival, Abraham Voyageur came into camp carrying a beaver and a rabbit and laid them down beside the woodpile. (They were immediately picked up by the children, who carried the dead animals around, playing with them, for quite a while before their mother took charge of them.) The people had almost come to the end of their beaver trapping, and were now primarily engaged in preparing the skins for Speers to pick up the following week. But in addition there was always something interesting happening inside the lodge. Either Ronnie would be fashioning a snow shovel or repairing a handmade sled, or Sam would be making a new harness for his dog, or Mrs. Blacksmith would be cleaning out the entrails of a beaver, or Sam cutting young Abraham's hair, or Philomen and Mary skinning rabbits, or Louise tending the beaver as it hung from the roof alongside the hot stove, spinning slowly all day roasting—incidentally performing the function of adding to the fragrance in the lodge.

These people had had an extremely hard winter. Cree life in the bush, though it provides many satisfactions, can be tough. Sam seemed tired. Since we left in October the men had spent only ten nights in the lodge with the women. The rest of the time they were out on long trips, away from the lodge sometimes as long as two weeks at a time, moving their tents every two or three days, traveling always by foot and on snowshoe, gathering the beaver. Because of the early arrival of the winter they had not had time to locate all of the beavers, and in these circum stances the job of finding a beaver lodge, and then pinpointing the entrance under the ice, and opening it, and setting a trap in the entrance, could take almost a whole day. "Things do not always go well when we are trapping and hunting," said Sam, "and that is how it has been for us. We have had to try very hard to survive." Sam had trapped sixty of his quota of ninety beaver, and Ronnie and his family had taken seventy of their quota of 130: not a particularly good result. They had not found any big game except one bear asleep, so for the most part the hunters had had to feed their families on small game, by keeping up a constant supply of rabbits, partridges, porcupines and fish. None of these was easy to catch. The work demanded remarkable capacities. Abraham Voyageur would walk off across the lake in the morning and, as often as not, return after three or four hours carrying a bag full of rabbits or partridges: he would enter the lodge with the bag, place it at the feet of his wife as she squatted on the floor in their corner of the lodge, and then take one or other of the small children on his knee as Shirley wordlessly opened the bag and emptied out the frozen rabbits one by one. No one would say anything in congratulation to the hunter, but the scene would be suffused with an air of enormous satisfaction. Similarly, if anybody returned without game, nothing would be said, no reproaches made, only the feeling would be different.

We began to get a glimmering of the extent to which the Cree people communicate by the transmission of waves of feeling rather than words. But it was still a mystery to us how sixteen people could have lived in this one-room lodge, have had such a hard winter and yet have come out of it so cheerful and emotionally intact. They had not had any quarrels, apparently, or any illnesses or accidents. Each family occupied a corner of the lodge and each hunter looked after his own family. But tradition demanded that a careful watch be kept all around so that no one should go without.

Even though the men felt that their winter had been hard and their hunting unsuccessful, their frustration probably had more to do with their failure to get big game than with any other single factor. Possibly their self-esteem was damaged by this failure, for it is an essential part of the hunting experience that meat should be sent back to the settlement to relatives, to keep them supplied with country food—especially the old men who are not allowed to go into the bush for health reasons. It cannot be said that, in spite of the hard winter, anyone in this camp was ever near to going hungry. I figured out that between the beginning of the beaver-trapping in mid-November and our return in mid-March, some thirty pounds of beaver meat had been entering the lodge every day. Although they are enormous meat-eaters, this amount of food, in addition to the many rabbits, birds and fish that the men and women caught, suggested a level of eating very far from starvation.

Nevertheless the men were somber in their account of their hunting. "It appears," said Abraham quietly, "that the land is less productive of beaver than it used to be."

"We look for all kinds of game," said Sam. "We wouldn't be able to find it if we didn't try. But sometimes it is just not there."

"Things are going much the same for me as for them," said Ronnie. "This is the way things happen on the land: it cannot always be productive. I have always been hunting and trapping, and in the past have observed the land and how things are going on the land. Even in an area which has been left unhunted, it happens that there is no game. I have known since three years back that the beaver is declining. I have been watching the beaver as they were born."

As soon as he had dropped us off at Lac Trefart, Marcel returned the 185 miles to Chibougamau to pick up the second planeload of our equipment. When we heard about their failure to find big game, we suggested to the men that they might like to take a quick flight around their territory when Marcel returned with the plane. They did so, and after about a twenty-minute flight came back with the news that they had spotted some moose in the southern part of Ronnie's territory, near Lake Paneman, about twenty-five miles to the northeast.

BONES, SKULLS AND FEET IN THE TREE

Outside the lodge a tree had been stripped and was now carrying a tremendous load of rabbit and beaver skulls and feet, tied there by the hunters, part of the religious observance of respect for the animals.

LE BEL [cross-examining]: In your testimony you refer to religious ceremonies. When you're talking about religious ceremonies, are you referring to the Anglican religion?

ADRIAN TANNER: I don't believe I used the word ceremony.

LE BEL: Or Catholic religion, or another religion, an Indian religion of some kind?

TANNER: I was using the word religion in the anthropological sense of general belief system.

LE BEL: What is religion in the anthropological sense?

TANNER: It's a system of values, beliefs and ideals encapsulated in ritual and legends, ceremonies.

LE BEL: It's a very general definition.

TANNER: Correct.

LE BEL: Those beliefs or attitude or worshipping, have they been greatly affected by the missionaries? Have they been greatly altered, modified, changed?

TANNER: I would say several additional items of religion have been added through the teaching of missionaries.

I asked the men what they thought of the Christian religion, and after some discussion with Philip they answered: "The religion that is now in Mistassini appeals to us. We think it is good."

QUESTION: But the Christian religion seems to suggest that man can control and change nature and man is made in God's image. What you people do seems completely different, since you regard yourselves as just one of the participants in nature.

SAM [after discussion with Ronnie]: It appeals to us because—well, we have not gone into the Christian religion so deeply to understand the whole idea of man in the central place. We know there is a Great Spirit who is watching over us all, and we want to be with him.

QUESTION: I wonder why the Christians are coming and stealing your land?

SAM [after discussion]: The religion we understand from Christianity establishes relations between men. It appeals to us because, for example, it tells you you should not steal, or do this or that. That's why our reaction to the James Bay project was that we were totally taken aback that they had not bothered to ask us, because they display themselves as hypocrites when they cannot practice what they preach. But Cree society has its faults, too. We have our own thieves. For example, somebody in Mistassini stole forty ounces from me. [Laughs uproariously.] But sometimes when some of our people have left stuff behind on their trapping grounds, others have taken off with it.

Apart from the first visit made by Speers in January (which usually lasts for half an hour at most) the people in the camp had had no contact with anyone outside except for a quick visit by plane from Larry Linton, the Baptist missionary. A thin, tight-mouthed fellow with a nasal drone, bearing every appearance of the religious fanatic, Linton in fact had learned a great respect for the Cree people in the many years he had been flying around their camps, and he had done his best to subsume the traditional Cree attitudes into an acceptable Christian format. He said that much of the revival of such Indian religious observances as the spirit pole or bone tree stemmed from a visit to Rupert House by an anthropologist who kept asking a lot of questions about the old religion and got people thinking about it again. "I believe we can co-exist with it," said Linton, "for there are other things more important in the Christian life," though "definitely it was an animistic religion, sort of satanism." If he knew an Indian who had made "a definite commitment to Christ" and who was also going in for this old worship too much, "I believe it would be in my place to say something about it. But for the normal run of the bunch that go to church two or three times a year, I don't see anything wrong with it at all, because it doesn't conflict with Christian principles, seeing that they aren't committed to Christ in the first place. I mean, that's one way I get around it myself. I believe that if it's something wrong, the Holy Spirit will speak to the person about it, if he's God's child. And this will clear itself up, so I would never bring myself to preach against it. I know ministers who do, but I don't think that it's my place as a minister, because I think there's many things that are more important."

Linton said that because Indians had been treated as an inferior race for so long, they had a tendency to agree with the instructions of their various bosses to their faces and then, when the boss went away, to return to their own way of doing things. "This carries over into religion, too. A minister is the preaching boss, the praying boss, instead of the store boss or the village boss." He thought that the people were happiest in the bush, that the environment they created in their bush camps was a fine one, ideally adapted to their capacities and temperament.

It was Mrs. Jolly, Ronnie by her side, who told us about the visit of the missionary. He flew in on his little Cessna and stayed the night. The men were away when he arrived, but by evening they came back. "How many partridges did you bring home that time?" she asked Ronnie.

"Six."

"You brought home six partridges and a porcupine," she said, "and then when night came the priest had a small service." Ronnie whispered to his wife, who said to us, "We thank him. And the children received toys from him also." Ronnie whispered again and Mrs. Jolly said, "We were happy he came." Yet again Ronnie whispered, and Mrs. Jolly said, "We appreciated his visit, although he had a different religion."

STRANGE INCIDENT

All winter the men had kept a beaver lodge for us, not far away, so that we could film the setting of the trap and the catching of the beaver, and on one rather windy, extremely cold day we set out in convoy to film Sam setting the trap. Ronnie, Abraham and Malick came along, too, and I had a taste of the sort of work involved in the hunting life when I hitched myself to a sled laden with heavy camera equipment and began to pull it over the hill to the nearby lake. Before long, despite the thirty-degree-below-zero temperature, perspiration was pouring off me. I took off the inner lining of my coat and was able to put it back on when we stopped, which prevented the sweat from freezing solid. But I had nevertheless broken one of the first rules—never move so fast that you sweat.

We stopped at a windy, open place on a small river leading into one of the lakes north of Lac Trefart. The men immediately began probing with sticks and then digging away the snow with their little handmade snow shovels. When they reached the ice, they broke it open, using the long poles with blunt metal ends that are among the commonest tools of the Canadian north. They dug in two places, opened two holes about a foot wide and two feet long. Sam took a curved stick that he had prepared and felt around gently with it under the ice until he found the beaver-lodge entrance. Ronnie felled and trimmed a couple of thin trees. Sam now measured the depth of the hole and then, calculating by eye, cut pointed ends on these trees so that when the rings of the beaver trap were fitted over the top, they jammed about a foot and a half from the bottom. Working with extreme care, he shoveled off the top of the water the small pieces of wood that had drifted up to the surface, presumably from the beaver's food supply, and then, everything having been cleaned up, he opened the trap and secured it with the catch, leaving a light piece of metal for the beaver to swim against as it passed through the trap, thus releasing the catch. The beaver would be caught across the back and drowned. The sticks on which the trap was supported were secured in a diagonal arrangement, and the whole thing was covered with brush and snow, leaving only a few sticks poking above the surface. It looked almost haphazard, but one of these sticks had been carefully arranged so that the moment the beaver trap sprung, this stick would move perhaps an eighth of an inch—not enough to loosen the trap, but enough to dislodge some tiny particles of snow at the surface. This movement, which, frankly, when the time came I could not discern at all, was all that the trappers needed to tell them whether a beaver had been caught in the trap.

All expert craftsmen are fascinating to watch, and these men were expert craftsmen as they went about setting the beaver trap. What was really piquant for me, however, was the contrast between their superb capacities—the delicacy and sureness with which they performed every tiny operation, their air of total mastery and command as they whizzed over the countryside on their snowshoes—and the caricature of their life that I had just heard the government's lawyers offering to the court in Montreal, a caricature that appeared to arise from an honestly mistaken common perception of the nature of the hunting life.

By the time this was over I was somewhat rested, and with a bit of help I got the heavy sled of camera equipment to the top of the hill above the little river and laboriously hauled it back—I guess it was no more than half or three quarters of a mile, but it seemed like miles—to the camp.

Everybody was tired when we returned, and Tony suggested that the occasion called for a restorative rum. We had brought more liquor with us this time so that we could have more of those delightful evenings at which we all shared a couple of shots from a single bottle. The men came into our tent and we soon warmed up with the rum. We were joined by the women and children, and everybody became talkative. We had been meaning to give Sam a bottle of rum as a gift, but I had learned my lesson last time, and had postponed handing it over. On our first visit to the camp we had given them all our gifts on our arrival, so that later we had nothing left with which to fulfill the custom of returning a small gift for each plate of fish or meat that they provided for us. But now I gave the bottle of rum to Sam.

We were anxious to photograph a moose hunt, but it presented what seemed to be insuperable difficulties, and we discussed it for quite a while over the rum. We had to carry heavy equipment everywhere we went, and could not keep our batteries outside in the cold for too long. The half-mile trip to the beaver trap, from which I had returned exhausted, had illustrated the difficulties. The men were ready to walk off to get the moose: nothing could be more natural for them than to do so. Indeed, that is what they would do, for they needed the meat, and ultimately we could not interfere with that imperative, whatever the demands of the film. Their struggle for survival took precedence over our film.

The men had grave doubts as to whether we would ever make a twenty-five-mile journey by foot: they were ready to leave the next morning, and we had to ask for a little more time before deciding. Since we had given them the flight on which they had spotted the moose, they were prepared to try to accommodate us.

Ronnie now began to talk about the fact that we had not paid Eddie for his part—that of a full-grown hunter—in the filming in the fall. We explained that we had agreed to pay each head of family and we did not realize that Eddie was a fully qualified man. "I have seen how the government spends and wastes money," said Ronnie, "and know that they have plenty of money, so that is why I wondered why he could not be paid." Eddie sat by, nodding. "I didn't get a release," he said, referring to the release form which it is customary to get people to sign after they have been filmed. It releases the copyright in the film to the film makers, and Eddie was confusing it with the check in payment.

Sam came and asked if the women might have a little more to drink, and I opened another bottle: the tent by this time was jammed—all twenty-two people in the camp were there—and we were suffused with good feeling. It was perhaps the warmest moment we had experienced with these people since we had met them. They seemed used to us now, and were entering into the spirit of the film as they had not done at the beginning.

Ronnie always got extremely cheerful when he drank—indeed, I had first noticed him grinning and laughing at a party in Mistassini, and with his charming open face he had seemed so sympathetic that I went around the next day to ask him if he would be interested in our filming him in the bush. Before we finished the rum he began to fall asleep. Sam and Ronnie—like Philip's father—belonged to a generation of men in their fifties who had never seen or heard of alcohol until they were well into their twenties. Sam now came over and asked if I thought they should drink the bottle I had given them, but I suggested maybe they should keep it for a later day. It was with hilarity and good spirits that they eventually sloped off to the lodge, and we gathered ourselves together to cook some dinner and go to sleep. By the early evening there seemed to be a great deal of noise in the lodge, which was usually preternaturally quiet. My twelve-year-old son Ben, whom I had brought with me on this trip, had gone off as usual to sit with Malick in the lodge, and he now returned to say that they were arguing and shouting a lot. He disappeared again, fascinated. Philip was unconcerned, but Tony, always apprehensive about giving liquor to Indians—to that extent he carried the white man's burden with him into every relationship with an Indian—was uneasy. It was obvious that I had dished out too much liquor—I had not intended to do so, but Sam had asked for it, and I had got so used to Sam making wise decisions that it never occurred to me to question him.

Finally Mrs. Blacksmith appeared and asked Philip to come to the lodge, for they were fighting. We all went over—my own apprehension was as great as Tony's by this time—and found Eddie stripped to the waist and hauling Rosie about the lodge by her hair. She had called him a son-of-a-bitch, and he obviously didn't know what he was doing. The women were quarrelling bitterly, apparently about their children and the way they were always wandering into the other family's corner of the lodge. We got Eddie outside into the cold: he was raving on about not having had a release and, as a result, having risked and nearly lost his life. Tony started to talk non-stop to him: he'd been doing a great job for us, he had done so in the fall, and was doing so now. We hadn't realized he was such a skilled man, but if he was a man, he had to forget about attacking Rosie, he had to come to the tent and calm down and accept our assurance that we would put everything right.

Rosie came over to our tent to get away from the quarrelling parents: she sat hugging her knees, with her back to all of us, assuring us that she was all right. Eventually she went back to the lodge. Eddie stayed with us, and everybody went to sleep. When I got up at 7:30 the next morning, Sam was already walking around on his snowshoes, none the worse for wear, and clearly not at all disturbed by what had happened.

But it was an extraordinary incident to happen in those circumstances in that place. For we had not consumed much liquor, four bottles of rum among twelve adults. Apart from opening up a moral dilemma for Tony and myself, the incident seemed to confirm some of the preconceptions that white men have about Indians being unable to hold their liquor. Tony always thought it better to leave well enough alone and not give liquor to Indians. Philip was totally undisturbed by the incident. What most surprised me was that the invariable wisdom of Sam's decisions as boss of the camp broke down and he kept drinking when he could have stopped. But even more astonishing was to see the destruction of that wordless tolerance, the invariable friendliness, the extraordinary social self-control that enabled them all to live together in harmony for so long.

Yet by the next morning when they woke up, all of the social self-restraints were re-imposed. Rosie and Eddie got along like the others for the rest of the winter in harmony and peace, and Rosie, though she clearly wasn't fond of Eddie, couldn't be got to say a critical word of him. A year later I tried to find out from Shirley Voyageur, an intelligent and sensitive woman, exactly how they managed to live together so amicably. Did they have smoldering resentments that they sublimated? No, they didn't, she said. They really liked each other. Did they not have things they wished to say to each other that they kept back? Not often, she said. It was done by each family living in its corner of the lodge and telling their children not to break up things in the other areas. That way, they did not quarrel. Had they never, in the whole of the winter, I asked, had quarrels? Only, she said, when you gave us the liquor. She and her husband liked Sammy Blacksmith, she said, but not when he was drinking. I tried to get the same information from Rosie, but she was no more informative. Their getting along together is not just a question of holding back. They really feel friendly toward each other, she said. But, she added, it does happen that families cannot get along, and "then they just split." It had happened once to Abraham and Shirley that Shirley had been unable to get along with the wife of Reuben, Abraham's brother, so they had broken up their camp in mid-season.

Apart from that one reference by Shirley a year later, no one ever referred to the drinking bout again.

EDDIE GETS LOST

The rationale for Eddie's worry about not having been given a release was an interesting comment on the tenacity of the sharing ethic. After each of the hunters had collected his $325 in the fall, Eddie had felt that he had fallen behind by that amount in the collective effort to ensure survival and he must make some special effort to compensate for this failure. He thereupon redoubled his work as a trapper and hunter, went out on longer and tougher trips than usual and once wandered so far into unfamiliar territory that he got lost. "I risked my life," he said over and over again, "and nearly died." He went so far that he wandered south right off Sam's territory. While the snow fell all night, covering his tracks, Ronnie set out to look for him, knowing which way to go only from the smell of smoke from the fires set by Eddie from time to time as he moved south. Each time he arrived at a fire, Ronnie found that Eddie had moved on: not knowing where he was, Eddie was moving away from the lodge. His father did not catch up with him until he was some twenty-five miles to the south.

MALICK GETS SOME CARIBOU

In all their travels around the territory the men never saw moose until the plane trip, and only once did anyone see any caribou. The women had been working on the boughs while the men were away, and late in the evening, just as they were lighting the candles, some caribou passed right across Lac Trefart within sight of the lodge. The animals were already moving away when the women saw them. "We tried to get them," said Shirley Voyageur, "but it was already too dark, and we had only .22 guns in the camp. The men had not left any .3030s, so we couldn't reach them with our guns."

"They never seem to pass when the men are at home," said Mrs. Blacksmith, "but eventually someone will catch them."

Her prophecy came true later in the year, after our departure, when only the Blacksmiths were left at Lac Trefart: they had returned from a trip around Sam's territory and were staying in a tent near the lodge. Sam had got up and gone off hunting at three a.m.—later in the year he had to move early, before the sun got at the snow and made it damp and mushy—and at around six in the morning Mrs. Blacksmith saw some caribou cross the lake. She woke Malick, who floundered out in his underwear to try to shoot them with a .22. He shot one, and with the dogs barking on the other side of them, the caribou kept going round in circles, so that he was able finally to get a second one, a big success for a fourteen-year-old boy. In all the flurry little Abraham was left alone by the tent, and when a wounded caribou headed for the tent, he took fright and headed off into the bush. At that time of the morning the snow was too hard to leave tracks, and it took the people a long time to find the little fellow, who was by this time almost a mile away. When he returned to the tent, the caribou was lying dead. He kicked the carcass and said, "You were the one coming into my tent!"

ALEXANDER BANFIELD [in evidence]: Caribou have a trait which is perhaps strange to relate, but very useful. They what we call loaf, they loaf on lakes in April when the sun starts to get bright, they go out and they bed down on the frozen lakes and they spend a good deal of time, we think, suntanning. They're ruminating, chewing cud in the afternoons, and they loaf there and then you can see them gradually wander back to the shores to feed. This is why we use that to count them because you can count on a great percentage of the herd being on the lakes, and every caribou report always has pictures of hundreds of caribou bedded down on the lakes. So this would be a hazard. The collapse of the ice on the shores [of the new reservoirs] when they are drawn down would make it hazardous to get on and off this shore, and the caribou might well give up that area to loaf. . . . Another aspect is predators: they can sense a wolf coming on to a lake a lot easier than if they're bedded in the woods. . . . I think they would probably have difficulty too in travelling on the reservoirs in the spring and fall. The Caniapiscau river valley is a great migration route up towards Chimo. We don't know why they're attracted to that area, aside from assuming that the food is good there, and that there's excellent range, but this suggests that it's a special area, so it would have a greater impact on the caribou herds than if we simply removed less desirable range. We would suspect that this area is somewhere near the carrying capacity at present, so if it is removed from caribou winter range it is not possible to say they can go elsewhere, because the whole area may be pretty close to utilization, and you can't put more cows in the field than you can normally manage or feed.

ROBERT LITVACK: Now over the years, have you had occasion to study the caribou in relation to their importance to the native people in the region?

BANFIELD: Yes, in my preliminary caribou work I actually lived with native families and travelled with them, and I noted the use made of caribou in their daily lives, in their daily occupations, and the caribou is a very important resource, particularly in the interior. Natives living on the coast can use marine resources, fish, marine mammals, seals, walrus to a great extent. But in the interior the caribou is really important. It's excellent food, even Europeans agree it's heartily excellent, it's much tastier than beef, but it's delivered to the camp, you see, beef isn't delivered, but this is, you have a resource that literally walks into camp, more or less, so it does have a very big advantage from that point of view, and it's very important for meat. But also it provides a number of other supplies, particularly winter clothing. It's extremely light and extremely warm. . . . It also goes for boots and

shoes and for beds, of course, the traditional bed in the north country is an old bull hide, and there's a great lore and knowledge built
up as to what time of year, and what sex and age, class, you use for
clothing. For socks, mitts, for bedding, for instance. You use the
forehead of an old bull for the soles of your shoes, because it's the
heaviest part. And when the store resources run out, it's simply
wonderful to have this native knowledge and this native ability to use
resources. You can make thread out of the fibres, you cut the skin
into strips and make bindings out of it, it can be used for heating or
a little wavering light, so it's a survival kit for people who live in this
area.

AFTER THE BEAVER

We had reconciled ourselves to not seeing many animals in our film, for
there was a richness in the detailed activities around the camp that more
than compensated for the lack of animals. As March drew on, the weather
became glorious: I took to getting up early, so that I was usually walking
around as the sun rose over the lake, casting a magenta shade over the
whole landscape. Two days after we had set the beaver trap, I saw Sam
come steaming down the hill on his snowshoes at an early hour. He had
been over to check the trap, and he thought that he had caught two
beavers. This time Sam hitched his dog, Coffee, a large and powerful
mongrel, to the camera sled. Each family had a dog, and they were tied
up all the time. Even though they were not huskies, they loved to work,
and as soon as they were hitched to the sled they went off pulling and
barking joyously. I set off behind Sam on my snowshoes, thinking that I
would be able to keep up with them, but after a couple of hundred yards
they were out of sight. As Malick watched, Sam opened up the beaver
trap, picked the trap out of the water with the hapless frozen carcass
caught in it, carefully unhitched the trap and then slapped the beaver
back and forth on the snow to dry it off. He had caught a male and a
female beaver. He said he had known they were there since he had seen
them as he paddled around in his canoe in the fall, but he had kept them
until now so that we could film them. But he did want us to know that
it was not always as easy as it seemed here. Sometimes the hunters had
to check traps three or four times and still they would not have caught
a beaver, because there could be another entrance to the lodge they did
not know about. Sam drove a small stick through the beaver's nose,
attached the ornamental string to it and pulled the beaver on its back to
the lodge.

GEORGES EMERY: Would it be possible to tell us what proportion of animals are caught on the traplines?

GARRETT C. CLOUGH: . . . I think beaver could be harvested at 25 percent of production. . . . All animals overproduce, that's the general law of Darwin. If man takes that overproduction and uses it . . . he is not doing the animals any harm. The Indians take about one out of five. A 20 percent harvest would be sensible. If you took 30 or 40 percent then the beaver might produce more young. Animals have a way of compensating for that. When there's a need, when there's more space, they're not as stressed, there are not as many abortions. The female can have as many as six young, but if there's a lot of social stress going on, she may abort a few and have only two or three or she may have six, but those don't get enough food, or something gets them. It might be that the harvest could go higher than 25 percent and still maintain the stability.

THE DISTRICT COMMISSIONER

The visit of Speers was quite different from what we expected. The Indians have lived for a long time in a colony of the mind, and for an hour or so the camp was transformed into a physical colony receiving a visit from a powerful and distinguished district commissioner. He did not arrive on the expected day, for the weather was bad, but the following day he turned up about noon. They had made extraordinary preparations: a completely new floor of spruce boughs had been laid, and outside the entrance to the lodge was a fan-shaped carpet of balsam leaves like a welcoming mat. Inside, a tarpaulin was laid as carpet over the boughs, and a specially made table, used for no other purpose, was brought out and covered with a cloth. A trunk was used as a seat, covered by a pillow and an ornamental cloth. They had even brought up with them for the occasion from Mistassini a glass ash tray, which they now laid on the table for the manager's use.

Speers was dressed in high moccasins and a cap that covered his ears, and carried a heavy briefcase up to the lodge after shaking hands with everyone in the Indian party. He tried to service four camps a day, and worked quickly. Muttering in Cree, he handed out the family allowance checks, and then received them back, entering them in a ledger. He told the Indians the prices they had received for the beaver skins at the January sales, and entered in his books credit for each of the hunters in the amount of his checks.

As Speers worked over his books, making entries against each of the names, the atmosphere in the lodge was hushed, respectful and worried. Rosie was dressed in her best sweater for the occasion and looked most fetching, though no doubt Speers didn't notice such a detail. But Sam looked anxious as we had never seen him before: the confidence of the *auchimau* of the camp, the dominant figure giving orders and organizing these people as they fanned over the countryside, commanding their environment, was totally gone. He had the air of a supplicant. And even so strong a figure as Mrs. Jolly seemed as though she were in the presence of a respected schoolmaster. Speers took the four lynx skins they had caught, and paid them $75 for each—sums that he entered as credit against the names of the sellers. Philomen handed over a muskrat skin, for which he paid $1.50—she had expected $2.50, but took the price without demur.

As Abraham moved to the table to accept a slip, Tony asked him to stand closer to the company manager for the filming. "I was thinking," said Speers later, "as Abraham came closer and closer, Gee, hold it a bit, that's getting a bit intimate, isn't it?"

The following is an excerpt from our filmed interview with Glen Speers:

QUESTION: People in the south assume that the Indians are having a tough life in the bush. Is that true?

SPEERS: Well, they live a lot better than the average person can live in the city. Probably much better than anybody else unless they have a farm.

QUESTION: That statement would come as a revelation to most city people.

SPEERS: Well, some people come up here and they think, "Oh, the Indians are living in tents in the winter" and so on, but they're not tents as we know them. They make camps, split log teepees, they bank up the walls four or five feet with snow, they're quite warm, comfortable, and they're clean.

QUESTION: I suppose you know James Bay as well as any white man?

SPEERS: Well, I started in the early forties. I was two years in Great Whale River. And I learned to speak a bit of Cree there and Eskimo, and from there I went to Kanaaupscow post. The last year I was there we built the store that is there now. I went out west for a year, but they called me back and I was at Nemaska post for a year and a half. Another chap and myself built the store that's in there today. From there I went as manager at Rupert House for four and a half years. Then I went out in the Western Arctic for a while, and then I came back, and I've been here for the last sixteen years.

QUESTION: So you've seen it all?

SPEERS: Well, I've seen quite a lot of it. We used to travel on the Fort George [La Grande] river by canoe all the time. It was all by paddle. We had no outboard motors. We used to haul our supplies up to Kanaaupscow. The Indians used to live out there all summer, and that gave them quite good employment. They used to carry 2,000 pounds of freight in a twenty-three-foot canoe that was made in Rupert House.

QUESTION: How many portages?

SPEERS: Well, generally there were twenty-nine portages if you go by river, and if you go by the lakes I believe it was fifty-six. But we always went by river. And it's a wonderful river. I've seen a lot of Canadian rivers, and it's the nicest one. I always liked it. It's . . . ah, fascinating, and it's very nice to travel on, both ways. It has its dangerous spots, but the Indians were extremely proficient in rapids and so on. . . . I would imagine there is no one in the world that is better than they were. There must be some old-timers today who can tell you about that. Then, when I was in Nemaska, I made probably five or six trips from Nemaska down to Rupert and back. That's quite a nice river, there are not quite so many portages. I think there are nineteen. There's one that's quite long, a mile and a half maybe. That's a nice river, it's not nearly as large as the Fort George, but it's very nice. I have also traveled up the Great Whale River to the second falls. It's very pretty country.

QUESTION: It's one of the last wildernesses in North America, I guess. What leads a man like you to spend his life in a wilderness?

SPEERS: Well, I like it. I've always liked it. Uh . . . I think it's a good life. And I like working with the people. I like working with the Cree. I guess you won't find any people any better. And the Fort George men . . . I've heard a lot of people say that they are some of the finest men in the world.

QUESTION: Could you tell us something of the Indian attachment to the land?

SPEERS: Well, it's difficult for me to describe. They take a great pride in their land. Their ancestors owned ground that they have been handed down from generations. And they are natural conservationists. For instance, the trapper might kill, say, ten or twenty mink on his land in a particular winter, but he'll stop because he wants to leave enough so that he can get that particular animal every year. Most of them are quite careful like that. For instance, the Nichicun people . . . There are caribou there in the winter. When we fly in there we see caribou on the lakes all the time. Not tremendous concentrations, because the main concentration is east of Scheffer-ville. But quite often there are caribou within half a mile of their camps, and they are not touching them. If they've got what they need, they leave them.

QUESTION: When you see Indians in the bush, it's quite different from seeing them around the city?

SPEERS: Well, you see, you get a better impression of the person when he is in his own environment, and living the way he is skilled to live. All the Indians in Mistassini are not skilled to live in the bush. Some work right in the village. But the ones that do live in the bush, inland, you get a very favorable impression of them.

"I thought I had better tell you," said Speers later, "that it's not exactly usual for them to put out a table for me like that. It's only seven or eight camps do that. Usually I just sit on a box." As the people rummaged around on the ice among the trade goods he had brought in —flour, coffee, jackets, sugar, biscuits, sweets—they would choose what they wanted and he would mark it down against their account, so that the whole system, almost as in Anderson's day, was a book operation in which money hardly ever changed hands. Speers paused long enough to be interviewed laconically about how difficult the flying had been because of the glare ice, and how far he was falling behind schedule. Ironically, though we were the first people to whom he had ever granted a filmed interview, we were not able to use it in the films we made: his flat, laconic style, his persistent habit of understatement, did not lend itself to film. He had promised us about fifteen minutes to shoot his visit. We delayed him by maybe half an hour. But about an hour after arriving he was shaking hands unemotionally and was in the air, leaving the people to read their mail, to enjoy the few luxuries he had brought with him and to get back to the business of surviving.

I remembered something he'd told us: "When you're in business, you can't afford to be too personal." But after seeing him at work in the camp, I thought I understood the attraction of the job for him. He perhaps more than anybody was responsible for the arrangements which made the Mistassini people one of the greatest hunting communities left in North America. He was fulfilling with dignity and decency an essential function for a fine group of people. It certainly beat being an ordinary shopkeeper.

Two boys

My son Ben and Sam's son Malick got along splendidly together, though they were so dissimilar. Ben was a typical city kid, self-centered, idle around the house, disliking anything that remotely resembled work, and used to life in a suburb where in five years he had never got to know the names of the kids across the street. Malick was used to living three families in one lodge, and was capable of making the accommodations to other people's personalities that such a situation demanded. Ben had seen quite a lot of the world, and already had a somewhat jaundiced view of it; whereas Malick's world was more or less restricted to Mistassini. Malick was supremely competent for his age: though he was still learning from his father, he could already handle himself in the bush, could snowshoe for hours and keep up with the men, could recognize the tracks of animals, set snares and shoot game. Ben was competent, really, in nothing whatsoever. Without constant parental bugging he wasn't capable even of keeping himself washed and brushed, and he could not possibly—as Malick did the following year—have taken a job and maintained himself in it. He was a small boy, just beginning to grow up; Malick, though only two years older, was already finished with his schooling and was ready to make his way in the world. Yet Ben could adjust admirably and with generosity of spirit to the strange circumstances in which he found himself, whereas Malick could not face making a trip to Montreal and, had he done so, would have been lost and anxious to get back home. Ben spent most of his time in the lodge, reading comics, absorbing the atmosphere, observing the manners and habits of these people whom he grew to like. And Malick spent a lot of time with us, reading the girlie magazines, picking up a little of the glamour of city life that we carried into the bush with us. Ben went out into the bush with him once, when he shot some partridges, but it was a long walk and he never really pressed Malick to go again. In the end it was hard to know how much of value they had rubbed off each other.

How it's done

White people who later saw our film about the hunting camp, with sixteen people of all ages living in one room, often asked one particular question: How do they manage to do it? What do they do to make love? It was not something we were able to take up with the families, but I knew that Indians generally had a free, natural attitude toward sex, one of those functions of nature which they fulfilled without fuss. Philip assured us that no girls of fifteen in Mistassini were still virgins, which indicated that these charming young girls in our camp must be very much more open and accessible in their manners within Cree society than they were toward people from outside. A couple of years later I was able to ask a hunter from Fort George how it was managed. "We are not like white people," he said. "We do not use beds. We do it right there on the boughs. Secondly, we try to do it when we think everybody is asleep. And, thirdly, the white man likes to take a lot of time with the breasts and all that. The Indian people are not like that. They believe in getting down to the basic thing without any delay."

WE SHOOT FOUR MOOSE

We had discussed the problem of the moose hunt again and decided that the only way we could combine the killing and filming of the moose was to hire a plane so that the cameraman could have at least some chance of being on the spot for the event. We had to acknowledge reluctantly that none of us was capable of keeping up with the men on snowshoes for twenty-five miles (or twenty-five yards, for that matter). So Speers arranged for Dolbeau Air Services to send a small plane from Chibouga-mau, and after the customary day's delay for bad weather it arrived early one morning. Tony took off with Sam, Ronnie, Eddie, Malick, some sleds and snowshoes and a camera.

They flew low over the hills and down into the valleys, all the way up to Lake Paneman. Eventually in one valley they found moose tracks everywhere. The plane swooped and dived, soared and dipped as the hunters calmly sorted out the location and identified which of the many tracks was the freshest. They soon realized that the moose were in a certain patch of woods, and asked to be let down at a nearby lake. The men assigned Malick to look after Tony, who, before they disappeared, had time to ask if, when they got to the moose, they could wait for him to catch up before shooting, though he understood that the killing of the animals was the first priority. The men made a long detour so as to approach the animals from downwind, and Tony struggled along after them, up and down a lot of difficult ridges and hills, trying to protect his camera, but every now and again going over on his ankle, the snowshoe harness slipping off and his leg disappearing into three feet of snow, from which Malick would pull him. By the time they caught up, the hunters had shot four moose.

Sam and Ronnie immediately started to dig a great hole in the snow around the moose, right down to ground level, while Eddie got a fire going. The two men rolled the carcasses into the hole and were able to take off their snowshoes and get to work. They had a short time before the plane would be back, and they had to skin, quarter, dress and haul out these four huge carcasses. Though Sam was fifty-eight, even after all that trekking he still seemed to have a bottomless well of energy, and the two men worked over the moose together in perfect rhythm.

One of the moose was a female. The men dug around in the steaming carcass and lifted out the womb, which they then slit open so that they could remove the twin foetuses. They laid them on the ground; then Sam cut flesh off the mother and popped some into the mouths of the dead foetuses. The explanation they gave us later was that since the mother had been unable to fulfill her maternal function, a little bit of her was given to the dead calves to ensure that the moose would continue to flourish. "They treated it so very respectfully," said Tony. "They performed this act without any remorse at all, as a natural thing, as if it were just something that happened from time to time."

Finally, after their exhausting afternoon's work, the two men still had the energy to load the meat on sleds, hook cords over their shoulders and haul it a mile and a half to the lake where the plane landed for them.

While they were away I had persuaded Philomen and Rosie to walk with Ben and me to the abandoned lodge nearby. We set out together, but the girls left me a mile behind, and had it not been for the tracks in the snow I would have got completely lost. We were coming back across the lake when the plane appeared. Even Rosie, though apparently alienated from Cree traditions, immediately betrayed intense excitement about the prospects of the moose hunt, and the girls broke into a run on their snowshoes and raced toward the plane. "They've not got anything," cried Rosie, "or they wouldn't be back so soon." But she was wrong, and there was great joy on the ice around the plane as the pieces of moose were unloaded and we learned that they had shot four. They pulled the moose up to the lodge, where, in an atmosphere of great satisfaction, the children brushed off the fur, picked up the limbs, shouted and laughed and finally pitched in with everyone else to carry the meat inside. Ben decided he would like to photograph the four moose heads together, and asked if he could take them outside; but Mrs. Blacksmith said no—once in the lodge, the moose must not be further disturbed. The moose hunt had cost us $600, but we figured that at current meat prices, its value to the three families was close to $3,000.

The next day was our last full day in camp. Now we had cause for a real feast. The day was occupied in preparing the meat: Sam and Ronnie squatted on the spruce boughs of the lodge and methodically and with tremendous expertise (they were qualified butchers, along with all their other talents) carved up the four heads, of which every piece except, curiously, the brain is eaten. That night, the preparations completed, we had the feast. It happened to be my birthday, and when we entered for the feast we found that the girls had cooked round slabs of bannock and had stuck a candle into the center of each in my honor. (Thus our ethnographic film, authentic in all its detail, has this curious aspect, which no one has ever remarked, of the candles standing up in the bannock during the moose feast!) The meat was delicious and we ate for quite a long time. We had a judicious amount of liquor left and when we sat around drinking it after the meal we felt that finally we had been accepted as friends. Philip had gone away on the moose-hunt plane so that he could get to a meeting of friendship centers in Sault Ste. Marie, and without his restraining influence we were able to ask Sam and Ronnie if we could film them playing the drum (something that Philip said they would never do, and which we knew he would never have asked them to do).

After they had played the drum, we thought it polite to leave, not wishing to outstay our welcome. They had been pleased with their experience, they said, had enjoyed having us on their land and showing us their lives. And we said again that for us it had been an honor and a privilege.

We went back to our tent and had just got into bed when the girls arrived and said that the feast could not end without a dance and we must return to take part in it. We dressed again and returned to the lodge, where those who had been sleepy were now wide awake. We danced around the lodge Indian-style a few times, and then Rosie turned on her tapes of country music and she and her father showed themselves considerable experts at the traditional jig. Along with the sacred drum, we improvised drums from cans and kettles. Our sound man, Jean-Guy Normandin, never very sensitive to Indian traditions, played "Frère Jacques" on the sacred drum, grabbing the intestines of the moose which were hanging from the roof and using them as an improvised microphone. But his outrage had no bad effect: the more outrageous he became, the more the Indians seemed to enjoy his performance. We opened our last half-bottle of liquor and the jigging became uproarious.

Oddly enough, we had an emotional leave-taking the next day. We packed up early, and the girls came to our empty tent to help us wash dishes and clean up. For an hour or so we made a great game of wanting a goodbye kiss from each of them, a demand that was greeted by that strange, inimitable groan of amusement and pleasure that I have never heard anywhere in the world except from Philomen, Mary and Rosie together. Finally, as the plane was coming in, they all relented, offered their cheeks to us (turning their faces as far away as possible) and consented to be kissed. We ran down to the plane. Everyone gathered on the ice. The parents were smiling and happy, the girls pulling at Ben, asking him to stay. We all knew that as we climbed into the plane, lots of company for the young people was disappearing, and at least one of them was, in her mind, climbing in with us, soaring over the trees away from the camp, anywhere as long as it was away. As the plane turned around, the youngsters waved and waved, and ran a little after the plane and stood waving. I was very moved.

The next time we met them in Mistassini, they were all super-cool, nodding at us as if they had seen us only five minutes before, and seemed surprised when we ran after them asking how they were.

13

ROSIE

Rosie was determined to go to Montreal in the summer. Her mother promised she could go if she would stay an extra couple of months in the bush camp. I suggested that in Montreal she should stay with my family, since the big city chews up and spits out girls like Rosie two a minute. After she returned to Mistassini in June she wrote a polite little note asking if she might spend the summer with us, and we wrote telling her to come. But we heard no more, and after a couple of phone calls to Mistassini I was told by Smally that the parents wouldn't permit her to travel alone.

For months we worked over the film in a small, dark room at the National Film Board and by August were ready to take it to Mistassini to show to the families so that they might exercise their editing prerogative. Philip Awashish was living for the summer in a small cabin five or six miles out of Chibougamau on the road to Mistassini. When we stopped off there they told us they had heard that Rosie was to be married on August 17, and Philip's wife, Anne-Marie, said she had the impression it was not so much a case of Rosie getting married as being married off.

The Voyageurs were out of Mistassini working in tourist camps, but we showed the film to the Jolly and Blacksmith families, telling them that they could see it as many times as they wished. The families laughed all the way through the first screening, since they had never seen themselves on film before. But the second time they saw it they were more sober and appeared to enjoy it.

Rosie, as usual, hung back in a bedroom when we met the families again, but after we had shown the film she stayed on and cooked some food for us, and as I drove her back to her parents' house she said, "Will you have room for me on the way back to Montreal?"

I said yes.

"I really want to come," she said, the most vigorous expression of opinion I had ever heard her make. She said she didn't want to get married, and since her parents had told her before that she could go to Montreal, she was going to talk to them about it again. I said Malick was invited, too, but she said Malick didn't know much about the city and didn't seem keen to go.

That afternoon, as we were recording one or two additional comments from Sam for our sound track, he added: "I cannot expect my children to take my land. They have always lived in Mistassini. But I am hoping to solve the problem by finding a son-in-law to take it over. There is a young man who is capable, and we want him to marry Rosie. But there is only one problem: there is some resistance to the idea from Rosie. My wife and I feel that Rosie should stay with us and settle down."

If she would marry this young man in the summer, then the winter after next (for next winter Sam's land would be resting) Sam would take his new son-in-law into the bush and begin to teach him everything he knew. That would be an ideal solution, for it would ensure the handing on of his land to a capable hunter, and Sam and his wife would have someone to live with as they grew older.

Expressed in those terms, Rosie's problem took on an added dimension. If she was right in wishing to avoid being forced into marriage, Sam was also right, within the context of the cultural attitudes he had inherited, in wishing to safeguard the future of his land, the sacred duty that lay upon him for the benefit of future generations. It was one of those situations, so dearly beloved of moral philosophers, in which both sides were right. But there was no doubt who would win. The very fact that the confrontation with his daughter was occurring showed how Sam's cultural and moral base had been withered away by outside forces. Like the arranged marriage in India, Sam's solution could work only so long as the cultural context remained intact.

It was as if 300 years of assimilative processes, grinding away, had come to a ghastly climax in the misunderstanding and incomprehension that now existed between this man and his daughter. Neither was in control of anything. If Sam succeeded in securing the future of his land, it would be at a heavy personal cost for someone else, and there was no guarantee, given the white man's political and economic drive into the north, that the solution would last long. If Rosie succeeded in getting away, it would probably lead to disaster. For though our society had alienated her from her roots, she was hopelessly ill-equipped to make her way through life alone. She had no education worth speaking of, and would never get one, so that in any white man's social context she was destined to be a servant. She had only a gentle nature and a wonderful grace: physically, she could have become one of those exotic models who, from time to time, become the rage of London and Paris and appear on the cover of *Vogue*. But no such glittering future could be foreseen for Rosie: her tragedy was that, though she yearned to get away to the city, within a couple of months she would have become the mute victim of the urban jungle.

After seeing the film the second time the families said there was nothing they wanted to add or subtract, and they wished the whole community to see it. Abraham and Shirley now arrived from the camp where Abraham was working as a guide for $28 a day and Shirley was doing the washing for $10 (in such a way did contact with the wage economy reduce these people to the level of servants). We had sent a message to say that we wanted them to come and see the film. They too liked it, and when we showed it to everyone in the village that evening, though it was still in a rough and rudimentary state, the reaction was lively.

The next day Mrs. Blacksmith told me that they had decided Rosie could go to Montreal, and they would send for her when they wanted her. "Don't let her be too independent," she said. As we were driving around the village that morning we met Sam leaning against the wall of a house, finishing off a bottle of beer. He had been drinking ever since the screening of the film the previous evening. He put the bottle in his mouth, zipped up his fly and then, spotting us in the car, lurched over to us and cheerfully said, *"Agaday,"* over and over again, laughing each time. He climbed in and we drove him to his house. He wanted to go with us to Chibougamau, but in the background Mrs. Blacksmith, smiling, was skillfully steering him back into the house. Rosie drove with us to Chibougamau and back, maintaining her usual silence.

Before we took off for the airport the next day we filmed a rather strange scene: Rosie walked out of her house carrying some bags, put them on the ground, went expressionlessly around shaking hands with members of her family and then picked up her bags and carried them to a car which drove off with her. We had in mind that the dilemma of Rosie, caught between the two worlds, might dramatize the political and social points we were hoping to make. But there was something very final about the farewell scene we filmed, and I wondered what the parents thought was happening, for ostensibly Rosie was going away only for a couple of weeks' holiday.

We gave a lift to Nancy Gunner, a strong-willed and capable Mistassini girl who had lived in Montreal for some years and had a job at the Department of National Health. On the way to Chibougamau, Rosie's tongue was finally unleashed—but in Cree. She chatted and laughed without stopping, all the way. The contrast between her tongue-tied shyness in English and her volubility in Cree was strange to the point of being bizarre: she did not care whether the Cree language lived or died, and wished to escape from the Cree world. Yet outside of her own language she was crippled as a personality. If Cree died, so did she.

At the airport, before I went to buy my ticket I asked her if she had money to pay for her fare.

"How much is that?"

"Forty-five dollars."

"No."

I sat with her in the plane, and tried unsuccessfully to penetrate that curtain of reticence. But it was mostly, as it had been in the bush, a monologue. She had left home with $15, the assumption apparently being that I had more or less adopted her.

"You know," I said, "a city like Montreal can be rough for a girl like you. So, whatever happens, I want you to promise to keep in touch with us." She nodded. I mentioned the name of a Mistassini girl who had gone to the city a few months before, leaving her baby behind. A quiet, pleasant girl when she left, she was already deep into hard drugs, was painted up like a whore and could be seen almost any night hanging around the bars on Crescent Street waiting to be picked up (within a couple of months she was in jail for a long term). "That can happen to anybody," I said. "And there's another thing. All you need to do is have a few friendly beers at the Boiler Room"—(the Indians' favorite bar) —"go off home with one of the young Indians you will meet there and you can wake up with a baby. I don't want that to happen while you're down here, so the first thing you should do is get some contraceptive advice." She nodded expressionlessly.

For the next three weeks we had this oddly convolute girl living in our house. We found her to be totally Cree in her behavior. The state had taken her away from home and educated her, but it had failed to make her over into the image it required. Though she had spent years in the white man's schools, she was lost in the white man's world. Her education had been, quite literally, a work of terrible destruction. We were living

during these three weeks in a small house on a lake forty miles north of
Montreal. The presence of bush, water and a canoe were good for her,
but it was far from the city life that she was apparently yearning for. She
shared a small room with my four-year-old daughter, and she spent a lot
of time in it, sleeping and occasionally reading the children's comic books
and smoking a lot (all Indians seem to smoke a lot). But she would emerge
from time to time, busy herself around the stove, wash some dishes, clean
up around the place, or wash her small collection of jeans and sweaters.
She was always neat and extremely clean, and, as in the bush, she often
washed her hair and constantly combed it. But for the most part she was
just a wordless presence around the house and it was almost impossible
to discover if she was happy or not.

For us, in a way, it was like having a bomb in the house, for we were
anxious to please her, and we could never get any idea whether we were
doing so. It was probably the first time she had ever lived in a white man's
house, and she was embarrassed. She would often pick up my daughter
and remark that she reminded her of little Abraham, the kid in the bush
camp. And then occasionally, while standing around, she would volunteer
some remark to my wife and even get into a considerable conversation
with her, which would end as abruptly and apparently purposelessly as
it had begun.

We asked if she would like a bathing suit so she could swim in the
lake, and she said she would. So we took her to a little shop in the
French-Canadian village of St. Sauveur. When the shop assistant asked
if she could help she was rewarded by a view of Rosie's back. When my
wife tried to help her to choose, she agreed with every suggestion, but
adopted none. She wouldn't try on any suit, but finally chose one that was
certainly too big for her. And then of course she never wore it.

For two days she came to the Film Board and worked with us on
refining the translations for the film. She clearly had an excellent grasp
of the Cree that the people were speaking in the bush, and was most
helpful. Tony, with his customary kindness, arranged for her to be paid
full translator's rates, so her take from the two days of work was $240,
an unimaginable sum for which her father would have to kill half his
beaver quota. She took the news of this windfall utterly without expression, though presumably it pleased her, for it made her financially independent at least for a week or two. Yet though she worked so well for
us, occasionally in the cutting room she would suddenly lapse into that
curious inability to speak that had characterized her ever since we had
met, and we would have to ask her a question three or four times before
she could give us an answer.

Sometimes I would ask if she was happy with us, and she would say abruptly, "Yes." Once she replied, "Well, I'm missing little Abraham, that's all." She was very much attached to her parents, and when I asked how her father got on in the village, she launched into reminiscences about his drinking which were more affectionate than critical.

"Last year he would go two or three weekends without drinking," she said, "but this year he is drinking every weekend." If he visited Chibougamau he would usually come home sober and start drinking when he got home. But Mrs. Blacksmith was equal to him. One time when Rosie and her mother were shopping at the Bay in Chibougamau and Sam was at the Waconichi, he sent Malick over to the store with $40 to give to his wife. "What's this for?" asked Mrs. Blacksmith, and Malick said Sam had sent it over because he was going to be in the tavern all night. "We can't have this," said Mrs. Blacksmith. She went over to the tavern, hauled him out (leaving all the beer behind) and shoved him into a taxi (loudly protesting his desperate need for a beer). He was soon asleep. Rosie said he was very funny when he was drunk. He would tell them the same story over and over again until Malick would say, "You told us that one already. Can't you tell us something different?"

"How about the boy they want you to marry?" I asked. "Do you like him?" She shook her head. "Is he a good hunter?" She nodded. "Is he a nice fellow?"

"He's a big drunk," she said with unaccustomed vehemence. "He's already been in hospital for two and a half weeks through drinking. And when he got out he was already drunk again by the time he took the bus home."

Rosie herself proved to be the best drinker in the family: one night when we were having a drink, she downed six Scotches without any noticeable effect except about a 5-percent increase in her willingness to initiate a conversation.

On an evening when we sat far into the night with some neighbors round a fire on the shore of the lake, she became quite talkative and didn't have to be coaxed to utter every sentence as she spoke about the life in Mistassini. But another time when we finished a tramp through the bush at a friend's house, she took off along the road for our place without a word to anyone. We discovered her gone ten minutes later, and when we picked her up and asked why she had walked, she said only, "I was hungry."

She was anxious to go to the city, and after staying overnight in Montreal with a friend of ours, she rang and said she would stay the weekend with Nancy Gunner. This time she was on her own: we could only hope she knew how to handle it. For all we knew, Indians being independent and capricious in their behavior in the city, we might never see her again.

"She never said a word when she was sitting around the house," said my friend, "but when we went for a walk along Ste. Catherine Street she was so afraid to lose me, she held my hand like a child and kept talking all the time and saying 'Don't go so fast, wait for me, I will get lost, there are too many people.' For me it was very strange to be with this Indian girl, and she so afraid, so out of her element. But she was so happy. She didn't want to go back to the country. Later she walked by herself to Alexis Nihon Plaza, a place with three stories of immense balconies around a central well, and of course she'd never seen anything like that, and then she came home and said, 'That's enough for today.'"

She stayed one more night, but by next afternoon was on the phone asking if she might "come home." In forty-eight hours the city had proven too much for her. She was to have stayed that second night with Nancy Gunner, but at a late hour had found herself stranded at the Boiler Room. The Mistassini girl who was heavily into drugs was there, urging Rosie to turn on, alternately boasting and crying piteously, but Rosie found some other young Crees who took her home for the night. She returned to my friend's house the next day, penniless, but she had gumption enough to phone my friend's sister and borrow money for her bus fare. When we picked her up at the bus stop in the country, she declared, "I was already bored with the city."

"I had the impression that for her it was like going back into a hole," said my friend. "Something had happened in the city that had made her very unhappy." But we felt she had been frightened by the impersonality of her experience. It seemed that, in spite of her dreams, the city really was no place for her.

Whether she was to escape permanently from Mistassini or not was now for her to decide. She said she wanted to go back to school, but at the age of nineteen, with only grade six behind her, she was not in a good position from which to further her education. Fortunately, a new college for Indian students had recently been opened at an abandoned Bomarc missile base in the forests and lakes of the countryside about 100 miles north of Montreal. They were offering exactly the sort of educational upgrading programs that many young Indians like Rosie needed. I asked if she would like to enroll for the following September: she said she would. Her parents had told her she could go back to school if it could be arranged. One day we went to the downtown office of the college. Here, among other Indians, I expected that she would be somewhat thawed out. But again the experience was too much for her: she was tongue-tied when confronted with the simple questions demanded of her, and I had to do most of the talking.

She was no sooner enrolled than she decided that by the end of the week she should return to Mistassini "to talk it over with my parents." She would come back, she said, in the last week of August, ready for the beginning of school.

She had been with us for three weeks. We drove her to the airport, the air heavy with her embarrassed silence, since she no doubt knew that some kind of farewell would be expected of her. When the time came, she could not get the words out. "Where do I go?" she asked, after she'd checked in. We pointed the way down a set of stairs. She turned her back and without a glance at us or a word, she walked off and never looked back.

Two weeks later I had a phone call from her. She was at Chibouga-mau airport. "I am on my way," she said. I suggested that since the school did not start for another two weeks, and she had to complete arrangements for her bursary with her regional Indian Affairs Office, she should stay another couple of weeks with her parents and then come down. "All right," she said.

But she never came. She did not turn up at the college, though three weeks after it had started I rang and asked if she still wanted to come and she said she did. Her mother had been sick, she said, so she had been unable to leave. She soon found a better, wiser solution. Within a few months she was married, not to the prospective inheritor of the hunting territory, but to a boy who worked for Glen Speers at the Hudson's Bay store. Soon she gave birth to a baby daughter. She became a Cree mother, and forgot about escaping from the only culture she knew.

I saw her in Mistassini the following March. She was sitting with her parents and husband in the village hall. I came up behind her. "I've brought you a wedding gift," I said. She didn't turn around. "Thanks," she said. And nothing more.

JUDGMENT

On November 15, 1973, after nearly five months of deliberation, Mr. Justice Malouf issued his judgment. To the astonishment of everyone, he declared a ringing verdict in favor of the Cree and Inuit and issued an order to the James Bay Development Corporation, the James Bay Energy Corporation, Hydro-Quebec and twenty-two contractors,

> their officers, employees, agents, servants and those acting under their authority and pursuant to their instruction
>
> (a) to immediately cease, desist and refrain from carrying out works, operations, and projects in the territory described in the schedule of Bill 50, including the building of roads, dams, dykes, bridges and connected works
>
> (b) to cease, desist and refrain from interfering in any way with petitioners' rights, from trespassing in the said territory and from causing damage to the environment and the natural resources of the said territory.

The judgment created consternation in government and establishment circles in Quebec, though it came as an unexpected breath of hope to the many people who persisted, in spite of everything, in believing in decency in human affairs. That the Crown corporations had produced a ramshackle case in court was confirmed by the scant regard the judge showed for their evidence and most of their arguments. In his 170-page opinion Mr. Justice Malouf based himself on the nature and history of the

Indian rights. He first quoted Lord Haldane, from a 1921 Southern Nigerian case, to warn against any assumption that all concepts of property should be identical with those of the white man:

> Their Lordships make the preliminary observation that in interpreting the native title to land, not only in Southern Nigeria, but other parts of the British Empire, much caution is essential. There is a tendency, operating at times unconsciously, to render that title conceptually in terms which are appropriate only to systems which have grown up under English law. But this tendency has to be held in check closely. As a rule, in the various systems of native jurisprudence throughout the Empire, there is no such full division between property and possession as English lawyers are familiar with. A very usual form of native title is that of a usufructuary right, which is a mere qualification of or burden on the radical or final title of the Sovereign where that exists. In such cases, the title of the Sovereign is a pure legal estate, in which beneficial rights may or may not be attached. . . . On the other hand, there are indigenous peoples whose legal conceptions, though differently developed, are hardly less precise than our own. When once they have been studied and understood they are no less enforceable than rights arising under English law.

The judge then ran through the statutes which, he said, showed that until the present day the Canadian authorities had consistently undertaken the obligation to alienate Indian title to the land before opening it up for white settlement. And he said Quebec had undertaken a similar obligation in 1912, the nature of which was fully understood and spelled out in a federal order-in-council (dated January 17, 1910). The judge said he was not called upon to decide the exact nature and extent of the Indian title, but merely to satisfy himself that the petitioners had made out a *prima facie* proof that they had clear rights sufficient to ensure that they had a substantial case to be considered by the court in the final hearing, and that he found.

Malouf totally accepted the credibility of the native witnesses and the scientists supporting them. In spite of the masses of statistics and the long parade of witnesses produced on the corporations' side to show that Indians were just like white people, the judge paid little attention to their testimony, preferring the testimony given by the native people themselves. He dismissed the evidence of Rita Marsolais, the Hydro-Quebec economist who claimed that Fort George families were making $10,167 a year.

She included as revenue all the sums paid by the federal government authorities for health, the economic development of Indians,
the administration of programs, the maintenance of roads and the
treatment of refuse. . . . Why should such calculations be made to
apply to the Cree Indian and Eskimo population when such a calcula-
tion has never, to my knowledge, been made with respect to the
revenue of any other individual in this country?

The evidence, he wrote, led to the following conclusions:

(a) The Cree Indian and Inuit population occupying the territory and the lands adjacent thereto have been hunting, trapping and
fishing therein since time immemorial.

(b) They have been exercising these rights in a very large part
of the territory and the lands adjacent thereto including their traplines, the lakes, the rivers and the streams.

(c) These pursuits are still of great importance to them and
constitute a way of life for a very great number of them.

(d) Their diet is dependent, at least in part, on the animals which
they hunt and trap, and on the fish which they catch.

(e) The sale of fur-bearing animals represents a source of revenue for them, and the animals which they trap and hunt and the fish
which they catch represent, if measured in dollars, an additional form
of revenue.

(f) The hides of certain animals are used as clothing.

(g) They have a unique concept of the land, make use of all its
fruits and produce, including all animal life therein, and any interference therewith compromises their very existence as a people.

(h) They wish to continue their way of life.

The judge found that the native petitioners "are justified in their
apprehension of injury to the rights which they have been exercising." In
his description of the likely effects of the project on the environment, he
drew heavily on the evidence of the scientists supporting the Indians, and
noted that whereas many of the corporation's witnesses could not foresee
any drastic effects, they mostly admitted under cross-examination that
they did not know what the effects of the work would be, and some
admitted the effects would be considerable.

After noting (in a detailed seventy-page summary of the evidence) the
likely effects on the animals, fish and birds, the wetland habitat and the
balance and integrity of a delicate ecosystem in which plant regeneration
is unusually slow, the judge declared: "In view of the dependence of the

indigenous population on the animals, fish and vegetation in the territory, the works will have devastating and far-reaching effects on the Cree Indians and the Inuits living in the territory and the lands adjacent thereto."

Mr. Justice Malouf decided that the Indian rights were so clear that he was not even called upon to consider the question of balance of inconvenience, but he analyzed the question just the same, because of the importance placed on this argument by the corporations. He said he found it difficult to compare the monetary losses claimed by the corporation (in the event that work should be stopped by the courts) to the damages that the native people would suffer. "The right of petitioners to pursue their way of life in the lands subject to dispute far outweighs any consideration that can be given to such monetary damages," he wrote. "It is my opinion that greater damage would be caused to petitioners by [my] refusing to grant an injuction in the event that the final judgment maintains the rights of petitioners than to respondents by [my] granting the injunction in the event that the court in considering the final application concluded that the injunction should not issue."

The events which followed the Malouf judgment were remarkable. The Quebec government and the corporations, with the willing cooperation of the newspapers, swung into a beautifully orchestrated campaign to prove how much the people of Quebec were paying for stoppage of work in James Bay. The James Bay corporation's immediate response to the judge's order to cease trespassing was to seize the territory, seal it off from outside observers and continue the work. The order to stop was issued on Wednesday. The following Saturday, in defiance of the law, the work was still going on. Pilots who were hired to fly inquiring reporters north to the work sites were intercepted at Matagami and told by officials that if they proceeded into the territory the officials would see to it that their licenses were canceled and they would never fly again. Reporters who managed to get aboard the scheduled flights which left Montreal as usual were hustled into a trailer at the new airport built five miles from the LG-2 camp (on the trapline of Josie Sam's brother) and told that they would not be allowed on the site. They talked to one or two workers waiting to take the plane south, and then were told that if they didn't get back on the plane and return to Montreal they would have nowhere to sleep that night.

"The work is still going on," said one worker, quoted in the Montreal *Gazette,* "but prove it. Try to get one of the men to say it. He'll tear you apart."

For days the newspapers carried enormous headlines over stories issued by the James Bay corporation to the effect that they were losing $500,000 a day through the stoppage. This would have amounted to a loss of $182 million a year. Even in court the corporation had claimed likely losses of only $87 million a year, and Malouf, in his analysis of that figure, had decided that the more accurate loss would be around $9 million.

The day after the Malouf judgment, the corporation entered two appeals to the Quebec Court of Appeals. One was against the merits of the Malouf judgment, and one was an application that the effect of the Malouf judgment should be suspended pending the hearing, at a later date, of the appeal. The Quebec Court of Appeals sat on the case less than a week later. The atmosphere in court was unusual, to say the least. Normally a party which enters an appeal is called upon to justify the grounds for the appeal. But no sooner had the three judges, Lucien Tremblay, Jean Turgeon and P. C. Casey, settled into their seats than they directed a surprising question at James O'Reilly, the Indians' lawyer. "Now, Maître O'Reilly, what do you have to say?" O'Reilly, unexpectedly called upon to give reason why the works should not be re-started, staggered to his feet and began to pull his case together. But the judges did not want to hear anything about Indian rights, and whenever he mentioned them, they pulled him up sharply, saying that Indian rights were irrelevant to the application for a suspension of the injunction on the grounds of public interest.

The hostility of the judges and the fact that at no time were the government lawyers asked to justify their appeal indicated that the judges had made up their minds before they entered the court to suspend the Malouf injunction. On the morning of the second day, exactly a week after Malouf brought down his judgment, the Appeal Court swept it aside, without a nod in the direction of appearances.

They had spent five hours considering the case (compared with the seventy-eight days of hearing before Malouf), and in less than three double-spaced pages (compared with Malouf's 170) they now simplified the whole matter:

> The works of which the suspension was ordered are carried out in virtue of chapter 34 of the laws of Quebec, 1971, and certain orders-in-council adopted in virtue of this law. This law created the James Bay Development Corporation, and by article four gave it the objective of undertaking the development and exploitation of the natural resources found in the territory, of carrying out the adminis-

tration and management of the territory, and of giving priority to Quebec interests. This law was adopted by the National Assembly elected by the people of Quebec, and so long as it has not been declared unconstitutional, it ought to be applied, save in absolutely exceptional circumstances. . . . It is, then, the general and public interest of the people of Quebec which is opposed to the interests of about 2,000 of its inhabitants. We believe that these two interests cannot be compared at this stage of the proceedings.

The judgment made no reference to the Indians' rights, which were the subject of the whole proceeding. The Appeal Court based itself on Article 4 of Bill 50, mentioning the interests of Quebec, but gave no consideration to Article 43, ensuring the protection of the rights of Indian communities, an article which Malouf had found was being violated by the corporation. So the works were allowed to continue.

The Indians now decided to go to the Supreme Court of Canada in Ottawa to ask leave to appeal against the suspension of the Malouf judgment. In Montreal a handful of us decided to march to Ottawa to protest against the quality of the justice being handed out to the Indians. A couple of days before the Supreme Court hearing, 500 or so, mostly Indians, marched around Parliament behind a banner marked "INJUSTICE IN JAMES BAY," and holding up signs such as "KANGAROO COURT OF APPEALS," "QUEBEC APPEAL COURT IN CONTEMPT OF THE PEOPLE," "JUSTICE IN JAMES BAY BY BULLDOZER, DYNAMITE AND HELICOPTER," "JAMES BAY IS INDIAN LAND." It did no good (apart from relieving our frustrations), for though the Supreme Court was extremely careful to give the impression that justice was being done (five justices sat on the appeal and occupied a day and a half of hearings for a matter usually disposed of in fifteen minutes), they decided—again without any reference to the Indian rights —that since the Quebec Court of Appeals had yet to hear the appeal against the merits of the Malouf judgment, the higher court could not be seized of the matter at the same time.

Both courts said that the appeal would be dealt with expeditiously, in a comparatively short time. But though the appeal *against* the Indian victory came to court within a week, it was eight months before the same court could get around to hearing the next stage of the appeal. During these eight months the work proceeded.

Yet the project kept running into trouble. The government seemed to be short of money to build it, and then, in the depth of winter, most of the work had to be abandoned because of the intense cold. When the work was resumed in the early spring—still extremely cold up there, with

nightly temperatures down to thirty degrees below zero—the project ran
into a jurisdictional dispute between rival union groups, at war in the
construction industry all over the province. O'Reilly kept hammering
away with appearances in court on minor matters, escalating the issue as
much as possible and keeping it alive during that winter, but his efforts
were as nothing compared with the anger of the construction workers.
After a lot of tension on the site, some workers exploded: on an intensely
cold day they seized some bulldozers and giant shovels and ran them into
the power plant. They then set fire to the camp, destroying part of it.

The irony of this event was striking: an area, the second rapids on
the La Grande, which had been occupied for countless generations by a
people living harmoniously with nature, had been handed over to a
supposedly more civilized and advanced form of humanity, whose aliena-
tion and hatred was so great that they took their tools in hand and
smashed the power plant which was the only source of heat capable of
guarding them against the sub-zero weather.

Some 1,400 men were hurriedly flown out of James Bay; damage was
estimated at $2 million and work was delayed for some months. The
ringleader of the men who caused the damage was identified and given
a ten-year jail term. The government took the opportunity to set up an
inquiry into the recurring problem of violence in the Quebec construction industry, an inquiry whose scandalous revelations of corruption and
violence at James Bay titillated the Quebec public in the coming months.

Once again the work had been stopped. When a reporter phoned
one of the lawyers for the native people to ask for his comment on the
destruction of the site by the workers, he said: "If you don't quote me,
I'll tell you: it sure as hell beats an injunction."

15

NEGOTIATING TABLE

For the Indians the importance of the Malouf judgment was that it confirmed their rights. The principle that these rights could and should be alienated to the government by agreement or treaty was accepted on both sides, and the Malouf judgment created a favorable political situation for such an agreement. The Quebec government was anxious to get rid of the Indian problem so that it could proceed unencumbered with the James Bay project, and had every reason to seek a quick settlement out of court. The Indians, too, had reasons to get down to negotiations quickly. They were in a strong position so long as their rights were recognized, but they knew they could expect little sympathy from the Quebec Court of Appeals when it considered the merits of the Malouf judgment, for this court had shown that, like many other institutions in this French-speaking society, it was determined to interpret events in a highly nationalistic way.

At the same time, the James Bay Indians were not really ready for negotiations. For two years their energies had gone into the creation of a political cohesion with which to defend themselves against the project. They had not had time to work out what they wanted over the long term. Now suddenly they came under intense pressure to make a settlement. Within a day or two of the Malouf judgment, the federal Minister of Indian Affairs, Jean Chrétien, appeared in O'Reilly's office in Montreal to urge that the Indians withdraw the case from court and trust to a political settlement. And within the next two weeks the Quebec government made an eleven-point offer to the Indians and Inuit. Though the government had delayed sixty-one years since undertaking in 1912 the obligation to treat with the Indians, they expected an answer to their offer in a couple of weeks. The young political leaders of the Cree refused to answer without consulting their people, who were at this time mostly in the bush hunting and trapping.

Meantime, on the Indian side there were still divisions and misunderstandings. The officers of the Indians of Quebec Association, who had been petitioners in the case, had been dismissed from it by Malouf on the grounds that they did not live in James Bay and therefore had no status to take action, and they were now worried that they might lose their mandate to speak and negotiate for the Cree. The Cree chiefs, at a hurried meeting, renewed the association's mandate to negotiate, but appointed Philip Awashish, twenty-five, and Ted Moses, twenty-four (who had been the chief interpreter during the court hearings), to join the negotiating committee. They also decided to keep the court action going.

When the Indians and Inuit had not replied to the government's offer within two months, the Quebec premier, in a fit of impatience, made public his proposals. He offered them $100 million. Of this, $40 million was to be a cash payment, while $60 million would come later in royalties. He offered to set up a development corporation of native people to handle this money and run economic development and social and training programs to ensure that native people were involved in running the tourist industry in the area. He offered a guaranteed income for trappers; hunting, fishing and trapping rights throughout unoccupied Crown land; and exclusive trapping rights in certain areas to be designated by agreement between the two sides. He offered 2,000 square miles of reserves for the Indians and Inuit of northern Quebec, to be allocated on the classic and traditional basis of one square mile per family of five.

Premier Bourassa also threw in something that had not been in his original offer—a number of modifications to the project, designed, he said, to take account of Indian and Inuit fears. "It was not a great offer," said Philip Awashish, "but it was not really bad. We decided to reject it because the major concerns that the people had expressed were land, hunting, fishing and trapping rights, and environmental protection to guarantee the continuation of the tradition." The young leaders, with their lawyers, made a quick tour around the James Bay villages. Not a single question was asked them by any Indian about the monetary aspects of the offer. All the questions were about the future of the land.

The leaders were confirmed in their intention not to accept anything until they had had a chance to consult the hunters and trappers who would not be returning from the bush until the spring. Though the Minister for Indian Affairs was pressing them to accept (since he said the Quebec offer was by far the best ever made to Indians in Canada), they decided to reject it formally. At a January press conference in Montreal, Chief Billy Diamond said the Cree people would continue to oppose the James Bay project. "Abraham Martinhunter, of Fort George," he said, in a quote that hit the headlines of every Canadian newspaper, "has stated that the Indian lands are not for sale, not for millions and millions of dollars." The country was impressed: the Indians were spurning $100 million. The governments, of course, were furious, and the federal minister, Mr. Chrétien, overstepped the mark: if the Indians remained unreasonable, he said, he would consider cutting off the funds with which they were pursuing the court case. That threat was too much for even the newspapers to stomach. The Indians were faced with an appeal against their victory in the courts, and if Chrétien withdrew their money he would make it impossible for them to contest the appeal, an intolerable decision. Under heavy attack, he quickly said he had been misinterpreted.

The Indians' advisers believed it was still possible to achieve important modifications in the LG-2 dam if they could get down to negotiations quickly, for irrevocable decisions about the design and extent of this dam and reservoir had not yet been taken. They could not negotiate, however, because the Indians of Quebec Association was still hoping to use the James Bay issue as a lever for a package political settlement of Indian problems in the whole province. For three months the negotiating committee to which Philip and Teddy had been appointed did not have a meeting. Finally, the Cree chiefs decided to withdraw the mandate they had given the association to act on their behalf. First they placed negotiations in the hands of a three-man committee made up of Philip, Ted and Chief Robert Kanatewat of Fort George, and eventually decided to go it as Cree, entirely alone, setting up a Grand Council of the Cree to conduct all future negotiations. Billy Diamond became the Grand Chief. The federal government agreed to fund the Grand Council directly, cutting out the association as intermediary. On the day on which the Cree solemnly set up the first political organization in the 5,000 years of their history, Chief Matthew Shanush of Eastmain said: "Our forefathers were never faced with a crisis like this. They never had to sit at a round table, as we are doing here, to resolve such a crisis, because they were well content with their hunting life and nothing was bothering them."

A CITIZEN'S SUGGESTION

Chrétien's crude threat to undermine the Indian resistance to the project and the prevailing Quebec public opinion that the Indians should accept the money and get out of the way of progress indicated to me that there was a serious misunderstanding of the moral and ethical basis of the Cree hunting culture in northern Quebec. To try to draw attention to the underlying issue, I suggested to the Quebec government that they should offer to the Indians and Inuit the ownership of all the wild animals of northern Quebec. The implications of such an action, if followed through, would be considerable. For it would mean that no sports hunting would be permitted in the region except through the intermediary of Indian bands and Inuit villages, and the native people themselves would be in a position to decide how much of the animal resources they needed for subsistence and how much they could afford to sell, as a cash crop, to sports hunters. The evidence given in court had made one thing abundantly clear: the real threat to the native culture was not just the James Bay project, but the imminent prospect of the arrival in force, all over the James Bay area, of the white man's civilization. Up to now, the Indians had simply made way; as Harvey Feit had shown, as soon as the white man arrived, even the animal resources were handed over without question to him. Deprived of their economic base, the Indians were then faced with a pace of change that they could not handle. It seemed to me that if the means could be found to establish native control over the animals, the Indians and Inuit could themselves decide, in the next fifteen years or so, how much and how fast they wished to integrate into the white man's wage economy and technological culture. Although I had little time to put this proposal together, I was able to get considerable publicity for it in Ontario and Quebec over one weekend, indicative perhaps of a changing climate of opinion toward the Indians. The James Bay chiefs endorsed the idea and accepted it as a guideline for negotiations.

THE HUNTERS MEET

As soon as the Cree and Inuit themselves took over negotiations, the bargaining with government became intensive, and it became urgent for the young men to consult the hunters about the government offer.

Though they knew that compromise was the essence of negotiation, the young men were unsure how far they could go in compromise without fatally undermining the integrity of Cree life. The Quebec government had always been notably indifferent to Indians and to the environment, and the tone of the government in court did not incline the young men to believe in the sincerity of the men they were negotiating with. In two things especially they were bothered by government insincerity: when Bourassa announced the modifications in the project, he said they were to meet Indian objections. But on examination these changes proved to *increase* the size of the project, adding the Eastmain River to those which would be diverted, while the Great Whale was now to be left alone. The two reservoirs of Delorme and Caniapiscau were to be combined into one. These were all changes obviously dictated by engineering considerations, and the Indians were disturbed at the attempt to pass them off as meeting Indian objections. Secondly, the Indians had uncovered a sneaky act during that winter which filled them with misgivings about the government's real intentions. For years the beaver quotas "given" to the hunters by the provincial government were based upon their being allowed to kill one beaver for every occupied beaver lodge they had observed in their territories in the previous season. In other words, they virtually decided their quotas themselves by the reports they handed in to government agents, though, formally, the government, having received the reports, would send back to the settlements a statement of the quotas for the following year. Following the court case, the hunters discovered that when the annual statement came back from the government, the beaver quotas had been almost doubled throughout James Bay. In Mistassini the quota was lifted from 7,000 to 11,000. It appeared plain, all over James Bay, that the government was using this underhand method to persuade hunters to kill off too many beavers, and so further weaken the validity of the hunting life.

It was against the background of these misgivings that in March 1974 the hunting families were flown out of the bush camps so that they could hear about the government's offer, made five months before, and express their opinion of it. Never before had there been an issue so solemn and urgent as to demand that they leave their bush camps. It was not an easy thing for them to do; each man had to be warned by a flying messenger so that he could prepare his camp and his traps for his absence, and a few days later, all over the huge area, everyone had to be picked up. I went to the meeting in Mistassini. Weather-beaten and ragged from their months of hard work in the bush, the hunters gathered in the school auditorium, and when they got to their feet they spoke of only one thing, their land. They spoke with the passion, feeling and perception of poets. They talked about the purpose that the Creator had when he created the earth and put the animals on it and gave them to the Indians to survive on. They talked of how they had worked and suffered for the land, and of how the animals and the land had helped them survive. They talked about the white man, and his thoughtless ways, his failure to ask their permission before he invaded their lands, the things they had silently observed him do over the last two decades. Over and over again they declared their affection for the land and their knowledge that its destruction meant their destruction.

At the end of the first day of that meeting we showed our film *Cree Hunters of Mistassini*, a chronicle of a season in the bush at Sam Blacksmith's camp, which we had now completed. Every white audience that had seen it seemed to have enjoyed it, but nothing had prepared us for the experience of watching the film in the presence of men who had only the day before left their bush camps. The atmosphere was electric from beginning to end. They discovered pleasures and jokes in the film which we did not know existed, and the auditorium was abuzz to such an extent that much of the English commentary was completely drowned. An absorbed silence or roars of pleasure and amusement, however, met the long sections in Cree. It was the first time these people had ever seen a film which had anything to do with their own lives. "We will see that film a hundred times," a young hunter told me.

ISAIAH SPEAKS

On the second day Isaiah Awashish spoke, an extraordinary sight in his usual old maroon windbreaker, slightly stooped, his hair and thick eyebrows jet black in spite of his advancing age, his copper-colored face lined and marked by the years of travail he had endured for his land.

"The people from Montreal," he said, in a gesture of reconciliation toward those young men whose ways had become so mysterious in recent years, but who were now speaking so well for their fathers in Montreal, "have come to help us, they have come to hear what we have to say, and we should tell the lawyers what we think. They want to hear from us what we feel about the land, how we love our land, and what we hope for our children on the land in the future." A person did not have to bring things to this meeting in order to be believed: he had only to talk and he would be believed. Everyone should be confident in what he said. For all had come together to get help from each other. The money they had been offered, "we must realize what it is," said Isaiah. "The money is really nothing. The land is the most important thing of all. It is what everyone here has survived on, and we cannot sell it. We cannot exchange money for our land. That way cannot be. In ten years the money, maybe, will all be gone."

The young men who now had to sit day after day in wearying meetings with bureaucrats knew that the purpose of the negotiations was to alienate the Indian rights in the huge territory of northern Quebec so that the white man could be free to develop it. They knew they would have to negotiate about money, about economic development, about management skills and job training, things that were going to be forced on their people by the inexorable movement of history. But after the March meetings the main thing for them was to be faithful to the trust their elders placed in them. In their negotiations they were determined to try to create the conditions—through land grants, a hunting regime and environmental controls—which would enable the culture to survive.

16

BACK TO COURT

The negotiations for a political settlement had continued through the summer of 1974, when the two sides were called back into court. After eight months the Quebec Court of Appeals was ready to consider the appeal of the corporations against the merits of the Malouf judgment. At the negotiating table the Indians were experiencing the velvet glove of the government's intentions; the arguments the corporations now produced in court and the response they elicited from the Appeal Court were the mailed fist within. Outside, the government was recognizing the reality of Indian rights and of its obligation to deal with them. In court it now denied that there was a subsistence hunting culture in James Bay, argued that Indian rights were totally non-existent, insisted on the need of the Cree people for a brutal shock, and claimed that the project was the essential instrument for the transformation and survival of Cree culture.

Five Appeal Court judges were put on the case this time, two of whom had made previous judgments against the Indians. One of these, Mr. Justice Jean Turgeon, had been a member of the three-man bench which swept the Malouf judgment peremptorily aside eight months before. By now the case had moved into an atmosphere as remote as it possibly could be from the hunting life with which it was concerned. The five judges sat in the handsome paneled courtroom, each with a small library behind him of the ninety-eight volumes of evidence, legal argument, plans, documents and exhibits which had been prepared for the higher court. The court was full of soberly clad legal figures whose combined salaries for the day's argument must have been very much more than either Job Bearskin or Sam Blacksmith earned from a year of hunting and trapping.

For those with a feeling for such things, the cultural paradoxes which had always underlain this case seemed, if anything, to have been heightened. The James Bay hunters, in town for a meeting, filed into the courtroom and sat down, clad in the same ill-fitting and well-worn jackets, the same rudely tied boots, their faces the same ruddy color, their expressions still open and wondering. Headed by O'Reilly, the lawyers for the Indians, English-speaking and younger than their adversaries, seemed to stand a little outside that atmosphere of solid bourgeois respectability uniting the appellants and the judges. The corporation lawyers were there in force: Emery, joky and saturnine; LeBel, plump and bushy, still looking not quite on top of things; oil-company agent Boulanger, peering suspiciously from behind his thick lenses as if trying to penetrate the hoax being visited on Quebec society by these Indians; Thibodeau, elegant, detached, far above this particular battle. These lawyers, encouraged by the community of feeling they must have felt with the three judges of the Appeal Court at their last hearing, now decided to brazen it out with their original arguments and to concentrate their attack on the waywardness and capriciousness of Malouf's judgment. For the sake of appearances, this time there were two English-speaking judges on the five-man panel, and one of them, Mr. Justice George Owen, presided. But when, five months later, the court came to write its judgment, it was left to Mr. Justice Turgeon to expand to forty pages the four-page rejection of Malouf and all his works that he had written in November 1973. The other judges agreed with him.

ONSLAUGHT ON MALOUF

The heart of the Appeal Court exercise lay in the four-volume factum which each side had prepared, in effect two 500-page books written by the lawyers on either side making their arguments from every possible angle. It was in this factum that the corporations now, for the first time, brought their case into a coherent form. If their denial of the existence of a subsistence hunting culture was to carry any conviction, they had to discredit Mr. Justice Malouf. They were not sparing in their efforts to do so. They accused him of ignoring testimony given by their witnesses as to the effects on the animals, and, when he had taken it into account, of paying attention only to testimony given under cross-examination. He made "numerous errors," they claimed, in his account of the effects of the project on sedimentation and erosion, he made "a grave error" in his account of the effects on the beaver, he gave "a false account" of the testimony of one of their major witnesses, he was "guilty of manifest exaggeration" in saying that the work would adversely affect the birds, and in other sections of his judgment made "false interpretations," "many mistakes," showed "lack of comprehension" of their arguments, "lack of objectivity" in his summary of evidence and "illogicality."

FRONTIER OF THE QUEBECOIS

The corporations' view of James Bay, now produced to the Appeal Court, was not of an Indian homeland, but of a Quebec frontier. The Bill 50 area, they said, had a population of 27,000, of whom 21,000 were white. The Indian population was only 5,309, and "it is interesting to note" that 3,808 of them, or 72 percent, were less than thirty years of age. The territory already had a considerable infra-structure, with 300 miles of main roads, 700 miles of secondary and forest roads and nearly 150 miles of railway track. There were several small airports, twenty companies exploiting the forest in the southern part of the territory, seven sawmills and one pulp-and-paper mill, and these companies spent $8 million a year on salaries and provided jobs for 620 people. Nine mines extracted 30 million tons of copper, zinc, gold and silver a year, including 40 percent of the province's total production of copper and 12 percent of zinc. These nine mines, with 3,465 employees, produced minerals worth $378 million a year. Another 2,500 persons were employed in hotels, garages, hospitals and government services, and all of these activities in the Bill 50 area represented an investment of $1.2 billion.

Curiously enough, in his judgment Mr. Justice Turgeon used exactly the same description of the territory, including even the parenthetical note of interest in the numbers of Cree people who were less than thirty years of age.

JUST LIKE WHITES

The corporation factum gave a similarly statistical and bureaucratic view of Indian life. Trapping had already been completely institutionalized by the government when it stepped in following the disappearance of the beaver in 1932 and created the beaver reserves. There were now 273 traplines, of which 203 would be completely untouched by the work undertaken on the La Grande River and the building of the roads. Of the seventy which would be affected, forty-four would be affected over only

0.1 percent of their area, and another one over only 0.5 percent of its area. Of the sixty-two traplines which would be permanently affected, only fifteen would be affected over more than 5 percent of their surface. Besides, the number of trappers was diminishing, while the number of Indians in work was increasing, having already reached 267 out of 268 families in Fort George. The Indians were using a very wide range of services such as health, education and housing, provided by the government at a cost of $20,631,444 between 1968 and 1973, an expenditure which reached $12 million in a single year by 1973. Because of all these activities there had been "considerable modifications" in the way of life of Indians, who now had many Skidoos, and houses with electricity, refrigerators, radios, beds, furniture, dishes and even telephones.

Mr. Justice Turgeon for the Appeal Court had no difficulty in accepting any of this. He quoted all the facts and figures as interpreted in the corporation's factum, and added some pithy summaries: "A considerable number of Indians occupy interesting jobs, and do not give themselves over to hunting and fishing except as recreation. . . . For means of transportation on lakes and rivers they use canoes with outboard motors, and no longer paddle. . . . In summary, the Indians and Inuit have abandoned the way of life of their ancestors and have adopted that of the whites. These facts are the reality, and I apologize for displeasing those who take pleasure in speaking of the question . . . with emotion and romanticism."

The corporation's factum contained page after page of the evidence of Therese Pageau and half a dozen other witnesses, who testified that the Indians bought between 75 and 80 percent of their food from the store, that, indeed, they spent $24 per family a week on store-bought food. The judge had made an error, said the factum, in generalizing from the figures of Feit and Tanner, which were based on very small groups, and which, in any case, covered only the winter season and did not take into account the Indians' increased dependency on store food during the summer.

The Appeal Court accepted this: the figures of Tanner and Feit, they agreed, were based on "very small samples and incomplete information," whereas the witnesses produced by the corporations had "lived in the territory for many years and in various settlements . . . have travelled a great deal in the territory, and have lived closely with the native people. . . . These witnesses . . . furnished a preponderant proof with respect to the lack of importance of country food in the diet of the Indians," wrote Mr. Justice Turgeon, with the agreement of the four other judges. "The native people eat as do people inhabiting the urban centres. . . .

They have moved rapidly towards a way of life similar to that of all Quebecers."

The court even accepted the famous estimate of $10,167 a year as a *bona fide* figure for the income of families in Fort George, making no reference to Malouf's critical analysis. The corporation concluded that "it is difficult to find in the proof anything justifying a great dependence by the autochthones on the land . . . the proof shows that [hunting and trapping] are of little importance and constitute perhaps a way of life for a very small number of them . . . in any case the works and projects will not interfere with these occupations." The sentiment was echoed, in as many words, by the appellate judges: "It appears that the works undertaken will not adversely affect the . . . activities of respondents."

A SALUTARY SHOCK

Anthropologist Paul Bertrand's evidence provided the main intellectual justification for the corporations' attitude to the Indians. Bertrand, said the factum, had grasped that the Cree could no longer live in a closed society but could "progress" only by suffering the assaults and impacts of change.

"In fact the James Bay project represents for the Indian culture its main cohesive tool, and the salutary shock that will permit it to rediscover its identity and its personality. . . . The development of the James Bay territory, then, from the cultural point of view, has a doubly positive aspect: on the one hand it will create a salutary shock to the autochthones which will allow them to take cognisance again of their originality and stop the slow and continual withering that their culture has suffered since generations; and on the other hand, it will bring with it men and knowledge that can help in the elaboration of the necessary policies of transformation. These aspects, which are essential for the future of the Indians, should have been noticed by the judge, who stopped at the notion of conflicts without trying to see in the proof all that these conflicts could bring of a positive nature."

STARTLING NEW EVIDENCE

So assiduous were the corporation lawyers to prove that the Indian culture was of doubtful value that they now reached far back into the past to prove that the Cree and Eskimos were actually a warlike, belligerent people. A writer in the *Jesuit Relations* (1633–1677), Gabriel Marest, had written that "the Kriqs are . . . lively, always in action, always dancing or singing. They are brave and love war." And in 1720 a writer published in Amsterdam, Nicholas J. Lamontagne, had written of the Eskimos: "They have this in common with the country that they occupy, that they are so wild and so intractable that no one has been able up to now to attract them to any commerce. They make war on all their neighbors, and when they kill or capture some of their enemies, they eat them raw and drink the blood." The lawyers solemnly produced these revelations to the Appeal Court as new evidence.

GOOD FOR THE ENVIRONMENT

The arguments of the opposing sides produced a confusion of statistics. O'Reilly's factum said the La Grande complex, by flooding an additional 3,407 square miles of land, would increase the water area of the entire drainage basin from the present 15 percent to 20 percent, an increase of a third. Moreover, since the wetland habitats on which the animals depend are near water bodies, a high percentage of them would be destroyed by flooding: 63 percent in the LG-1 area, 46 percent in LG-2 and 54 percent in LG-3. Some 50 percent of all wetland habitat now available along the La Grande River and its tributaries would be flooded. These facts led the judge to speak of the "devastating" effects of the project on the animals and Indians. But they came out differently in the corporation factum. Of the 135,228 square miles of the Bill 50 area, 14.8 percent was naturally covered with water, to which the project would add only 1.85 percent.

Once again the Appeal Court had no difficulty in accepting the corporation's version of the figures, and they agreed, too, that Malouf's opinion that the flooded lands were of extreme importance to the Indians was totally without foundation. Indeed, Mr. Justice Turgeon went much further. The scientific witnesses for the Indians had mostly testified "on matters that were outside their field," whereas those for the corporations had been experts of great experience who had spoken about what had happened in other reservoirs in Quebec and Russia. The adverse effects of the project on the environment, he wrote, had been greatly exaggerated. The corporation claimed that its proof demonstrated that beaver, caribou, seals, fish and birds would be little affected by the works. But in Mr. Justice Turgeon's opinion, the dams and reservoirs would actually increase beaver and fish populations, the reduction in the spring floods of the rivers would be beneficial, the speed of regeneration would be aided because forest fires would be brought under control and would decline in number, very little wetland habitat would be flooded, the caribou, geese and ducks would not be adversely affected and, in general, "there is no positive proof that the management of the James Bay territory will not render the ecological modifications beneficial as a whole."

The Appeal Court again quoted the figures produced by the corporation: only twenty-four trappers had reported fur sales in the fifteen traplines which would be affected over more than 5 percent of their area. Against this tiny number of negligibly affected people, one must weigh the costs of stopping the project, placed by the corporation at anywhere between $265 million and $701 million, as well as the need for electricity of 6 million Quebec people.

THE DEADLY WORK OF A KING

On the question of Indian rights the corporation factum returned to the initial arguments in the first days of the case: in 1670, when King Charles II gave the charter to the Hudson's Bay Company, no reference was made to Indian rights, and the company obtained the territory ceded by the King "without restriction, and in full proprietorship." Thus if any Indian right had existed, it was surrendered, legally speaking, from that moment, "whether by discovery, occupation, conquest or royal decision." Again Mr. Justice Turgeon concurred: "All aboriginal right in the territory, if it ever existed, was extinguished by the decision of the King." After quoting aspects of the corporation's historical argument, he observed that "I do not wish to bind the judge who will hear the case on the merits. I will content myself with expressing serious doubt concerning the rights of [the Indians]."

The same arguments held true of the Inuit, only more so, concluded the Appeal Court. The government of Canada had always considered them different from Indians, and had treated them differently. And they could not claim any rights pursuant to various acts or orders-in-council, resolutions, addresses, legislation or jurisprudence.

The four supporting judges, in their brief opinions, had little to add: Mr. Justice George Owen said he found the Indian rights in the area far from clear and the inconvenience (to the corporations) resulting from a stoppage of the work would be far greater than the inconvenience (to the Indians) in the event that work on the project was allowed to continue. Mr. Justice Marcel Crete found the Indian right doubtful and, in any case, not that of a proprietor but merely a right to live on the territory by trapping, hunting and fishing. Considering the size of the territory, twice as big as England, 60 percent the size of France, and the fact that the works would occupy only 1.85 percent of it, he was not convinced that the inconvenience to the Indians was on the same scale as the "growing need for energy of all of Quebec." Mr. Justice Francois Lajoie maintained that the *status quo* which Malouf had wanted to maintain by his injunction was the opposite of what he supposed—the *status quo* was the application of the law under which the James Bay work was being executed, not the suspension of its effects by the judicial branch. And Mr. Justice Fred Kaufman, like Owen, said the rights of the Indians were "far from clear," the flooding was still four years away and the corporations should be allowed to continue their work in the hope that the basic issues in the case would have been decided before the flooding occurred. It was as if those months of testimony in Malouf's courtroom had never been heard.

17

SETTLEMENT

The Crees felt they had no alternative. The project was already being built in spite of their protests. The Quebec Court of Appeals was quite evidently set against them. And on October 14, 1974, after months of wearying negotiation with the Quebec government, they were confronted with an ultimatum, a final offer they must accept or reject by November 10.

In that month of decision, they hired helicopters to fly the hunters out of their bush camps once again so they could ratify an agreement-in-principle. At all of the earlier meetings the hunters had protested that the land settlement was too small to permit the traditional life to be protected. But now, each of the eight Cree communities agreed that the Grand Council of the Crees should be authorized to sign the Agreement. The wording followed closely the classic renunciations made in Canadian treaties of the past: ". . . in consideration of the rights and benefits herein set forth . . . the James Bay Crees and the Inuit of Quebec undertake to cede, surrender, and convey all their claims, rights, titles and interests in and to land in Quebec, whatever they may be. . . ."

Chief Billy Diamond, announcing the settlement at a press conference in Montreal, hovered between jubilation and regret. It was a "big victory," he said, for the six thousand Crees against the six million Quebecers. Yet the Cree people had been "very reluctant" to sign the Agreement giving them this big victory.

"The Cree people all understand that the province must be allowed to build the hydro-electric project in the James Bay area," he said sadly. "We realize that many of the friends we have made during our opposition to the project will label us sellouts. . . . I hope you can all understand our feelings, that it has been a tough fight, and our people are still very much opposed to the project, but they realize that they must share the resources."

A complex agreement

It took another year of hard negotiation before the James Bay and Northern Quebec Agreement (JBNQA) was formally signed on November 15, 1975.

The Agreement was certainly the most complicated and far-reaching ever signed with a native group in Canada. Billy Diamond, however, was correct when he prophesized that the Crees would be labelled as "sellouts" for signing it. Objections of other native people in the country centered on the surrender by the Crees of their rights in their traditional lands, according to the formula used in earlier treaties.

Other natives felt that this would jeopardize their own efforts to negotiate native self-government on the basis of their aboriginal rights. The Agreement was signed just as native groups across Canada were beginning to insist that their aboriginal rights stem from their pre-European occupancy and ownership of the land, and are thus inalienable to any non-native government or authority. Fifteen years later they have begun to get some powerful encouragement in that view from decisions of the Supreme Court of Canada. These decisions have moved native opinion even further from acceptance of the surrender provisions of the James Bay Agreement.

Because of the surrender provisions, of course, the Canadian government hailed the James Bay Agreement as the desirable model for all future land agreements with native people; and in the following decade and a half have acted on that assumption.

The hunters return to the bush

Having made the decision to sign, the hunters went back into the bush in the winter of 1974–75, leaving their young political leaders to cross the t's and dot the i's of the Agreement. Deep in the wilderness, out of contact with the rest of the world, they resumed the trapping of beaver for fur, the killing of game for food.

Up in Fort George, Job Bearskin, who had never missed a winter on the La Grande River, went as usual to join the hunting camp of Clifford Bearskin at Kanaaupscow, blissfully unaware that his evidence before Mr.

Justice Malouf had rapidly been accepted as a classic statement of an indigenous people's love for the land. "I would rather think about the land and the children. . . . The money means nothing." While Job tramped the land he knew so intimately, and his wife Mary prepared the food he brought back to the lodge, Job's words were going around the world. His warnings of the folly of destroying the land were quoted to the Stockholm Conference on the Human Environment, and elsewhere. But, ironically, they were having no effect at home: he would have few winters left before the land at Kanaaupscow was drowned under seventy feet of water — and with it the fabled story rock of the Crees, on which they had for generations gathered to tell stories and watch for their colleagues returning along the river from the interior.

Sam Blacksmith returned to Lac Trefart after having allowed his land to rest for one year. This year he invited a different man, George Brien, to hunt with him. The year in the village had not been a good one for Sam. He had drunk a lot (he gave up drinking a year or two later), and failed to make arrangements for the succession of his land. But since there was no work available in Mistassini during the winter, there was nothing to keep him there a second year. He felt he was coming to the end of his hunting life, but he returned to the bush with a feeling of relief: "A man will return to the land, even when things are hard. This is the man who truly respects the land."

Ronnie Jolly had not stayed around Mistassini in the winter following our visits to Lac Trefart. He had gone to his own land, farther north, but had found that in spite of its two-year rest the land was still not rich in beaver. Ronnie's perception that the beaver was entering a downswing in its cycle was confirmed. This year he accepted the invitation of a hunter called Joseph Ottereyes to join him in his land near Lake Evans, 130 miles west of Mistassini, in territory usually occupied by the Waswanipi band. Ottereyes had not trapped this land for some years, and Ronnie was hoping for a better result than he would obtain on his own territory. (Lake Evans emerged as a central component of a revived Nottaway–Broadback–Rupert hydro-project, fifteen years later). Eddie Jolly, now a confirmed hunter, went along with his father, after working as usual as a guide during the summer at a tourist camp near Mistassini.

Isaiah Awashish, that proud, strong and lonely figure, chose for almost the first time in his life not to return to his own land in the headwaters of the Rupert River. His brother Sam, to whom he had given his drum, was now to be the owner, or *auchimau* of the Awashish territory, and Isaiah wanted his brother to exercise his new responsibilities alone. Isaiah had never been invited to trap on the land of any other Mistassini hunter, an omission that must have had something to do with his somewhat stern and self-contained personality. But now Isaiah, too, was asked to join Joseph Ottereyes and Ronnie Jolly at Lake Evans, taking along with him his fourth son Kenny, fourteen, who left school to join the hunting life.

Two months before the hunters left for the bush, the whole Mistassini community was saddened by the accidental death of Isaiah's splendid hunting son Willie, who had ranged with his father over the family's hunting territory from the time he was eleven years of age. Willie had never gone back to school, but under his father's tutelage had grown into a young man with unusual qualities of physical and spiritual strength, and of commanding personality and character. A little more than a year after he'd left his father, walking in mid-summer with some friends across a shallow, muddy lake near Mistassini, he was sucked into a sump hole. His friends could not pull him out.

At the time his father was fishing at the head of Lake Mistassini. People in the village, conscious of the special bond between the two, asked the traditional chief, old Isaac Schecapio, to go up the lake to break the news to Isaiah of the death of his third son, who the father had dreamed would be a hunter. Willie was the eighth of Isaiah's twelve children to die.

Isaiah, though deeply distressed, took charge of the funeral, spoke Cree prayers for his son, and sang a traditional hymn for him.

"My father has accepted it," said Philip. "I am very proud of my father. He is a strong man."

Making the best of it

Meantime, back in contact with the government, the young Cree leaders realized that their toughest job lay ahead, to make the best of the Agreement that had been forced on them, to insist that the governments implement it fairly, honestly, and completely.

"We believe that the Agreement is the best way to protect our land from white man's intrusions," said Billy Diamond. "We have always said that we wanted to maintain our way of life and pass on our land to the children. We believe this Agreement guarantees the future of our children, and also that we can continue to live in harmony with nature."

Fifteen years later he said he could read the desperation of the Cree people in the haunted eyes of the young and the elders, and added: "If I had known in 1975 what I know now about the way solemn commitments become twisted and interpreted, I would have refused to sign the Agreement."

The Crees were "embarrassed" about having signed the Agreement, he said. Hydro-Quebec, the governments of Canada and Quebec, had broken their solemn treaty, and in the 1990s proposed to destroy the Crees and "utterly erase us from northern Quebec."

"We are standing up," he said. "We will not be destroyed. We will not let them do it."

A MODERN INDIAN TREATY

The major provisions of the James Bay and Northern Quebec Agreement signed in 1975 are:

MONEY

A total cash settlement of $225,000,000, payable over twenty years, the Crees to receive $90 million in cash and $45 million in Quebec bonds; the Inuit $60 million in cash and $30 million in bonds. This money was to come in three payments of $75 million, the first over 10 years, the second from the James Bay Energy Corporation as royalties on electricity production, and the third in the form of Quebec government debentures as compensation for fifty years of mining royalties.

To administer these funds a Board of Compensation was established under provincial legislation setting up a Cree Regional Authority. A number of restrictions were put in place to ensure that the compensation money be used for the benefit of the Cree community as a whole, and not for the personal benefit of individual Crees.

By 1990 some $137 million had been paid to the Crees under these provisions, although not without serious disagreements, threatened lawsuits, committees of inquiry and the like. These controversies became so intense that within five years of the signing of the Agreement, the Canadian government was forced to carry out a major review of how the Agreement was being implemented. The review concluded that because of the pressure-cooker atmosphere in which the Agreement was signed, "many provisions are vague, ambiguous and open to widely varying interpretations."

For instance, the Crees and Inuit both claimed that once the Agreement was signed Canada began to eliminate or reduce services that the native people were entitled to in housing, community infrastructure, and health and education. The natives also said Canada had failed to promote economic development in the communities. As a result of this review, some $32.4 million of additional federal money was made available to the Crees, and a similar amount to the Inuit.

To these sums was added $110 million ($15 million in cash, $95 million in Hydro-Quebec bonds) when the Crees agreed in 1986 to the construction of four additional power houses on the La Grande River. To manage these extra funds, the James Bay Eeyou Corporation was established.

In 1979, $25.5 million was paid to the people of the coastal village of Wemindji, to compensate for additional flooding around Sakami Lake.

The original Agreement provided a guaranteed income for Cree hunters and trappers who spent a minimum of 120 days a year hunting and trapping, of which 90 days were away from the settlement.

By 1989 this Income Security Programme for Cree Hunters and Trappers had paid out $100 million in benefits, covering almost exactly a third of the entire Cree population. The original limit of 120 days was later doubled to 240 days; and the global limit of 150,000 person-days for all Cree trappers and hunters (set according to the practices of the early 1970s) was expanded until by the end of the 1980s the program was paying out on 340,000 person-days a year.

LAND

Four categories of land were established under the Agreement.

Category IA:

Crees: 1,274 square miles of land in and around existing communities were designated for the exclusive use and benefit of the Crees. Title to the land was retained by the province, but administration, management and control were held under federal jurisdiction. These lands were subject to the Indian Act, but not as "Indian reserves" within the meaning of that Act.

Inuit: 3,130 square miles were to be selected for requirements of existing Inuit communities, title to be held by Inuit municipal corporations.

Category IB:

Crees: 884 square miles were designated to be owned by Cree communities. These lands are to be alienable only to the province, which can expropriate, but must pay compensation. Mineral rights are to be retained by the province, but their development is dependent on consent of Cree authorities.

This division of Category I lands apparently arose from the extreme reluctance of Quebec to agree to establishment of areas of federal jurisdiction within the province.

In 1978, 420 Naskapi living near Schefferville were, in effect, written into the James Bay and Northern Quebec Agreement, and in 1984 the Canadian government passed the Cree-Naskapi (of Quebec) Act, establishing a municipal form of government covering the Category IA lands. This Act replaces the Indian Act, and removes the Crees and Naskapi from the direct control of the federal Minister of Indian Affairs.

The Act established a Commission charged with reviewing implementation of the Act; within a year or two the Crees were complaining that the federal government, though handing over many duties and powers to the Cree communities, had refused to make available adequate finances with which to do the work. The Commission, headed by a Quebec judge, Mr. Justice Rejean Paul, in its 1988 annual report agreed with the Cree contention that they had been forced to use compensation monies to pay for rights and benefits that all other citizens, and all other Indians of Canada are entitled to. "Cree financial statements clearly demonstrate that compensation money has been used to meet what the Crees regard as obligations of the government of Canada," commented the Commission.

Category II:

For the Crees, 28,130 square miles, and for the Inuit, 35,000 square miles were set aside exclusively for Cree and Inuit hunting, fishing and trapping. These lands are available to the province for development without the consent of the natives, but subject to replacement or compensation. Non-native people are allowed to enter for development, scientific or administrative purposes, but may not hunt.

These lands are subject to public servitude without compensation, and the native people must agree to grant forestry concessions, timber limits or equivalent wood rights. Fifteen years later this provision of the Agreement poses a major threat to the Cree way of life. Logging companies have begun intensive clear-cutting operations on many traplines in the southern part of the territory, and the Crees have no authority to stop them, even on Category II lands that appear on modern maps as Indian lands.

The province can take for development without compensation 5,000 square miles from the Inuit lands.

The total of Category I and II lands is 41,793,000 acres, compared with the 6,000,000 acres of all other Indian reserves in Canada. But the rights granted in most of these lands are heavily circumscribed. Under the settlement, the Mistassini hunting territory has shrunk from 100,000 square miles to 7,000 square miles, and even in that area they are granted exclusive rights only "so long as it is physically possible and does not conflict with other physical activity or public safety".

Category III:

The rest of the lands of northern Quebec — 350,000 square miles — are surrendered by the native people, and are available to the province for development.

ADMINISTRATION

In addition to a complex structure of village and municipal government, the Agreement has given rise to four separate political regimes involving Canada, Quebec and the Crees:

1. The land regime, already described.
2. The hunting, fishing and trapping regime
This covers the whole territory, including Category III lands, and is governed by a co-ordinating committee of eight native and eight government representatives.

The decisions of this committee, according to the Agreement, are to be governed by the principle that native harvesting of wild animals for subsistence has priority over non-native sports hunting.

The natives have the exclusive right to trap a wide range of wild game in all of northern Quebec: mink, ermine, weasel, marten, fisher, otter, skunk, wolverine, beaver, lynx, bobcat, fox, coyote, squirrel, polar bear, raccoon, muskrat, marmot, musk-ox, porcupine and lemming among the animals, whitefish, sturgeon, sucker, barbot, mooneye and goldeye among the fish.

The major animals excluded from this list are moose and caribou, bear and wolves, and, among the fishes, trout.

The committee has the right to control the number of non-natives permitted to hunt and fish in Category III land, but effective control has proven to be almost non-existent. By 1990 the Crees estimated that 40,000 non-native hunters were using what was once their exclusive hunting territory every year.

Native people were also granted under this regime exclusive right to open and operate hunting camps for tourists in Category I and II lands, and first option for twenty years in Category III lands. Crees have complained that they have not been given first option, and that in fact the James Bay Development Corporation has recently begun overtly to favor Quebec businessmen in the awarding of contracts, business opportunities, and so on. This led in 1989 to a court challenge.

3. The environmental and social protection regime
This is governed by an Environmental Advisory Committee, on which the Crees have four representatives, the governments, eight. This committee is supposed to review and oversee the environmental regime established by the Agreement and advise Quebec on forestry management, but in effect, the committee has been inoperative during most of the fifteen years since the Agreement was signed.

Sections 22 and 23 of the Agreement provide for environmental and social impact assessments for development projects in the whole territory. This system is backed by an Evaluating Committee (two Crees, four government representatives), responsible for recommending on the extent, orientation and content of any necessary environmental impact study, and a Review Panel (two Crees, three government representatives) to study and analyze the impact report.

Confronted with the prospect of an expanded hydro-electric project in the late eighties, Cree leaders began to realize that the provisions for the environmental regime are so weak that they leave Hydro-Quebec virtually free to do its own assessments, without public hearings or independent review.

4. An economic development regime

This body is governed by tripartite committees to coordinate federal and provincial support for the Crees and Inuit in the areas, among others, of renewable resources, arts, and crafts. This regime has also been almost completely inactive.

As can be glimpsed from this list, a huge bureaucracy, conceived in terms of non-native norms, has been imposed on the few thousand people of northern Quebec by the Agreement. The major new administrative bodies set up for the 6,000 Crees include:

The Cree Regional Authority, to coordinate the work of village band councils and other regional Cree organizations. Though this is an administrative body, it has been treated by the Crees as identical with their political organization, the Grand Council of the Crees;

Eight *Cree village corporations* to exercise powers over Category IA lands, under the jurisdiction of the Cree-Naskapi Act; and, separately, over Category IB lands under the Quebec Cities and Towns Act;

Cree Landholding Corporations, to take full ownership of Category IB lands;

The James Bay Regional Zone Council, to exercise municipal powers over Category II lands;

The Cree School Board, to run all Cree schools, funded by the Quebec Education Ministry;

The Cree Regional Board of Health and Social Services, funded by the Quebec Ministry of Social Affairs;

James Bay Native Development Committee;

Cree Trappers' Association;

Cree Native Arts and Crafts Association;

Joint Economic and Community Development Committee;

Board of Compensation;

James Bay Eeyou Corporation.

This immense bureaucratic structure, as outlined in the Agreement, created more than forty separate legal entities to govern the eight Cree villages, a structure that now provides an elected, appointed or salaried role for almost half the working population of the Cree nation.

EPILOGUE 1991:

So progress has come to the land of the Cree hunters.

John Ciaccia, the Quebec provincial minister who negotiated the Agreement in 1975, stated clearly that the objective of the Agreement was to open the Quebec north to development, and that development has occurred.

The first power station of the hydro-electric project, which was already being built in the early 1970s as the Crees pursued their fruitless court battle, was completed in 1979 and opened to the accompaniment of a huge public relations fiesta, organized by the James Bay Development Corporation. Journalists were flown in from all over the world, and the television networks willingly set aside hour after hour for programs extolling the gigantic achievement.

The Canadian Broadcasting Corporation, the publicly owned network that in 1973 had refused to broadcast the film Job's Garden in which Job Bearskin expressed his attachment to the land and fears for the future, now made a new film whose opening words were almost identical to those used by Quebec Premier Robert Bourassa when he launched the project in 1971. The new film portrayed James Bay as a grim wilderness, empty, and virtually useless except for its power and resource potential. It was as if the Crees had never spoken of their land and their attachment to it.

Premier Bourassa himself was ejected from office by the electorate in November, 1976, leaving behind, it was generally agreed, the James Bay project as his major accomplishment and indelible monument. From its inception, the project had been wrapped in the fleur-de-lis, the flag of Quebec nationalism: more than any other project Quebecers had undertaken, this veritable marvel of high technology and industrial power embodied the 1970s slogan: "Quebec sait faire" — Quebec knows how. The nationalist pride in the project was reinforced by the provincial party that succeeded Bourassa in office, the Parti Quebecois, dedicated to the independence of Quebec. Performing the official opening of James Bay I, the new Premier, René Levesque, appeared full of nationalist pride, almost as if he had conceived the project himself.

INTO THE BUREAUCRATIC MAZE

But in the villages of James Bay, and in the boardrooms of Montreal and Quebec City, the young Crees who had mobilized their people for the fight now had to confront the complex task of giving flesh to the intricate and bewildering Agreement they had negotiated. To say that they found the federal and provincial governments slippery as they tried to pin them down to the letter and spirit of what had been agreed would be an understatement. Yet under the determined and sometimes bombastic leadership of Grand Chief Billy Diamond, the James Bay Crees became established in the minds of Canadians and Quebecers as a vigorous entity, a united people, and a community (a nation, in their own words), that was handling with skill and aplomb the transition from a subsistence economy and lifestyle to a successful integration with the Canadian mainstream. Whether reality in the villages of James Bay has ever matched that euphoric impression does not really worry the Canadian or Quebec public.

To have created this Cree Nation of James Bay has been, indeed, a heroic achievement by these young men and women, but one for which many of them have paid a great price in personal terms. They had not been educated for this life of politics and meetings; it was all worlds away from the life they grew into as children. The strain of trying to follow Canada's governmental structure through endless tunnels of evasion, prevarication and deception led many of these young leaders into alcoholism, drug addiction, family dissolution, divorce, and, for some, detoxification, reconciliation and rehabilitation, in what seemed like a pitiless maze of crisis and response.

These are problems held close to the chest within the Cree communities; yet the Grand Chief himself, Billy Diamond, lifted a corner on all this with extraordinary frankness in describing his personal traumas to his biographer Roy McGregor in 1989. (*Chief: the Fearless Vision of Billy Diamond*, Penguin Books.)

Beginning as a young village leader, one of the few in his community capable of relating to the white man's structures, Chief Diamond emerged eventually as politician, businessman, and fire-and-brimstone Pentecostal preacher, undergoing a personal evolution parallel with that of his anguished people, who in the fifteen years after the signing of the Agreement converted from the Anglican church to the Pentecostal in enormous numbers.

Billy Diamond became a sort of Lee Iacocca of the North, president of the Crees' own airline, Air Creebec, initiator of a groundbreaking deal with Yamaha of Japan for canoe construction, persistent scourge of governmental backsliding: a colorful, outspoken figure, plump, smiling, a man easy for Canadians to identify with as he thundered his imprecations against those who would hold back his people from attaining their promised dream.

CREE VILLAGES TRANSFORMED

By 1989 and 1990, when I returned to the James Bay villages after sixteen years to update the films I made in the early 1970s, certainly a great transformation had been wrought. Along the road from Chibougamau to Senneterre, right above the Waswanipi River, where the Rev. Muller introduced me to the four Gull brothers in 1969, a whole new village has arisen where the Waswanipi people have been able to come together at last, ending their two decades of dispersal. When the whites arrived with mines and new towns in the 1950s and 1960s, the Cree community on Waswanipi Lake had disintegrated, scattering its people over a wide area, plunging them into confusion and dismay.

Though far north of Montreal, the Waswanipi people are still some five hundred miles south of the James Bay project, and have not been affected by its reservoirs, dams and dikes. Yet the general opening up of the James Bay territory in the last two decades has confronted the Waswanipi people with even more difficult choices than those of the Crees further north. The roads have brought in the logging companies that have begun to hack the Waswanipi forests to pieces, and sports hunters who compete with the Crees for the fish and game.

Only 40 percent of the Waswanipi people are now dependent on traditional pursuits; the other 60 percent require paid work, wage employment, and the challenge is to meet the needs of both groups, defending the culture and traditions for everyone while finding ways to integrate into the wage economy. The two influences often seem to be on a collision course.

A BUSINESS-MINDED CHIEF

The new Waswanipi village cannot escape an intimate relationship with outsiders, even if the people wished it, which they do not. They now live on a main road, no longer gravelled and rough, but paved all the way, joining them to the outside world of neighboring French-speaking towns. For almost a decade they have followed the lead of a business-minded chief, Abel Kitchen, who has encouraged the villagers to create their own enterprises, hoping to compete with the surrounding towns on their own terms.

Thus, confronted with clear-cut logging, the Waswanipi people have established their own logging company, clear-cutting their own Category I lands. Their cutting is governed by principles of sustainability and takes account of the needs of the Cree trappers and hunters as well as wildlife habitat. Waswanipi has probably gone further than any other native community in Canada in identifying what the aboriginals who live and work in the forest need if they are to coexist with logging. Waswanipi's environmental coordinator, Sam Gull, has an office full of maps on which he has marked for every trapline the areas essential to the survival of moose, beaver, otter, waterfowl, spawning fish, and Cree hunting camps. He has drawn up these maps after detailed consultation with the trappers. He has then approached the Quebec government, asking that these areas be protected, and telling them that if forests are to be preserved for a variety of uses, then aboriginal use should have a high priority. In every case, he has been turned down: the companies have told the government, and the government has accepted, that such protections for aboriginal use, such respect for wildlife needs, would cost the logging companies too much.

So Waswanipi people, foiled in their effort to protect the forest for trappers, have begun to talk with a major logging company, Kruger, about a joint forestry enterprise, a proposal that arouses a certain nervousness among other Crees, further north, who are not so directly in the path of the logging monster.

Confronted with outside sportsmen fishing out their lakes, Waswanipi has created its own commercial fishing operation, first studying productivity patterns in some thirty lakes before establishing quotas and fishing plans designed to maintain production forever.

Waswanipi is closer to the surrounding French-speaking life of Quebec than are the other Cree villages. And though most of the Waswanipi Crees speak English as a second language, the Waswanipi schools are conducted 80 percent in French. This move into French is no doubt necessary if their people are to coexist with the Quebec culture; but the confusion of three languages within almost every Waswanipi family has added another level of tension for people already trying to cope with enormous changes.

Not surprisingly, perhaps, Waswanipi is a place that now boasts its own Lions Club, whose Cree members solemnly elect such office-holders as Tail Twister and Lion Tamer, raise money for good community works through the facetious Lions system of fines and jokes, and sponsor walka-thons to buy books for the school library. Is this what the future holds for all Cree villages?

AN EXPANDED, BUSTLING MISTISSINI

Waswanipi village has been built from scratch, and is a rather cheerless place, to tell the truth, with all the trees cleared to make way for row after row of suburban-type housing surrounded by sand; but one hundred miles further north, past Chibougamau, Mistassini (now, by decision of the band council given a new spelling, Mistissini, and a new pronunci-ation) has been almost completely rebuilt.

The summer meeting place on the shore of Mistissini lake that I first saw in 1969, with its crude log cabins, its Hudson's Bay store, its one restaurant, its small green shack housing the chief and band manager, has blossomed into a thriving town where Cree people at last enjoy the comfort and convenience of neat homes with "all mod cons": running water, electricity, central heating, an efficient telephone system, televi-sions and VCRs.

The band office shack has been replaced by a splendid new adminis-tration building, where Cree civil servants scurry back and forth with files under their arms as they organize all the services of what looks almost like an ordinary Canadian small town. The transformed houses, largely fi-nanced by government programs for low-income people that are avail-able to all Canadians, were built by an entity known as Cree Housing and are now being handed over to the ownership of the band council. The council also owns a company known as Mistissini Enterprises, which runs the village's own Waasheskum Airline (maintained primarily to transport hunters back and forth from their traplines), a network of tourist camps, the village radio station, and so on.

Charlie Brien, who ran a tiny restaurant when I first visited Mistissini in 1969, has now handed that over to his daughter, and has himself taken to spending most of his time in the bush on his family land, a rather surprising turn for the first entrepreneur that Mistissini ever produced.

There are now two restaurants, and the Hudson's Bay (now known as Northern Stores) has serious competition at last. No longer are the Cree villages trapped in what, twenty years ago, was a sort of Indian disease: the money that came into the village went straight into the coffers of the Hudson's Bay Company and so directly out of the village. Twenty years ago, Mistissini operated almost entirely on credit provided by the Hudson's Bay manager. He advanced hunters the money to pay for their flights into the bush; he brought the planes in from outside; he collected the furs; he checked off the fur prices against the credit he had advanced the trappers. Now, they service their own trapping and hunting season, using their own airline; their camps are organized by their own Cree Trappers' Association, which provides them with two-way radios, keeps in touch with them almost daily, and flies them back and forth between trapline and village far more frequently than was ever possible in the past.

THE HIGH PRICE OF PROGRESS

There is, however, a price to be paid for this creation of a normal Canadian village out of a gathering place for hunters and trappers. The new housing, the improved services, the running water and electricity, do not come cheap. Every family that occupies a house is now confronted with rather large payments — rents or, in some cases, mortgages — and bills at the end of each month, just as are people in other Canadian towns.

A town like Mistissini continues to be the center of a hunting culture, and in spite of the Income Security Programme for Hunters and Trappers, this is a culture still based mostly on subsistence harvesting of animals for food. There has been a great deal of paid work available as village infrastructures, such as houses and public buildings, have been put in place, and this work has given many who would prefer to be on the land an incentive to stay in the village, hammer in hand, earning the money to cover the month's bills. It is not altogether clear how these people will earn this kind of money when the building or re-building of the villages has been completed.

From one point of view, this new economic structure can seem like giving with one hand merely to take away with another. A hunter like Sam Blacksmith, recipient of the Income Security payments, says the payments make no difference, for though they give the hunters more money, that money is absorbed by the higher costs they must pay to live in the rebuilt villages. Many other hunters told me the same thing. Left out of this equation, however, is the fact that the money buys considerably better living standards than in the old days, and these better living standards have virtually eliminated the major diseases of two decades ago. Mistissini's main health problems now are diabetes, connected with the changeover from a diet of country food to store-bought food, and obesity (especially among women), caused by careless eating and lack of exercise.

MANY OLD FRIENDS

In Mistissini I was able to meet many old friends. Sam and Nancy Blacksmith are now living near their daughter Rosie in the new village of Nemaska. We met them by chance in the streets of Chibougamau, and I was amazed to find that even at age seventy-four, and after the hard life he has lived as a hunter and trapper, Sam seemed as vigorous as ever, and, if anything, even more handsome. Rosie, no longer shy and inarticulate but a trained social worker, the mother of three children, and a friendly, talkative character, does not now accompany her parents into the bush. But Sam and his wife have never missed a hunting season since we filmed them.

They had suffered a tragic loss seven years after our filming when their son Malick, who was fourteen when we were at Lac Trefart, was accidentally drowned. He was by then the father of three children. But Sam and Nancy had continued to bring up in their family the little boy Abraham, son of François Mianscum, who had been in the bush camp as a five-year-old in the year we made the film. Recently Sam had solved the problem of what to do with his land by handing responsibility for it over to Abraham, who, at the youthful age of twenty-four, becomes a tallyman and recognized "proprietor" of a hunting territory. This confirms a traditional Cree pattern for passing on the hunting territories. It has never been simply a question of handing it from father to son, or even necessarily within the family. The hunter has always handed on his land to the person he believes best suited to look after it.

SAD FATE OF THE VOYAGEURS

Malick's was not the only tragedy to have struck the families with whom we filmed in the 1970s. Abraham and Shirley Voyageur, their daughter Louise and one other daughter were drowned in a freak storm that blew up unexpectedly on Lake Mistissini. There was a particularly cruel irony in this: Abraham was one of the most competent and reliable persons I had ever met, the sort of man to whom one would have happily entrusted one's life. He appears in our film, running the rapids in his canoe at the very moment that the commentary observes that few Cree hunters learn to swim, but because of their supreme skills, Cree drownings are almost unheard of.

Ronnie Jolly died a year or two ago, but Mrs. Jolly remains in good health, still going into the bush every year because, as she told me when I met her in the village, "I just love it so much."

Among younger people, Philip Awashish, indispensible to all my Cree contacts as translator, interpreter and facilitator, had become one of the half dozen leading figures among the James Bay Cree, and had now returned to Mistissini, where he was once again a member of the band council and a political consultant. But though the Agreement and its implementation had come to be the dominating influence in Philip's life, his father Isaiah had not changed his style of life to accommodate the new realities. We planned, at Philip's invitation, to visit his hunting camp, but the September weather in James Bay is capricious, and we could not fly on that day. I particularly regretted that, because it was on Isaiah's land many years before that I had first experienced the unique ambience of a Cree hunting lodge.

JIMMY MIANSCUM STILL WAITS

The transformation of the Cree villages is not yet quite complete. The people on whose land the towns of Chibougamau and Chapais were built, during my first visit in 1969 were living in rude tents in Doré Lake (Chapter 4), or in makeshift shacks scattered along the main road; and in 1990 many of them were still living in similar conditions, in little wooden cabins along the main road between the two white towns. Over the years many of these people have tried to keep together in small, rough settlements, but they have been moved six times to make way for various mines or other developments imposed by incoming white settlers. Now known as the people of Ouge-Bougamau, they were, somehow, left out of the James Bay Agreement. Later the Canadian and Quebec governments recognized they had an obligation to these people. But the federal government, in particular, resisted for years fulfilling its obligation. Only in the last few days of 1990, as the federal government scurried to assuage a growing public concern about its incompetent handling of Indian affairs, did they agree to provide thirty million dollars for a new Ouge-Bougamau village, at a beautiful spot overlooking a lake not far from the town of Chapais.

Jimmy Mianscum, whose moving testament about the injustice of having his meat seized by game wardens so impressed me in 1969, when he was a resident of Doré Lake, is now the leading elder of the Ouge-

Bougamau. No longer a full-time drinker (indeed, like so many of the
Crees, no longer a drinker at all), he is still living in a small cabin along the
road — still wondering why, though an airport, transmission lines, roads
and forestry operations have invaded his family land, he has never received any compensation of any kind. It is still a good question, as it was
twenty years ago.

THE DRAMATIC EFFECTS OF THE PROJECT

One has to travel north to Eastmain, on the James Bay coast, before
observing the direct effects of the hydro-electric project. During the long
negotiations in the mid-seventies, the Quebec government managed to
get the Crees to agree that the Eastmain River be included in the La
Grande project; so it has been blocked off, its waters turned north into the
La Grande reservoir, where they pass through the LG2 power house, and
out to James Bay. This diversion of what was one of the great, raging
rivers of James Bay (and indeed of Canada) has had a devastating effect
on the hunters who for so long depended on the animals that, in turn,
depended on the wetlands around this river.

Naturally, the mouth of this once-great river, deprived of its fresh
water from the land, has become a broad saltwater estuary. Along the
many miles of this estuary, the beaver that were once plentiful have
disappeared, the supply of drinking water has dried up, the traveling
animals are no longer so easily found.

The hunters of Eastmain thought that the James Bay Agreement to
which they gave their assent in 1974 was designed, above all, to protect the
Cree way of life, the hunting culture, and the Cree language. Now they
feel that it has failed to do what they expected of it. They cannot understand why money paid to the Crees has been spent on buying an airline
and on other large capital expenditures when, year by year, they see the
viability of their Cree hunting culture diminishing. An old hunter like
Abraham Weapinacappo now says, "I am one of those who went to Quebec City to put my signature on the James Bay Agreement. I should not
have gone."

When we arrived in Eastmain with our cameras, the words burst out
of him like a torrent. He was wounded, hurt, he said, hammering his
hands against his chest: it was like he had been punched, when he saw
what had happened to the land and to the animals. With his friend
Charlie Mayappo, he marched on a bitterly cold September day across
the broad rocks that once were covered by the water that poured over the
first rapids of the Eastmain River, thirty miles or so above the village, and

he shouted at us against the wind, pounding one hand into another to make clear to us his abiding sense of betrayal, his pain at what the land he so loves has had to endure. It was, he said, the first time he had had a chance to say what he has to say. He talked for forty minutes, and afterwards insisted that in our film we must use all of what he had said.

Sadly, much of the dismay felt by these older men seems an inevitable consequence of the bind in which the Crees were trapped when confronted by the multi-billion dollar plans of the Quebec government. Only the young men, who could speak English, were able to talk directly to the government. They certainly tried their best to speak for their elders, to consult with them, and carry them along as they edged towards an agreement.

Years later it can be seen that the arrival of industrial civilization with such overpowering force, in the middle of a primeval wilderness and a culture of subsistence hunters, has placed that culture under pressures that threaten its survival. So it is hardly to be wondered at that those who embody the values of that ancient culture should feel betrayed and even abandoned.

One feels melancholy hearing a splendid couple like Mr. and Mrs. Willie Moses, of Eastmain, now in their seventies, assure us that if we return among them twenty years from now, we will find that the Cree hunting culture is no more. They feel their way of life is in decline. They see all the signs around them, in the village, where many of the younger people know little of the bush, and on the land, where the animals are under intolerable stress from the ever-increasing presence of Canadians and their works. Yet can it really be said that their culture has been betrayed by their leaders? For even within their own community, even among their own children — within one generation — the modern world has remorselessly intruded its values, and no one could hold it at bay, no matter how hard one tried.

One of their sons, Ted Moses, a former Grand Chief of the Crees, represents his people at the United Nations Social and Economic Council in Geneva. Now the Eastmain chief, he is almost as often in Geneva, putting the case for Canada's native people to other governments, as he is in his tiny village of four hundred people on the shores of James Bay. Similarly, his sister Bella Petawabano, the community health worker in Mistissini, is involved day after day in negotiations about budgets, about relations with provincial and federal governments, as she and others work to improve the welfare of their people. Their world is very different from that of their parents.

From Bella in Mistissini we carried a large bag of food to her parents in Eastmain. Farther north, when we met her brother Joe, working as a public relations man for Hydro-Quebec at the new LG1 dam being built on the La Grande River, he could not get rid of us quickly enough so that he could head south to meet with his parents for a weekend of hunting moose. Though his father seems old and walks with a cane, says Joe, he is still a hard man to keep up with in the bush.

A number of Crees have told Joe that he should not be working for Hydro-Quebec. But most accept, with some fatalism, that everyone, especially a family man with children, must have work, and must take it where he can find it. Carefully, however, Joe refused to be filmed while giving us his company spiel about the new dam and powerhouse. Many Crees have faced this same dilemma. Within a month or two of winning the court case in the early seventies, the Crees were given a contract to clear the forest from the proposed reservoirs. They have been taking similar work ever since.

Later in Chisasibi, Larry House, one of the leaders of a youth movement there, who was only five when the James Bay project was first announced, told us how as a young married man with a family he had no alternative but to take work clearing the forest perimeter of the LG1 reservoir that would flood some of the land that has been in his family for many generations.

Though they want to have nothing to do with the agency that they feel is destroying their way of life, their options are limited: they must pay the bills for their new homes, heating, electricity and the like. Having been sent away to school at a young age, they do not have the skills necessary to feed themselves in the bush. But Larry House had recently decided to refuse work with the project, and had turned down one job. Other young people seem to be moving in the same direction as they realize that only by root-and-branch opposition will they be able to fend off the enormous dangers posed to them by the Quebec government's new plans.

THE TENACITY OF THE CREE CULTURE

The Moses family, like many others in James Bay, spans within its own experience a cultural change of oceanic proportions. In this it is probably typical of most Cree families. Yet any easy assumption that the culture is dead should be resisted: for appearances are not always what they seem, and few Cree people have been really detached from the hunting culture and the values it embodies. Throughout North America, the Indians have been treated as a disappearing people; yet nowhere have they disappeared. Somehow, they are still around, and more vigorous than ever.

"Mistissini is not an ordinary Canadian small town," insists Bella Petawabano. It remains the center of a hunting culture, and essential to its health is the perception of the land as the foundation of everything, the knowledge that it will always be there, that it can be returned to at any time. The link between the land and the Crees remains unbroken, and ensuring the survival of that link, she says, should be the aim of all policy.

"Ours is a hunting culture," says even the business-minded former chief of Waswanipi, Abel Kitchen, who is now a partner in a large super-market that faces the main road. "Those are the values we must find a way to protect, because without those values, everything else means nothing."

"We have a clear sense of who we still are as a people," said Philip Awashish, when I met him in the comfortable new house he occupies in Mistissini. "We are a people with a special relationship with the land, a relationship that still ensures our survival. In that sense we know our place in nature. We are still basically a people who hunt and fish and trap and depend on the land."

He has never been a hunter himself, and has spent twenty years as a negotiator with governments. But he regards the position of the tallyman, the individual trapper, as central to the whole governing structure that has been built by the Crees.

The authority of the tallymen, he said, flowing from the traditional laws of the Cree nation, has been recognized by the state without the state determining what that authority is. He believes that in this way the Crees retain authority even over the land they have surrendered, though the nature of this authority may not be clearly defined.

This is more than just a lot of idle talk. No matter how acculturated the younger generation of Crees, including their leaders, may seem, a large number of them spend a lot of time in the bush, as much in search of spiritual reinforcement and emotional security as in the hunt for game. For this reason, statistics about the number of full-time hunters, the number of people in need of wage employment, the relative decline of the hunting economy, can be misleading.

ENDURING ATTACHMENT TO THE LAND

In the early seventies, on my first visit to Fort George, I had the impression that few people remained as full-time hunters, and within a few years the hunting life would be as good as dead in that village. Yet in the fall of 1990 when I was in Chisasibi, the new town that has replaced Fort George, very few people, whether young or old, remained in town. Most of them had fanned out across the countryside, and up and down the coast, into family goose camps.

Compared with two decades before, it was certainly hard to detect any diminution in the attachment of the Crees of that community to the hunting life. Young people in their twenties seemed more absorbed in the values of the culture than I expected them to be. I was surprised to find that people I had known years before as youths who were apparently already acculturated to the white lifestyle were now to be seen in the village only on infrequent visits from their camps in the countryside.

When I went out to a goose camp along the James Bay coast from the village, I found not only the middle-aged hunter Norman Sam, but also his three sons, young men who work from time to time at paid jobs, but who seem almost as expert as their father in the work of the goose hunt. Back in the tents and teepees, of course, was the whole family, from great-grandmother down to small children, rocking in traditional cradles, plucking the geese and ducks, cooking the bannock—learning the ways that their people have followed for generations.

Figures for the Income Security Programme for hunters and trappers of Chisasibi bear out this continuing attachment to the hunting life: more adults are covered by the program now than when it began in 1975. Though the percentage of the total population covered is somewhat lower, it has certainly not gone into a steep decline.

In short, rumors of the death of the Cree culture may be grossly exaggerated.

Nevertheless, the impact of the James Bay hydro-project on the people of Fort George-Chisasibi has been immense. The arrival in their region of up to twenty thousand workers to build the project, with all the amenities needed to sustain them, was a shock. From the beginning, liquor was a problem; and it remains so today, to such an extent that in 1989 the Crees of Chisasibi, with the support of their first woman chief, Violet Pachanos, established a permanent blockade on the road outside the town. They man the blockade twenty-four hours a day, check every incoming vehicle, and confiscate and destroy all liquor found. The Quebec police say the Crees have no legal right to carry out this blockade and make these confiscations, since the road itself is classified as Category III land and is therefore a public road. But they have done nothing to dismantle the blockade, and the villagers are unanimous that the road-block has greatly reduced the amount of liquor entering Chisasibi and has begun to bring drunkenness under control.

After my visit to Chisasibi in 1989 I wrote an article about the town in the magazine, Canadian Forum. "I was there for the weekend," I wrote.

"One youth tried to commit suicide. One girl took an overdose of pills. One young man was found passed out on the river shoreline, half in and half out of the water. Three teenagers ran a van into the river and had to be treated in the hospital. A woman who had been badly beaten in a drinking quarrel in a nearby house came and asked to be driven home. The bootleggers circled the town selling the beer from their vans. In the week after I left there were four more attempted suicides."

It was in response to this deteriorating, and alarming, environment that the roadblock was set up outside of town, and a month after that the new chief was elected.

THE LOSS OF THE FIRST RAPIDS

Other impacts, however, seem even more permanent and in the long term, more profound. In the original negotiations the people of Fort George had persuaded Hydro-Quebec to move the site of the proposed LG1 dam and powerhouse upstream from the first rapids, which are twenty-three miles above the village, to a site forty-seven miles farther upstream. As earlier chapters have indicated, the first rapids had been for millenia the most important fishing place for Fort George people. Even as late as 1972, the whitefish taken from these superb rapids were providing them with 20 percent of their entire diet.

This place, along these beautiful rapids, also had immense perceptual and spiritual meaning for the Cree people: for as long as any of them could remember, they had camped as children and adults on the broad slab of rocks that ran alongside the first rapids, the women in the family tents cooking up the fish that the men caught in the swirling waters beneath the rapids while their children played along the rocks, wandered in the bush, learned how to snare rabbits and shoot partridge or porcupine.

It was a shock to return fifteen years later to find that these wonderful first rapids, where Job Bearskin had so expertly negotiated the raging waters on our trip upriver in 1972, had already disappeared under the vast preliminary works of a multibillion dollar powerhouse, LG1. The magnificent rocks along the north shore had been blasted away to make a temporary channel for the river while the dam was being built.

It is an indication of the skill with which those who run our industrial machine manipulate others, that in 1986 the James Bay Energy Corporation had managed to buy its way back on to the first rapids, persuading the band council of Chisasibi to exchange this irreplaceable site for the money they needed to build their new village.

Of course, originally a new village was not contemplated. Fort George stood on Governor's Island in the mouth of the great river, and the original Agreement provided for remedial works to minimize the erosion that would be caused by the doubling of river flows, to provide a new village water supply, and to improve access by building a bridge from the mainland to the island.

The people of Fort George were still discussing the future problems of their village, still negotiating for the building of a new $30 million bridge to the island, when their intermediaries returned to tell them that the James Bay Energy Corporation was unwilling to finance a new bridge but was prepared to pay for the building of a whole new village on the mainland, if the people would agree that the LG1 dam should be located, as was originally intended, on the first rapids.

Though it was not made clear at the time, this generous offer arose from the engineering needs of the project. Hydro-Quebec had decided to add more powerhouses further inland, including a second underground power station at LG2. This would increase the discharge from the turbines, and create an even higher maximum flow downstream in the river. The banks of Governor's Island, already vulnerable to erosion, would probably be unable to withstand this additional water, so unless the village were moved, the utility would be unable to proceed with its planned expanded power stations. And so in 1978 the band council agreed to accept a proposal that the Corporation spend $50 million (later $60 million) to build a new village. The village was moved in 1980.

At the time, inlanders were predominant over coastal hunters in the band council; for them, easier access from a mainland village to the new inland road network was an attraction.

For a time there was a proposal that the coasters might stay on the island, living in the old village of Fort George while the inlanders moved, but the relocation authorities were determined to raze the old village to the ground.

Eventually they made an exception for the Anglican church, but most of the village was destroyed. There remain, however, a handful of families who never moved. In fact, nowadays the island, known as Old Fort George, is a favorite weekend destination when Chisasibi people want to get away from it all. Each year more families move back there permanently, and many others dream of doing so. In contrast to the sprawling, dispersed white man's village of Chisasibi, dominated by ten miles of internal roads and acres of unsightly sand, Old Fort George today is an oasis of beautiful waving grasses, mosses, and open canopy forest surrounding picturesque, weathered, often unused, wooden buildings. To step on to the island is to step into a ghost town that has never quite been abandoned. Many families left their houses and shacks standing, painting poignant signs such as "Do not touch" on the walls, against the day when they hope to return.

There are a few vehicles, and as they meander along the dirt roads, the drivers will always offer a stranger a lift from one end of the island to the other, no money needed. The vehicles are unlicensed; indeed, all of the supposedly indispensible accoutrements of civilization—electricity, telephone, running water, sewage systems, television—are missing from the island, which, as one wanders along its peaceful forest trails and roads, seems like a tiny area of silence and stillness in a crazy world.

Chisasibi, on the other hand, with the creation of a brand new village, immediately inherited all the social problems and insecurities that go along with white civilization. Dealing with those problems, trying to find a way to accommodate the ancient Cree values to the new influences from the south—without losing everything—has become the full-time job of the people who run the village.

Loss of a moral leader

My 1989 visit to Chisasibi was, alas, just too late: for Job Bearskin, one of the most impressive men I have ever met, died two weeks before I arrived. Long before his death he had been recognized by everyone, young and old, as a spiritual and moral leader in the community of Chisasibi, and it was comforting to know that in the film Job's Garden I had, almost inadvertently, produced a permanent record of a man whose gentle wisdom and unfailing openness to others continued to inspire everyone around him. Indeed, that film, however technically deficient it may be, acquired a new lease on life in Chisasibi eighteen years after being made, as a whole generation of younger people turned to the worries expressed about progress and technology, about flooding and survival, by the elders who appear in the film. One hundred fifty copies of the film were distributed around the James Bay villages, largely, I think, because of the message conveyed by the old men as they sat around in Job's teepee and heard his report on what he had seen inland. "I think we are only beginning to understand what those old men were telling us," a young man told me. Nearly twenty years after the film was made, when they came to name the new Chisasibi arena, they decided to call it Job's Memorial Garden.

All I could do was visit Job's grave, and his wife Mary, shattered by the loss of her life's companion.

THE MASSIVE EFFECTS OF THE PROJECT

Inland, the James Bay project is a network of reservoirs, extending some 500 miles from the coast, impounding under water an area half the size of Lake Ontario, retained within 9 dams and 206 dikes that run for 80 miles across the countryside. This vast complex today requires only about 325 people to run it, of whom only 5 are Cree.

Though the Cree may have benefited from the James Bay Agreement (and more and more of them are beginning to question that), it is undeniable that they have received no benefits from the James Bay project itself. It has not created work for them. It has opened up their hunting territory to thousands of outsiders. It has flooded much of their trapping land, has diminished the animal populations, damaged drinking water supplies, and in addition, because of the unexpected accumulation of methyl-mercury in the reservoirs, now denies them the use of most of the fish they once took from the La Grande River.

So far as concerns about the environment were expressed in the early seventies they concentrated on the likely effects of reservoirs, river diversions and the loss of terrestrial habitat through flooding. These concerns were expressed in rather general terms, because so little was known about the biology of the region, and no real biological assessment was ever done before the huge project was built.

The government's view, later put into action, was that any problems could be fixed through remedial works. The governments decided to leave all environmental monitoring to Hydro-Quebec, which, in the words of Billy Diamond, thereby became "proponent, evaluator, judge, and jury" of the project and its effects on the land and the Crees. Hydro-Quebec set up its own ecological monitoring unit; though it has produced much valuable information, it has confined itself to monitoring readily observed changes. Alan Penn, the Crees' expert environmental coordinator, argues that more profound questions should be answered about environmental impact. But these questions require an understanding of the structure and functioning of ecosystems. For this, a research agenda would be needed going well beyond the normal responsibilities of Hydro-Quebec. Such work should be done by the governments involved; but, as Penn has said, "the La Grande development has so far been distinguished by the almost complete absence of such involvement."

It is known, however, that there have been some important environmental consequences of the project, though what is known about them is far from profound. They include:

1. The bio-accumulation of mercury in and below reservoirs
This is the major environmental effect so far identified, and one that was
never foreseen in the 1970s. Some mercury is present in an inorganic form
in a natural boreal forest environment. What seems to happen when land
is flooded is that an organic form of mercury — methyl mercury — is created during the feeding of bacteria on the organic matter (such as
drowned trees and plant life) in a newly flooded reservoir. Though in a
stable lake this mercury is broken down by other bacteria, in a new
reservoir it is produced more rapidly than breakdown can occur. The
amounts of mercury involved are tiny: there is probably only one to two
hundred pounds of inorganic mercury in the LG2 reservoir, recombining
to form only ten to twenty pounds of methyl mercury.

As the fish feed, however, they accumulate this mercury in concentrations a million times greater than in the water; and through the fish,
the mercury is passed on to any animals or people, who, like the Crees, eat
these fish as part of their staple diets.

By 1985 Chisasibi Crees were found to have mercury levels far exceed-
ing acceptable limits established by the World Health Organization. Hydro-Quebec has argued that this will correct itself within thirty years; the
Crees and their scientists, however, say that no one has the information to
know how long the problem will last. It could be fifty, or even a hundred
years.

Meantime, generations of Crees are denied a major part of their diet.
Hydro-Quebec spokesmen admit the problem will worsen with the proposed building of more reservoirs in the coming decades. Yet they insou-
ciantly toss off their proposed solutions: the Cree could increase their
fishing in unaffected lakes whose productivity could be improved by fish
management; they could sell the inedible fish by turning them into live-
stock feed, household fertilizers and so on; or "an alternative diet based
on food resources other than predatory fish could also be contemplated."

Hydro-Quebec, while underplaying the importance of the mercury
problem in its public statements, has nevertheless created an agency, with
$18 million, to study possible solutions.

2. Erosion caused by changes in the flow of rivers
Though the bed of the La Grande River has apparently remained stable
under the impact of a mean flow of 5,200 cubic yards per second, it is not
known if it will remain so when the flow is increased to 7,800 cubic yards,
which will happen when the four power plants now under construction
are completed. But along the river banks, many enormous slides have
already occurred, and more may be expected as the flows increase.

3. The impact of flow changes on the survival and reproduction of anadromous whitefish and trout

These fish normally spawn in the low-flow months of the fall, and over-winter in the warm fresh water near the river mouth. "But our knowledge of fish adaptation and survival is quite fragmentary," says Alan Penn.

4. The injection of 45–60 cubic miles of additional fresh water into the coastal regions of James Bay in the winter when shelf ice is present

It appears that this augmented winter flow from the La Grande accumulates as a freshwater lake three to six feet deep, thus turning the coastal bays and estuaries, which in nature are delicately balanced between fresh and salt water, into freshwater environments. The biology of these coastal regions has proven to be quite complex, and the impact of these changes on the sea-grass beds vital to migrating waterfowl is not yet understood.

Norman Sam told me in his goose camp on the shores of James Bay that the additional freshwater pouring into the bay through the La Grande River has already damaged the eelgrass along the coast, and this has diminished the number of geese taking this migratory route.

While admitting that mercury is a genuine problem, Quebec government and Hydro-Quebec spokesmen repeatedly say that the James Bay project has been built without any permanent environmental consequences. This euphoric and self-serving view does not survive even cursory inspection, and certainly is not confirmed by the people who know the region best, the hunters and trappers of James Bay.

"The food I get from the land doesn't taste the same," Waswanipi elder Billy Cooper told me. "The partridge tastes different, the rabbit does too. The stuff that grows is subject to fallout, and the salt on the roads affects the partridge when he eats the trees, the willows, and the sand."

"There has been a decline in big game," says Sam Blacksmith. "There's a lot less than there was in the past . . . . Helicopters are flying over all the year round, and that scares off the animals."

"When I clean rabbits, I notice a difference in the way they look," says Mrs. Blacksmith.

"Exactly what I said was going to happen, happened," says François Mianscum. "Some people's hunting grounds are in a poor state. It's the same on my land due to deforestation. It's all going to be clear-cut, it's going to have the same effect on wildlife as the dams. There will be nothing there. . . . "

"A hydro line runs over our land," says François' son Abraham, who hunts with Sam Blacksmith. "At first I thought it was not going to have any effect; but now there are no animals within three miles of that line."

"The developers have polluted our river, which is where we got our water," says Willie Moses, of Eastmain. (This river has been subject to a number of Hydro-Quebec's cherished remedial works, of which the hunters have a poor opinion.)

"How much worse is it going to be for the people if they build more dams?" asks Abraham Weapinacappo. "We already know what happens and how much is destroyed. The Cree way of life will be totally destroyed."

Four new generating stations, under construction in 1990, are being added to the three stations, LG2, LG3, and LG4 built in the 1970s and early 1980s. These will raise the La Grande's generating capacity to 14,791 megawatts.

MORE DAMS, DIKES AND RESERVOIRS

But that ain't all, as they say. Unfortunately it seems that the Crees, having adjusted to the interruption in their lives caused by the hydro-project, will not be permitted to live happily ever after.

For in 1986 their nemesis Robert Bourassa returned to power in Quebec at the head of a Liberal government, and almost immediately announced his intention to finish the harnessing of the hydro-electric potential of the James Bay region.

His new scheme is on a scale far greater than anything that has so far been built. At an estimated cost of some $48 billion, it proposes two further phases: first, further north than the La Grande, by damming the Great Whale and its surrounding rivers; and second, by implementing the so-called NBR scheme involving the Nottaway, Broadback and Rupert rivers in the southern part of the territory—a scheme Bourassa originally announced in 1971.

Preliminary work has already begun on the Great Whale project, where the Great Whale, the Little Whale, the Coats, and the headwaters of the Nastapoka rivers will be dammed and diverted to create four reservoirs that will feed three new generating stations, GB1, 2, and 3, with an installed capacity of 3,060 megawatts. Altogether this will create reservoirs of more than eighteen hundred square miles in area, and will add some 540 square miles of flooded land to existing water bodies.

Construction is slated to begin in 1991, the first job being to build a three hundred mile road north from the site of the LG2 dam to Great Whale River.

The NBR project, in its new guise, calls for the diversion of the waters of the Rupert and Nottaway through the Broadback, the creation of seven large reservoirs, and construction of eleven generating stations with a total installed capacity of 9,100 megawatts. When it is all built more than a dozen great rivers will have been dammed, diverted or submerged; some twenty-one reservoirs will have been created; and twenty-four power stations built to feed the insatiable appetite for energy of Canadian and American society. It is unclear how much more land will be flooded, but in total the reservoirs will cover nearly eleven thousand square miles, an area almost as large as New Jersey.

The Crees are therefore confronted with the prospect of decades of massive construction projects in their lands, which including the four power stations now under construction, will add some 16,600 megawatts of generating capacity to the 10,800 of the original project, already built. It is evident that to build this system engineers will arrive in force, with exploration and construction camps, roads, airports, and so on, not just along a single river, as was the case before, but throughout the entire traditional hunting territory of the Cree people.

This is one occasion on which one might be forgiven for using the word colossal to describe a scheme. In fact, one wonders at the sanity of those who propose it. The NBR system, in particular, to be built in low-lying, swampy, and unstable land, has a certain megalomaniac quality. It will, for example, require that the Broadback River drainage basin be increased from 13,000 square miles to nearly 77,000 square miles, a factor of nearly six. The river's mean flow of about 520 cubic yards per second, will be increased to 3,000, and the mid-winter flow (now below 130 cubic yards per second) to about 5,200 cubic yards per second. The existing channel could not possibly handle such flows without a series of forebays and turbines, and even with them immense erosion is likely.

A HUGE INTERFERENCE WITH NATURE

The environmental consequences of the Great Whale and NBR projects are likely to be far-reaching. The country around the Great Whale River is close to the limit of the treeline and is very fragile from an ecological point of view. The rivers to be dammed and diverted not only have many fish species, which the native people have depended on for generations, but, like the La Grande, their freshwater maintains the acquatic balance along the Hudson Bay coast, where beluga whales, seals, polar bears and artic char are to be found.

The rare species of freshwater seals in some of the inland lakes will be affected in ways that cannot be determined, and further inland, the flooding of Lake Bienville will interfere with caribou calving grounds. In addition, of course, many traplines will be flooded.

"Devastating" is the word used by Chief Billy Diamond, whose village of Waskaganish will be in the heart of the NBR project, to describe the likely effects in his area. The habitat of sea-run stocks of trout, cisco and whitefish which spawn in the rivers of Rupert Bay will be completely transformed, and the injection of an additional nineteen cubic miles of freshwater into the shallow basin of Rupert Bay in winter will change the ecology of the bay. This will affect the millions of migrating birds that use Rupert Bay in the spring and fall, just as Norman Sam says has already happened further north along the James Bay coast.

Indeed, under this scheme the country south and inland of Rupert Bay would be virtually given over to the project, with almost all major rivers and lakes affected in some way or other. This country is covered by a network of lakes, rivers, streams, and wetlands that are important nesting grounds for many migratory birds.

In spite of its somewhat surrealistic nature, thinking about this scheme is so far advanced that Crees in the Waswanipi area claim that forestry companies are already clearing the land likely to be flooded by the NBR project. Already, a minor new bridge over the Waswanipi River has been built at an immense height, evidently because the level of that river is to be raised many years down the road.

As they watch this sort of thing being done, entirely without their consent, and without any pretense at consulting them, the southern Crees are fearful of a future in which their rivers will be eroded, their lakes silted, their community water supplies contaminated, their canoe routes ruined.

The proposed project is so large, in fact, that Alan Penn questions whether it can be designed so as to permit the development of the region for purposes other than energy production. The prospects, he says, are not encouraging. The same question, as to whether any other activity can coexist with energy production on this scale, can, he says, be asked of the Great Whale project, and should have been asked about the La Grande project.

The key to the project is Bourassa's hope that Hydro-Quebec can make firm contracts to sell 3,500 megawatts of power to the New England states for delivery during the 1990s and beyond. Alan Penn's reading of the Hydro-Quebec documents indicates, he says, that the utility is hoping it can increase these sales as the years go by, with the possibility of providing an additional 6,000 megawatts after the turn of the century, depending on the level of United States demand.

THIS TIME, A VIGOROUS OPPOSITION

Not unnaturally, this huge scheme, for which this one publicly owned utility plans to borrow tens of billions of dollars over the next ten years, has drawn vigorous opposition, not only in Quebec and other parts of Canada but also in the United States. Whereas the first version of the James Bay project was opposed mainly on environmental and native rights grounds, the new version has drawn fire from economists and energy analysts who argue either that it is economically infeasible, or that it is unnecessary, if the future economic trend is to be, as it should be, towards energy conservation.

The Crees, of course, have expressed their opposition to Phase II from the beginning. But this time they are receiving powerful and vocal support from the United States.

The Crees are no longer as defenseless as they were in the face of the first scheme. The Agreement has provided them with resources with which they can afford to mount sophisticated legal, media and public relations campaigns against the project.

Early in 1989 the Cree chiefs filed in federal court, claiming that they retain aboriginal rights over their traditional lands, have the right to exclusive use of these lands for hunting, trapping and fishing and that the James Bay mega-project violates and prejudices exercise of this right. They oppose the project and refuse to consent to it, they state, because of the many negative consquences, including river diversions, flooding, the proposed road to Great Whale, the airports, the likely effects on fish and marine mammals, on migratory birds, navigable waters, water quality, climate, and because of the likely accumulation of mercury in the reservoirs and rivers.

The Cree statement of claim says that the federal government has a fiduciary obligation to protect, preserve and enforce the Indian title and aboriginal rights of the Crees, but has "abrogated and neglected" that responsibilty, has consistently breached the James Bay and Northern Quebec Agreement, and "will in all likelihood continue" to do so in respect to the James Bay project.

They say the federal Environment Minister has failed to apply Section 22 of the Agreement (requiring an environmental impact assessment of development projects), nor has the project been assessed under the Federal Environmental Assessment guidelines. The Cree chiefs therefore ask for a declaration that the Great Whale River project be declared unconstitutional, illegal and beyond the authority of Hydro-Quebec and the government of Quebec.

This question of environmental assessment assumes enormous importance. Quebec accepts the need for an assessment, but believes it can be done within six months or so. To suggestions that an adequate assessment of such a huge intervention in nature could not be done in less than four or five years, Hydro-Quebec answers that it has already done 160 studies which none of the project's opponents have ever bothered to read. (In fact, these studies have not been published, have not been translated into English, the second language of the Crees, and are not easily available to outside investigators, at least at time of writing.) In addition, the studies have been done before the relevant questions to be addressed by an impact assessment procedure have even been decided.

CHALLENGE TO NEW ENVIRONMENT LAWS

The matter seems to be central to Premier Bourassa's hurry-up timetable. A genuine assessment, following the requirements of Canadian law, would require public hearings, a careful delineation of the key questions for study, and a long period of study, all of which would disrupt Hydro-Quebec's pre-arranged export sales. The problem for the Quebec government has been to find some way around these laws.

In November 1990, Hydro-Quebec began to talk about giving permission for the work to be started on the road from LG2 to Great Whale without waiting for an environmental assessment. Then under pressure the utility a few weeks later magically produced some of the many studies it has done over the last few years, now putting these forward as the assessment of the road impact that only a few weeks before they had said they needn't bother with.

All that remained, they now claimed, was for the provincial government's environment department to approve their studies as adequate. Development projects in the James Bay territory have been carefully excluded from the law on environmental impact assessments that makes public hearings mandatory in southern Quebec. And the James Bay Agreement, which governs such matters in northern Quebec, itself does not provide for public hearings on environmental assessment. Thus the way seemed to be clear for the Quebec government (if it is determined enough, which it undoubtedly is) to make an end run around all proposed delays.

In addition, if a panel appointed by the federal environment department were to give approval to the road building, there would be no statutory need for public hearings at that level, either.

This, of course, is to mock the entire legal structure for environmental assessment. Throughout Canada, a critical public expected the federal government to implement its own law. But this would set the two governments on a collision course, and by the end of the year it had begun to appear that, as in the early seventies, the federal government was unwilling to exacerbate anti-federal feeling in Quebec, and was preparing to acquiesce in Hydro-Quebec's sham assessment.

In April of 1990 the Cree chiefs instituted proceedings in the Quebec Superior Court, requesting an injunction to prevent the proposed Great Whale River project, and to restrain the Quebec government and Hydro-Quebec from proceeding with related works, specifically with the planned road from LG2 to Great Whale. As before, the government argued that if anything were to go wrong during and after construction, remedial works would do the trick.

By this time Hydro-Quebec had persuaded the Quebec government that environmental impact on the roads and infrastructure should be assessed without any reference to the huge hydro-project for which the roads were being built. This was the tactic most feared by those opposing the project: it would enable the roads to be approved, on the grounds that they did not cause unacceptable damage to Cree traplines, and would allow the utility to argue in any later court proceedings, that, billions having been spent on infra-structure, they should be permitted to finish the project.

If these strategies all work, as they seem to be doing at time of writing, the laws requiring environmental assessment will have been nullified by the governments that passed them, and, failing intervention by the courts, the way will be clear for the construction of James Bay II.

The Crees have been active in trying to uncover the secret deals made by Hydro-Quebec to sell the newly generated power to aluminum companies that have located along the St. Lawrence River. Here again, it is curious that on a matter of such public importance, the initiative should have been taken by a group like the Crees, rather than by investigators from within Quebec society itself. The suspicion here is that these are heavily subsidized energy contracts designed to improve the competitive position of Quebec aluminum: but the criticism, again, from members of the anti-James Bay Coalition, is that by such deals Quebec is locking itself into an almost third-world condition as a supplier of subsidized energy to an enormous, global, extractive industry.

OPPOSING ENERGY EXPORTS

Meantime, the Crees also intervened in the application made by Hydro-Quebec to the National Energy Board of Canada for permission to export power to Maine, Vermont and New York. This time, the Crees have been able to hire highly qualified people, such as energy economist William B. Marcus of Sacramento, California, and conservation engineer Ian Goodman, of Boston, to analyze Hydro-Quebec's plans and figures, and produce detailed criticism of them to the National Energy Board, the Quebec National Assembly's committee on the economy, and to public meetings.

Not since the James Bay court case in 1972–73 have the assumptions of Hydro-Quebec been subjected to such rigorous analysis. The American experts are accustomed to confronting major utilities before regulatory bodies and legislative committees: they found that Hydro-Quebec was not at all used to this kind of opposition, but had developed an immunity from scrutiny that in effect had made it a sort of dictator over the direction taken by the Quebec economy.

The utility's appearances before the committees of the Quebec National Assembly are usually more like triumphant processions of royalty before an audience of loyal subjects, than tough, parliamentary-style interrogations. Indeed, when the James Bay plans were outlined before such a committee early in 1990, the members of Quebec's Parliament left no doubt that they had already decided they would support the building of the project, before they heard any evidence at all.

The Crees told that committee that they would not agree to "the continued intrusion into our territory for the destruction of eight more rivers and the ecology of a vast area.

"We consented to the construction by Quebec of the La Grande Complex, though we never accepted the project. We have made substantial sacrifices to allow Quebec to complete that project. But enough is enough. We will not agree to the future destruction of the Great Whale River or Nottaway–Broadback–Rupert rivers. Let this be clear, that the Crees have never consented to these projects, neither in the James Bay and Northern Quebec Agreement, nor elsewhere. Quebec cannot build these projects."

In addition to the critique mounted by Marcus and Goodman, the Crees hired to testify on their behalf the Natural Resources Defense Council, of Washington, D.C., a non-profit organization with a staff of eighty lawyers, scientists, and resource specialists, whose focus is on issues of energy and environment. Their brief indicated that none of the proposed power stations would be needed for many years if Quebec would cancel its proposed exportation of electricity to the American Northeast and invest in a serious program of energy conservation, as is being done in various parts of the United States.

THE ADVANTAGES OF CONSERVATION

The Bonneville Power Administration in the Pacific Northwest, for in-stance, by spending more than $900 million on energy conservation over ten years, has achieved enough savings to defer indefinitely new large-scale power generation, and has done so at an average cost of only two cents a kilowatt hour.

Four California utilities have recently announced plans to invest more than $560 million over the next two years in such measures as improved commercial lighting, use of more efficient appliances, greater efficiency in space heating and cooling, and introduction of stringent building codes and energy-efficiency standards.

According to the Marcus-Goodman analysis, similar measures taken in Quebec, combined with a cancellation of the export contracts to Ver-mont and New York, would obviate entirely the need for any new gener-ating stations not already under construction until after the year 2020. They also claim that such a conservation program would create at least as much work, and at far lower cost, as the building of mega-projects.

The Hydro-Quebec development plan foresees 317,000 person-years of employment, at a cost of more than $102,000 per job-year. Of these job-years, 149,000 would be in direct installation of the projects, and 168,000 in the manufacture of materials. Comparable studies in New England and the Pacific Northwest show jobs in energy conservation being created for $52,000 per job-year. California studies indicate that conservation also frees up capital for investment elsewhere in the economy, and it has been calculated that this spending alone creates at least five times more jobs than does the building of a power plant.

Also to be taken into account is that increases in electricity rates levied on consumers to support the massive investments in James Bay will re-duce investment, and therefore employment, in other parts of the Que-bec economy. All this alternative spending would have the additional advantage of creating jobs distributed throughout the economy in south-ern Quebec, and thus avoid the many social and economic problems resulting from the creation of boom towns in the Quebec north.

Hydro-Quebec claims that it has budgetted $1.8 billion to be spent on conservation measures in the next decade. Proponents of the so-called "soft energy" path argue that such an amount, well managed and well spent, would make the Great Whale project unnecessary. They therefore question the sincerity of Hydro-Quebec's intentions, believing that conservation measures are designed to free power for export sales, not to reduce power generation.

A UTILITY BEYOND CRITICISM

It is remarkable that among Quebec economists, there has been virtually no serious examination of the impact on the Quebec economy of the James Bay spending, compared with what such spending would have achieved in other sectors. A notable exception has been Helene Lajambe-Connor, who began in the early seventies as an environmental opponent of James Bay, and has spent the time since equipping herself with academic skills as an economist which have made her into a formidable analyst of the Hydro-Quebec budgets and proposals.

This academic silence apparently arises from an unwillingness among Quebec academics to offend Hydro-Quebec, and is yet another proof of the unnatural power exercised by this utility over Quebec economic life. It funds much research; endows university chairs; has even made a substantial gift to the nationalist newspaper, Le Devoir. Yet the economic facts are somewhat astounding: the first phase of the James Bay project has already resulted in Hydro-Quebec contracting a debt of $23 billion, and this is likely to go up to $60 billion due to present plans. These figures are sufficiently dramatic, one would have thought, to stimulate at least some scrutiny from within the Quebec establishment.

It is ironic that francophone Quebecers, so nationalistic, so determined to defend their autonomy and language, have played so small a role in analyzing and criticizing their government's grandiose plans. The leading role has been taken by the Crees, naturally enough perhaps, since they are the people most immediately affected. But their most urgent support has come not from other Quebecers or Canadians, but from Americans.

ALARM IN THE UNITED STATES

The Audubon Society, having examined the latest Hydro-Quebec scheme, weighed in with a statement that the James Bay II project is a temperate zone equivalent of the destruction of the tropical rainforest, threatening an ecological catastrophe of global proportions. Though Quebecers these days have a tendency to discount interference by outsiders, especially when it is expressed in the English language, this intervention certainly added gravitas to the opposition arguments about James Bay II. "Just as Canada came down and told us to clean up our act on acid rain," Jan Beyea of the Audubon Society told a Canadian newspaper, "we think we have a role to play in James Bay. Sometimes an outsider's perspective can help. It can open some eyes."

The Sierra Club felt sufficiently alarmed to establish a task force, funded to the tune of $300,000 by the Sierra Club Foundation. With that vigor and enthusiasm so typical of Americans, the task force quickly pulled together a coalition composed of Quebec, Canadian and American environmentalists and energy experts, and the Cree Grand Council. At their first meeting in May 1990, this task force established the objective of stopping the James Bay II project. The main Quebec member of this coalition is an organization called Mouvement au Courant, an umbrella group dedicated to examining the Quebec energy future and claiming a million members.

In response Hydro-Quebec launched a $6 million advertising campaign through newspapers, television, radio and billboards, to whip up enthusiasm for the "clean" energy to be expected from Great Whale River.

The decision by energy utilities in Vermont, Maine and New York to sign agreements expressing their intention to purchase Hydro-Quebec power, brought about a coalition between Crees and American environmentalists from the beginning of 1989. The Crees went all out to persuade the American states to reject these proposed contracts. They sent their leaders to address public meetings throughout New England, and appeared before committees of state legislatures to urge conservation rather than importation of energy.

American conservationists argue that to import such huge quantities of electricity from Quebec would set back the many promising programs of energy conservation already underway in most of these states. "The Hydro-Quebec contracts undermine years of work to make Americans more aware of conservation," said Eric Washburn of the Natural Resources Defense Council. "We need to invest money in energy efficiency to combat the greenhouse effect. More power from Quebec just encourages wasteful behavior."

They say that in recent years some 70 percent of the increases in energy availability have been achieved through conservation measures, at an incomparably lower cost than through generation of new energy.

SOME PRELIMINARY SUCCESS

The Crees and their allies won an early victory in January 1990, when the Maine Public Utility Commission rejected the Hydro-Quebec deal. In

March, twenty New York state legislators introduced a bill into that state's assembly which would make mandatory a review of environmental impacts "at the site and in the region affected by the development of any generating source" of any electricity purchased from outside the United States.

Two months later, the Grand Council of the Crees joined with the Sierra Club and the Atlantic States Legal Foundation in filing a legal action in a New York state court demanding a full environmental review of the export contracts between Hydro-Quebec and New York, arguing that such environmental review should be mandatory for imported electricity, as well as for new electricity generated within the United States. They lost that challenge.

The Crees kept up the barrage throughout 1990. In March and April of that year Inuit and Crees from Great Whale River (Whapmagoostui) built a special canoe, which they named the Odeyak, and paddled it from Montreal down the Hudson to New York City, meeting environmental and citizen groups at every town along the way. They arrived in New York on April 20, and were there to take part in the massive celebration of Earth Day on April 22.

In September and October, 1990, contact was established between the Crees, the Inuit and the Kayapo Indians of Brazil, who also have been fighting immense hydro-projects in their territory. A publicity tour was arranged for a Kayapo chief, but he was urgently recalled from Montreal to Brazil before he could make his intended visit to the Cree territory.

Building their vigorous case against Vermont purchase of James Bay power, the New England Coalition for Energy Efficiency and the Environment has had less to say about impacts on the Crees or the Quebec environment, but has instead emphasized the undesirability for Vermont itself of turning back the conservation drive. Speaking for the New England Coalition, James J. Higgins claimed that "powerful corporate and political interests" most of them from out of state, were hoping to buy the power for profitable resale to other states. Another Vermont argument was that the future lay with small, local generating sources. In fact, Higgins said that when the Vermont department of public services put out a bid for energy alternatives, they received schemes from small producers offering more than 1,000 megawatts — more than twice the amount Vermont contracted to buy from Hydro-Quebec.

When Vermont officials announced their decision in October 1990, both sides claimed it as a victory. In a three hundred-page decision the Public Service Board agreed that Hydro-Quebec could export 340 megawatts, 110 megawatts fewer than the twenty-four Vermont utilities

had asked for, but added that the Canadian utility "must not use the sale
to justify new dam construction."

"Hydro-Quebec testified the sale would require them to build new
projects," said Bill Namagoose, executive director of the Cree Grand
Council. "So they are stuck with their evidence." But Hydro-Quebec said
they did not read the Vermont decision in that way.

Similarly, the National Energy Board of Canada approved exportation of electricity by Hydro-Quebec, but added the condition that it be
subject to strict environmental assessment. Environmentalists immediately began to pressure the federal government to ensure that the NEB insist
on fulfilment of this condition. This regulatory body has never concerned
itself with environmental matters before, and there is what would seem to
be a well-justified concern that the condition might simply be ignored.

At time of writing, at the end of 1990, the battle is well and truly
joined. Mr. Bourassa persists in his view, expressed in a book he wrote on
James Bay power, that "every day millions of potential kilowatt-hours
flow downhill and out to sea. What a waste!" This can hardly be said to be
an ecologicallly sensitive man: but at least he is consistent. He said the
same thing when he launched the project in 1971, and appears to have
learned nothing since.

For that matter, nor have the Crees changed their attitude in the
intervening twenty years, in spite of the millions of dollars they have
received, and the comforts this has brought them. When one of Bourassa's ministers, Lise Bacon, contemptuously said there was no need to
worry about the Crees, because they could be bought off for a billion
dollars, Bill Namagoose replied that the government could keep their
billion dollars: the Crees had followed that route once before, when they
had no alternative, but they were not ready to follow it again.

THE HUNTERS SAY IT AGAIN

In 1990, as twenty years before, the hunters of the north came south, once
again to express before the white society's decision-makers their deep
reverence and love for the land. And this, perhaps, is where we should
end this account, as we began it twenty years ago.

Chief Robbie Dick, of Great Whale River (Whapmagoostui), before the
National Energy Board, Ottawa:

QUESTION: Can money compensate for the loss of land, can it create other
        things for the people to do?.
DICK: As far as my people are concerned, they don't think that the land can

be replaced by anything. Once it is lost, it is lost. Money can't replace it. . . . One of the things that bothers me a lot is that my people do not understand this project at all. It is impossible for them to even think of the idea that the land they use will not be there. They cannot conceive that this can happen. If it happens, we'll lose part of our lives, because our land is our life. The whole river is going to go. . . . That river was used as a highway by our people, and still is. All our canoes down there on the river bank, where can they go? The river is going to be totally lost up to the first rapids, just a little stream will come down in the middle where the river used to be at low tide. . . . Once the river disappears, part of our culture disappears with it. . . .

QUESTION: How will the culture and language of the Crees be affected?

DICK: Already the children of families that don't hunt much anymore are losing their language. . . . The language comes from the things that happen on the land, the way we live there creates the words we use to express who we are, what we do, and if you don't have the language you can't practice the tradition properly. . . . Language is dependent on the resources of the land. If we begin to more or less speak English or French in the inland, then we will act like the white man, and do things the white man's way. But if you know the Cree way and use the Cree language in everything that you do, then you do what your ancestors did. Only then are you fully practicing the Cree way of life. We fear that the Hydro-Quebec project will destroy our culture and our language by destroying our land, and bringing in thousands of workers.

And at the same hearing John Petagumskum, senior, spoke with that unique Cree intimacy about the animals he knows so well:

QUESTION: Where can the various species of game be found?

PETAGUMSKUM: The beaver which inhabits the areas in and south of the Great Whale River does not travel in the winter. It remains and feeds out of its lodge. It travels in the summertime. If the normal water level should rise in the wintertime it will drown. The otter which roams all over the territory constantly travels on land in winter and summer. The marten and mink move on the land and live in a shelter in the ground. They roam in the forest. All these animals live in the river valleys. The hills are rock. Fish are to be found in the waters all over the territory. Many kinds of fish. The caribou migrate from the east to southwest reaching the Chisasibi hunting grounds. The herds then travel west until they reach the bay and then go northwards along the coast. There are calving grounds around Bienville Lake.

All waterfowl migrate from the south in the spring and nest around the waters all over the territory. Most species of waterfowl usually nest on small islands to protect their young from predators. There are more islands than you can count in the Bienville Lake area. If the water levels are dammed up, many of the islands will be flooded. If the water levels rise during the nesting period, this will destroy the young. The porcupine which inhabits the forests along the river valleys is another source of food for the Cree families. Should the water level rise, its only choice will be to climb a tree. When the waters from the dikes and dams rise above the trees in the river valleys, it will die. The animals will have nowhere to go if the river valleys are flooded, because up there, there is little vegetation useful to the animals outside the valleys.

He gave a remarkable account to the no doubt rather bewildered economists and accountants who make up the National Energy Board, of the Crees' respect for the freshwater seals, that unique species that will be destroyed by the proposed flooding.

PETAGUMSKUM: We hold the highest respect for the freshwater seal; because although it was hard to hunt, it saved the lives of entire families at times past when famine struck. When all else was not available, by miracle a hunter killed a freshwater seal. It was as if the Creator planned it this way. There are a number of Cree stories about the freshwater seal. . . . You see, not every hunter is gifted with ability to catch the seal easily. Such Creator-given ability was found only in some hunters. This belief is still strong nowadays. . . . There was, and still is, a belief that one has to show respect for the game that was given to him by the Creator for his very subsistence if he is to survive on it. . . . For example, if the seal could not be taken before it sinks, there was despair. You see, when a hunter leaves the community for a year-round hunt in the bush, everyone knows this, that the future is laid out for him as to what he is going to catch during that year. Even if his future looks good, and even if that person is healthy, chances are, if he lost a seal, it is as if the seal went down along with all else he could have caught during that year.

QUESTION: What will be in the impact of the project on the Crees of Whapmagoostui?

PETAGUMSKUM: It will no doubt be devastating to the Crees. . . . It will impact on the lands, animals, fish and birds that live in the river valleys and around the lakes. There will be disease that will strike all those animals because new things will be in the water, the water will not be the same after the flooding. This disease will eventually affect other game, one after another, because they live on each other, and the water. It will eventually affect humans, because we eat this food.

He added a remarkable elegy, of an eloquence and sincerity that is hard to ignore or resist:

> We have no other land, and this project is on the rivers and lakes where all things live. . . . The animals all need each other, and with the passing of the animals, the Crees of Whapmagoostui will lose their way of life.

That is the danger that John Petagumskum now faces, just as Job Bearskin faced it in 1972.

On their side the governments seem, if anything, more determined to push their projects ahead than they were twenty years ago.

But these two decades have brought us much knowledge about nature and how it works; much concern about the integrity of our environment; a renewed appreciation of the skills and insights of the native people; and a greatly broadened understanding, based on many years of experience in many, many countries, of the dangers of the mega-projects so enthusiastically undertaken by engineers and politicians.

The good news is that the Crees are no longer alone. More people than ever before now understand that the first time around, in defending their land, the Crees were fighting for all of us.

At time of writing an impressive number of people, people from far and wide, people of diverse interests, people of different nations, races, colors, creeds, are gearing up to use every means they can to turn back this latest onslaught on the Cree homeland.

NOTES

CHAPTERS TWO AND THREE

All extracts in this book from evidence given in the court case, including summaries of legal arguments, are taken from the transcribed daily joint record, which amounts to ninety-eight volumes. The Indian evidence lasted from early December 1972 until mid-February; the evidence for the corporations and the counterproof from early March until the end of May; and legal argument ended June 21, 1973.

CHAPTER FOUR

The major studies of the Waswanipi people were carried out in the late 1960s under the aegis of the McGill Cree Developmental Change project conducted by the Programme on the Anthropology of Development of the university's Anthropology Department. The leader of this project was Professor Norman A. Chance, and the major report, published in 1968, was *Developmental Change Among the Cree Indians of Quebec,* published by the Rural Development branch of the Department of Forestry and Rural Development, Ottawa, Canada. Several other papers by Professor Chance arising from this project were published by the Canadian Research Center for Anthropology, Ottawa, and by McGill University. Ignatius La Rusic's major contributions were *The New Auchimau,* published in 1968 as part of the project report, and *From Hunter to Proletarian,* published in 1969 as an annex to the project report. Harvey Feit's major contribution to this project was a paper called *Towards an Ecology of the Mistassini and Waswanipi Indians* (1967).

He also wrote *Mistassini Hunters of the Boreal Forest* as his M.A. thesis (1969) and two important articles, published in French translations in the *Recherches Amerindiennes au Quebec* in 1971, "The Ethno-ecology of the Waswanipi Crees" and "The Exploitation of Natural Resources in James Bay." On the educational side, Peter S. Sindell's *Some Discontinuities in the Enculturation of Mistassini Cree Children* was published as part of the McGill Cree project report, and in 1969 he also published an article, "How Going Away to School Changes Children." With Ronald M. Wintrob he wrote in 1969 *Cross-Cultural Education in the North and Its Implications for Personal Identity* (Arctic Institute of North America), and *Education and Identity Conflict Among Cree Youth,* an annex to the McGill project report. The poem about anthropologists by the Rev. Hugo Muller was in his collection, *For No One Knows Waswanipi,* published by the Bishop of Moosonee as the autumn 1973 issue of *The Northland.*

CHAPTER FIVE

The history of the Hudson's Bay Company is the history of northern Canada, and there are vast archives and considerable literature. The archives have only recently been transferred from London to the company's headquarters in Winnipeg. On the French side the early northern history is contained in the *Jesuit Relations* covering the seventeenth century. An interesting document is the long research monograph called *The Other Side of the Ledger,* written by Ron Dick for the National Film Board of Canada preparatory to the making of a film of that name, viewing the Hudson's Bay Company's history from the point of view of the Indians. For a more up-to-date account of the work of company officials, see Duncan Pryde's *Nunaga* (Walker Company, New York, 1972) and J. W. Anderson's *Fur Trader's Story* (Ryerson Press, Toronto, 1961). Anderson also published a number of articles, mostly in the *Beaver,* the Hudson's Bay Company journal, between 1934 and 1956. Of the witnesses quoted in this chapter, Edward S. Rogers has published a mass of articles and papers on the area between 1948 and 1971, the best-known of which is "The Material Culture of the Mistassini," *National Museum of Canada Bulletin,* No. 218, 1967, and Adrian Tanner has published the article "Are There Hunting Territories?" in the previously mentioned bulletin of *Recherches Amerindiennes,* 1971, and contributed to the McGill Cree project. James Clouston's journey is described in *Northern Quebec and Labrador Journals, 1819–35* by K. G. Davies (Hudson's Bay Record Society). The story of Peter McKenzie's visit to Mistassini is told in *Chibougamau Venture* by Larry Wilson (Chibougamau Publishing Company). The economic analysis of Mistassini was contained in a two-volume report, *A Survey of the Contemporary Indians of Canada,* published by the Indian Affairs Department of the Canadian government, 1967, and edited by H. B. Hawthorn.

CHAPTER SIX

The Indians of Quebec Association, largely at the urging of Max Gros-Louis, carried out the first legal research into Indian land claims in the province in 1967, and on the basis of that research entered a claim to the provincial government for $5 billion. It was not taken seriously.

CHAPTER SEVEN

The Bishop of Moosonee, James Watton, was called in evidence by the Indians to produce baptismal records for various Cree communities. He said some of the Anglican Church records go back to 1790, stemming from the activities of the Hudson's Bay Company, and their own records for the diocese of Rupert's Land went back to 1854. The anthropologist mentioned by the Rev. Linton as having stimulated interest among the Indians in their old religious observances was probably Richard Preston, who studied Rupert House and published many papers on the area between 1964 and 1972.

CHAPTER EIGHT

The book referred to by the witness Dr. Harold Hanson, which he wrote, is *The Giant Canada Goose* (Southern Illinois University Press, Carbondale, 1965). And the story of Maud Watt is told in *The Angel of Hudson Bay* by W. A. Anderson (Clark Irwin, Toronto, 1961).

CHAPTER NINE

More stories by Samson Nahacappo were published in 1973 editions of *Tawow*, a cultural publication of the Indian Affairs Department, Ottawa.

CHAPTER TEN

Of the sources mentioned by Tanner, A. P. Low was a famous Canadian geologist whose report, *The Mistassini Expedition, 1884–85*, was published by the Geological and Natural History Survey of Canada, which ten years later published a similar report of his explorations along the Eastmain, Koksoak, Hamilton, Manicouagan and other rivers on the Labrador Peninsula. Frank G. Speck, an American anthropologist, published a great number of papers on the peoples of the Labrador-Ungava Peninsula between 1909 and 1939, and Julius E. Lips' contributions, concentrated mostly on the Naskapi and Montagnais Indians farther east, were published between 1936 and 1947 in Sweden and the United States. Rolf Knight's 1962 M.A. thesis at the University of British Columbia was called *Changing Social and Economic Organization Among the Rupert House Cree.*

CHAPTER FOURTEEN

For a good account of the alienating education to which Rosie had been subjected, see Chapter Four in *A Survey of the Contemporary Indians of Canada*, Volume II, as well as the papers by Professors Sindell and Wintrob.

CHAPTER FIFTEEN

The Malouf judgment has been printed in French by Les Editions du Jour, Montreal; and by *Akwasasne Notes*, St. Regis Reserve, via Rooseveltown, N.Y., in English.

CHAPTER SEVENTEEN

Many of the transcripts quoted from, especially those of conversations with Mistassini Indians, were recorded during the making of the National Film Board productions *Cree Hunters of Mistassini* and *Our Land Is Our Life*, two one-hour 16-mm. documentaries, the first of which is ethnographic in nature, dealing with an Indian hunting camp, and the second socio-political, dealing with the politics of assimilation and the rising tide of Indian land claims across the Canadian north.

INDEX

aboriginal rights concept, 25–6; *see also* Indian rights

acculturative stress, 152–6, 195–6, 330–1, 336, 338

Air Creebec, 329

Alain, Father, 91

Alaska pipeline, 133

Albanel, Father Charles, 69–70

Alberta, reservoir projects in, 148

Amory, Lord, 64

Anderson, J. W., 68, 69, 70–2, 74, 77, 87, 110–11, 282

animals, 223–4
  reservoirs and, 173–5, 179, 182, 277, 358–9
  in settlement, 306, 326
  see also environment

Angel of Hudson Bay, The, 108

Anglican missionaries, *see* missionaries

anthropologists
  and acculturative stress, 152–6, 195–6
  and Cree concept of ownership, 219–23
  Rev. Muller vs., 55
  testimony of, 59, 69, 71, 72, 88, 94–5, 247–50, 269
  see also Feit, Harvey; La Rusic, Ignatius; Tanner, Adrian

Armstrong, 47–8

Arrow to the Moon, 230–2, 243

assimilation concept, 154–5

Atlantic States Legal Foundation, 355

Atomic Energy of Canada, 258

Aubé, Mr. and Mrs. Jacques, 227–8

Audubon Society, 354

Austin Airways, 113

Australian aborigines, 152

Awashish, Anne-Marie (Mrs. Philip), 11–12, 82, 288

Awashish, Isaiah, 4–17
  passim, 93–4, 110, 273, 320–1, 334
  and Indian Affairs Department, 198–200
  and negotiations, 308–9

Awashish, Kenny, 93, 321

Awashish, Philip, 10, 11, 15, 61, 69, 86, 121–2, 261, 264, 288, 334, 338
  and *Arrow to the Moon*, 230–2
  and Indian Affairs Department, 198–200
  vs. James Bay project, 80–84, 85, 87, 90, 92–4, 97, 102, 104
  and Lac Trefart, 201–10 *passim*, 217–19, 232, 238–40, 265–6, 269, 274, 284, 286
  and negotiations, 304–5, 319
  and Superior Court hearing, 85, 101, 103–4

Awashish, Sam, 13, 15, 17, 93–4, 320

Awashish, Solomon, 10, 15, 69

Awashish, Willie, 4–17 *passim*, 69, 81, 93–4, 239

death of, 321–2

Bacon, Lise, 357
Banfield, Alexander, 180–2, 212, 276–8
Beacon reservoir, 148
bears, 158–60, 197
Bearskin, Billy, 124–5, 128, 134, 136–52
    passim, 189–91, 195, 243
Bearskin, Clifford, 165, 319
Bearskin, George, 136–52 *passim*
Bearskin, Job, 116–17, 118, 124, 136–52
    passim, 157–97 *passim,* 216,
    220–1, 232, 243, 311, 319–20,
    340, 342, 359
  and Superior Court hearing, 118–21,
    246
Bearskin, Johnny, 179
Bearskin, Juliette, 185–7
Bearskin, Maggie, 243
Bearskin, Martha, 189
Bearskin, Mary, 134–52 *passim,* 179, 320
Beaudoin, Jacques, 108
beaver, 109–11, 169–74 *passim,* 187–8,
    251, 307, 312, 358
    see also fur trade
Bennett dam, 148
Bernard, Max, 30–1, 40, 92
Berry, John, 152–6, 195–6
Bertrand, Paul, 247–9, 314
Beyea, Jan, 354
Bienville Lake, 165, 181, 221, 347, 358
biologists, testimony of, 107–8, 113,
    131–2, 143–4, 169–71, 175,
    180–2, 210–11, 212, 276–8
Blackned, John, 106–7
Blacksmith, Billy, 224–5
Blacksmith, Malick, 207, 271, 274, 276,
    278, 283, 284, 289, 293, 333
Blacksmith, Rosie, 288–95, 333
  at Lac Trefart, 207–8, 214–15, 226,
    229, 232–4, 238, 266–7, 274–5,
    280, 285, 286–7
Blacksmith, Sam, 288–90, 293, 308, 311,
    320, 332–3, 345
  at Lac Trefart, 201–41 *passim,* 262,
    265–87 *passim*
Blacksmith family, 288
Bonneville Power Administration, 352
Borremans, Guy, 122, 137–52 *passim,*
    172

Bosum, Charlie, 99–100
Boulanger, Gaston, 244, 311
Bourassa, Robert, 34, 80–1, 242–3,
    327–8, 348, 350
  on energy needs, 346, 356
  and negotiations, 304–7 *passim,*
Bourbeau, Joseph, 256
Boyd, Robert,
Brantford, 227, 232
Brazeau reservoir, 148
Brien, Charlie, 331
Brien, George, 320
British Columbia
  and Indian rights, 36
  Indians of, 152–6 *passim*
  reservoir projects in, 148
British East India Company, 64
British North America Act, 36, 67
  Eskimos and, 114–15
Broadback River, 34, 55, 102
  and James Bay project, 21, 81, 112–13
  and James Bay II, 347, 346–55 *passim*
Brun, Henri, 92

Cache Lake, 54
Canadian Broadcasting Corporation,
    327
Canadian Forum, 339
Canadian International Paper Company,
    76, 261
Canadian National Line, 47
Caniapiscau Lake, 45–6, 70, 181
  reservoir, 307
Caniapiscau River, 45, 70, 181, 253, 277
  and James Bay project, 20–1
caribou, 180–2, 212, 276–8, 358
Caron, Monique, 29–30
Carrier Indians, 152–6
Casey, Mr. Justice P. C., 300
Caughnawaga Mohawk band, 83
Chabat, 261–2
Chapais, 49–50, 51,61, 263, 334
Chapleau, 75–6
Charles II, 63, 316
Chibougamau, 15, 39, 49–54, 61, 74, 80,
    85, 98–9, 241, 261, 329, 331,
    333–4
  and Corporation witnesses, 244
  Indian Friendship Center at, 10, 82,
    201–2, 265

and Mistassini, 198–203 *passim*
winter carnival, 264–5
Chibougamau Mining Commission, 74–5
Chibougamau Mining and Smelting
Company, 16
Chibougamau River, 56
Chief: The Fearless Vision of Billy Diamond,
328–9
Chisasibi, 337–42
see Fort George
Chrétien, Jean, 303–6 *passim; see also*
Indian Affairs Department
Churchill Lake, 108
Ciaccia, John, 327
Clough, Garrett C., 132–3, 169–71, 279
Clouston, James, 70
Code of Civil Procedure, 41
Confederation, and Hudson's Bay
Company Lands, 26
Consolidated Edison, 257
Coonishish, Abel, 208, 225
Coonishish, Charlie, 224
Coonishish, George, 208, 225
Cooper, Billy, 345
Cox, Samuel, 137–52 *passim*
Cree Hunters of Mistassini, 308; *see also* Lac
Trefart
Cree-Naskapi Act, 323
Cree Trappers' Association, 326, 332
Crete, Mr. Justice Marcel, 316

Darwin, Charles, 279
de Bellefeuille, Paul-Aimé, 243, 244
de Hoop, Rev. Tom, 91
Delorme Lake, 181
reservoir 307
Deschenes, Chief Justice Jules-J.,
Desmaraisville, 57
Deveney, Brian, 157, 164–5, 176
Diamond, Albert, 103, 118–21 *passim*
Diamond, Annie, 103, 105
Diamond, Chief Billy, 84, 92, 97, 102–5
passim, 109, 121, 328–9, 341,
343, 347
and *Arrow to the Moon*, 232
and settlement, 305, 318, 319, 321
and Superior Court hearing, 32, 33–5,
41–2, 85, 102, 103–4, 120
Diamond, Mrs. Hilda, 103

Diamond, Malcolm, 103, 104, 242–3
Dick, Clifford, 127
Dick family, 125–6, 130–1
Dick, Chief Robbie, 357–8
Dolbeau Air Services, 239, 284
Doré Lake, 50–4 *passim*, 92, 202, 221,
334
Dutch East India Company, 64

Earth Day, 1990, 355
Eastmain, 25, 33, 69, 101, 113, 129, 204,
220, 224, 305, 335–6
testimony from, 36–9, 42–3
Eastmain River, 70, 201, 202–3, 224
and James Bay project, 20–1, 81, 92,
206–7, 307, 335
Elizabeth II, Queen of England, 64
Emery, Georges, 103, 279, 311
energy needs, 255–9
Court of Appeals and, 316, 317
engineers, testimony of, 127–8, 147–8,
167–8, 191–2
English law, 27
and concept of property, 297
English Privy Council, and Indian rights,
25
environment, 95, 102, 123–4, 164,
245–6, 249, 250–5
animals and, 223–4; *see also* animals;
bears; beaver; caribou; fish;
geese; moose
Court of Appeals and, 315–16
effects of James Bay I, 340–5
and energy needs, 255–9; *see also*
energy needs
French-Canadians and, 22
James Bay Development Corporation
and government reports on,
103–6, 132–3, 157, 349–51
and Justice Malouf's judgment, 296,
298–9
and settlement, 304–9 *passim*, 340–5
see also forest fires; mining; power
stations; reservoirs
Eskimos
of Baffin Island, 152
and Supreme Court of Canada,
114–15
see also Inuit people
Evans Lake, 104, 320

Fecteau Transport Ltd, 54
federal government
and James Bay project, 83–4

multiculturalism policy of, 154
and negotiations, 303–6
and Nouveau Quebec, 113–15
see also Indian Affairs Department

Feit, Harvery, 110–11, 306
and Court of Appeals, 313–14
and Superior Court hearing, 59–60, 69, 249–50

Fenton, Melville Brock, 175

Fillion, Alfred, 244

Fire Lake, 225

fish
effects of James Bay I on, 343–4
reservoirs and, 254; *see also* reservoirs
sturgeon project of Indian Affairs Department, 198–200
see also environment

forest fires, 6, 37, 38, 110, 188, 253, 315
caribou and, 181, 182

Fort Chimo, 25, 45, 108, 253, 277

Fort George, 25, 110, 245, 249–50, 281, 284, 313, 338
acculturative stress in, 152–6, 195-6, 338–9
income in, 246–7, 297–8, 314, 339
vs. James Bay project, 113–97 *passim*, 242–3, 305
and Kanaaupscow, 165
testimony from, 41, 45–6, 118–21, 126–7, 135–6, 140, 177–9, 193–4

Fort George River; *see* La Grande River

Fort McKenzie, 45

Fournier, Jean-Pierre, 122, 243

France
and Hundson's Bay Company, 64–5, 69–71, 124
trading posts of, 14, 70–1, 124; *see also* fur trade

French and Indian War (1756–63), 25–6, 65

French-Canadians (Quebecois), 80, 195, 303
Indians' dislike of, 195
and Superior Court hearing, 21–4, 27

fur trade, 62–5
and beaver, 109–11
France and, 69–71, *see also* France, trading posts of
and Indian rights, 26
see also Hudson's Bay Company

Gatineau River, 77

geese, 193–4, 106–8, 113, 193–4

Geoffrion, C.-Antoine, 23–4, 27–8, 31

George III, King of England, 26

Gill, Don, 173–4

Giroux, Roland,

Goeland Lake, 56

Goodman, Ian, 351–2

Gouin Reservoir, 77

Governor's Island, 191–4, 340–1; *see also* Fort George

Grand Council of the Cree, 305–9, 354–6
and settlement, 318

Great Whale River, 25, 45, 128, 165, 181, 221, 280, 281, 358
and James Bay project, 20–1, 307
and James Bay II, 346–55 *passim*
environmental consequences, 347–8
testimony from, 42

Gros-Louis, Max, 83, 84

guaranteed annual income, 323

Gull, Jacob, 56

Gull, Jacob, II, 56–7, 329

Gull, Jacob, III, 39

Gull, Rev. John, 56, 329

Gull, Peter, 61

Gull, Sam, 330

Gull, William, 39–40, 57, 61, 329

Gunner, Nancy, 290, 293–4

Haldane, Lord, quoted, 297

Hanson, Harold, 107–8, 113

Hare, Kenneth, 123–4, 128–30

Herodier, Gilbert, 124, 125, 127, 134–5, 136–52 *passim*, 243

Higgins, James J., 356

Hourdis, Marcel, 240–1, 264–5, 268–9

House, Larry, 337

Hudson's Bay Company, 26, 32, 49, 140, 141, 223, 238, 264
and Corporation witnesses, 224–5
in Fort George, 190–1
France vs., 64–5, 69–71, 124
closings of, 53, 55, 111–12, 164–7
in Mistassini, 62, 65–74 *passim*, 86, 87, 331, 332; *see also* Speers, Glen
Royal Charter of, 62–5, 316
in Rupert House, 101, 109–12 *passim*
in Waswanipi, 55, 58

Hudson's Bay Mining and Smelting Company, 64

hunting camps, 54, 66, 93–4, 140–3, 261, 334

and settlement, 319–20, 325
songs in, 150–1
see also Bearskin, Job; Lac Trefart

Huron Indians, Loretteville band, 83

Hydro-Quebec, 10, 85, 104, 116, 186, 246, 321, 326, 336–7, 340–1, 343–5, 348–57 *passim*
and energy needs, 255–9 *passim*, 327–9, 350
and Justice Malouf's judgment, 296

Ikon mine, 67–8, 98

Income Security Programme, 323, 332, 339

Indian Act, 75, 323

Indian Affairs Department, 57, 67, 78–9, 133, 189, 195, 208, 233, 234, 260–1, 295
and Mistassini decision to oppose James Bay project, 83–4, 85, 122
sturgeon project of, 198–200

Indian bands, 75

Indian Friendship Center, 10, 82, 201–2, 265

Indian rights, 25–6, 92
and Court of Appeals, 300, 301, 310, 316–17
and Justice Malouf's judgment, 296–7, 299, 303
and settlement, 306, 309
Philip Awashish and, 81–4 *passim*, 121–2
and Superior Court hearing, 21, 23–30 *passim*, 36

Indian River, 71

Indians of Quebec Association, 18, 82–4, 85, 92, 104, 115, 121–2
O'Reilly and, 24–5
and settlement, 304–5

integration concept, 154

Inuit people
and settlement, 318, 322
see also Eskimos; Fort Chimo; Great Whale River; Quebec Court of Appeals; Quebec Superior Court

Iserhoff, Charles, 74

Iserhoff, John C., 68, 74

Iserhoff, Joseph, 74

Iserhoff, Samuel 74

James Bay Development Corporation
difficulties of, 301–2
and energy needs, 255–9 *passim*
hydro-electric project proposed by, 15, 20–1, 53, 56, 74

and James Bay I, 327, 342–3
impact on Fort George-Chisasibi, 339–343
and Justice Malouf's judgment, 296, 299–300
and settlement, 325
witnesses for, 243–59
see also environment; La Grande River; power stations; Quebec Court of Appeals; Quebec Superior Court; reservoirs

James Bay Energy Corporation, 23
and energy needs, 255–9 *passim*
and Justice Malouf's judgment, 296
and settlement, 322, 340

James Bay and Northern Quebec Agreement (JBNQA), 319–22 *passim*, 329, 334–5, 340, 342–3, 349–50, 352
major provisions, 322–6

Jimiken, Lawrence, 104–5, 111–12

Job's Garden, 118, 124, 243, 327, 342

John, Willie, 47–8

Jolly, Charlie, 224

Jolly, Eddie, 207–8, 213, 234–5, 262, 273, 274–6, 284–5, 320

Jolly, Johnny, 208, 226–9

Jolly, Kathleen, 208, 226–9

Jolly, Mary, 208, 226, 234, 238, 267, 286–7

Jolly, Philomen, 208, 267, 280, 285, 286–7

Jolly, Ronnie, 204–42 *passim*, 262, 267–87 *passim*, 320–1, 333

Jolly, Winnie, 208, 229–30

Jolly family, 288

Joseph, Alvin M., Jr., 115

Kanaaupscow, 71, 134, 140, 164–7, 171–3, 176, 204, 280, 281, 320

Kanaaupscow River, 157, 164–7 *passim*
and James Bay project, 167–8

Kanatewat, Chief Robert, 305

Kaufman, Mr. Justice Fred, 317

Kawapit, John, 42

Kayapo people, 355–6

Kellerhals, Ralph, 147–8, 167–8

Kentucky Fried Chicken Restaurant, 244

Kershaw, Jimmy, 92–3

Khazzoom, Dr. Daniel J., 255–9

Kitchen, Chief Abel, 330, 338

Knight, Rolf, 219

Koksook River, 45, 251

Kruger, 330

Kudluk, Thomas, 45

Kushkan, Sagan, 48

Kwandibens, Cecilia, 48

Labrador
  caribou and, 180–2
  Churchill Lake, 108
  coast, 45

Lac St. Jean area, 69, 74–5

Lac Trefart, 201–41, 264–87, 320, 333

La Grande (Fort George) River, 45, 118,
      178, 210, 281, 343
  and James Bay project, 20–1, 81,
      112–13, 143–74 *passim*, 149, 155,
      156–68 *passim*, 191–4, 206,
      242–3, 256–9, 302, 312–13, 322,
      335–6, 344, 348, 352
  LG–1, 127–8, 147, 336, 340–1
  LG–2, 128, 157, 159–64, 177–9, 181–2,
      191, 305, 315, 335, 341, 344–6,
      350
  LG–3, 315, 345
  LG–4, 345
  Richardson on, with Job Bearskin,
      124–52 *passim*, 216; see also *Job's
      Garden*

Lajambe-Connor, Helen, 353

Lajoie, Mr. Justice François, 317

lakes, *see proper names of lakes*

Lamontagne, Nicholas J., 315

land ownership
  Cree concept of, 218–23
  English law on, 297
  see also Indian rights

Lanzelo, Tony, 201, 205, 209, 215,
      237–8, 239, 272, 273, 280, 284,
      292

Larch River, 70, 181, 253

La Rusic, Ignatius
  and Indians of Quebec Association, 83
  and Mistassini, 85–7, 91–2, 97
  The New Auchimau, 55, 57–8, 59

La Tuque, 227, 229, 260, 261

Leaf River, 212

Le Bel, Jacques,
  and Court of Appeals, 311
  and native testimony, 34–46 *passim*,
      119–21, 135–6, 194
  and scientists' testimony, 71, 269
  and witnesses for government and
      Corporation, 243–4, 247–8

Lammert, Rev. J. Arthur, 263–4

Levesque, René, 328

Linton, Larry, 89–90, 263–4, 270–1

Lips, Julius E., 220

Litvack, Robert, 143–4, 147–8, 153–6,
      167–8, 180–2, 191–2, 195–6,
      210–12, 277

logging, 324, 329–30

Loon, John, 95–6, 225

Loon, Sidney, 29–30

Loon Point, 33

Loretteville Huron band, 83

Low, A. P., 219

Macmillan, Harold, 64

Maine Public Utility Commission, 355

Malouf, Mr. Justice Albert H., 28, 30–1,
      46, 103, 118–21, 132, 155, 243,
      319
  and Indians of Quebec Association,
      304
  judgment of, 296–301, 303
  and Court of Appeals, 300–1, 310–17
      passim

Manpower Department training course,
      51

Marcus, William B., 351–2

Marest, Gabriel, 314–15

marginality concept, 153

Marsolais, Rita, 246, 297–8

Martinhunter, Abraham, 305

Matagami, 182, 299

Matagami Lake, 56
  and James Bay project, 21

Matoush, George, 87–9

Matoush, William, 87–9

Mayappo, Charlie, 335

McGill University, 10, 61, 122, 255
  Cree Developmental change project,
      245

McGregor, Roy, 328

McKenzie, Peter, 74

methyl mercury, 343–5

Mianscum, Abraham, 207, 333, 345

Mianscum, François, 46, 207, 333, 345

Mianscum, Jimmy, 50, 52–4, 74, 80, 92,
      98–9, 334

Mianscum, Samuel, 225

Mica dam project, 148

Michiskun, 69, 70

Mile 17 Lake, 177–8

mining, 16, 64, 74–5
Ikon mine, 67–8, 98
see also Mianscum, Jimmy

Miquelon, 57

missionaries, 263–4
Anglican, 49, 52, 91, 263; *see also*
Muller, Rev. Hugo
Baptist, 89–90, 263–4, 270–1
Catholic, 91
Pentecostal, 263–4; *see also* Pentecostal
sect
and Superior Court hearing, 269–71

Mistassini, 7, 10–14 *passim*, 25, 36, 47,
51, 52, 55, 61–79, 139, 141, 151,
260–1, 282, 320, 331–4, 336–8
beaver preserves, 109, 307
Hudson's Bay Company and, 62,
65–74 *passim*, 86, 87, 331–2; *see
also* Speers, Glen
vs. James Bay project, 80–102 *passim*
and land ownership, 218–23
missionaries in, 263–4; *see also*
missionaries
and settlement, 324
subgroups of, 221; *see also* Doré Lake;
Nemaska; Nichicun testimony
from, 29–30, 40, 46, 50, 85,
103–4
see also Lac Trefart

Mistassini Lake, 36, 50, 62, 67–70 *passim*,
73, 74, 78, 212, 333
and James Bay project, 21, 81, 85

Mistissini Enterprises, 331

Mohawk Indians, Caughnawaga band,
83

Montreal *Gazette*, 299

moose, 211–12, 253, 284–7

Moose Factory, 227

Moosonee, 111

Moses family, 337

Moses, Ted, 36, 44, 46, 101, 336
and negotiations, 304–5

Moses, Mr. and Mrs. Willie, 336, 345

Mouvement au Courant, 354

Muller, Rev. Hugo, 49, 50–2, 55–6, 58,
61, 67, 91, 329

Murray, James, 26

Nahacappo, Samson, 179, 1082–4, 187–8

Namagoose, Bill, 356–7

Naskapi Indians, 45, 323

National Energy Board of Canada,
247–8, 351, 356–8

National Film Board (NFB), 201, 288,
292

National Indian Brotherhood, 260

National Resources Defense Council,
352, 355

NBR project, 320, *see also* Nottaway,
Broadback *and* Rupert rivers;
Great Whale River

Neeposh, Edna, 61–2, 82, 85, 87, 94, 225

Neeposh, John, 225

Neeposh, Matthew, 40, 61

Neeposh, Philip, 85

Nemaska, 8, 14, 69, 70, 71, 73, 90, 221,
280, 281, 333
Hudson's Bay Company and, 111–12

Nemaska Lake, 21, 104

Neoskweskau, 14, 70, 71

New Auchimau, The, 55, 57–8, 59

New England Coalition for Energy
Efficiency and the Environment,
356

Nichicun, 14, 70, 71, 221, 281–2
and Mistassini, 222–3

Nonsuch, 62

Normandin, Jean-Guy, 286

North Bay, 74

Northwest Company, 65, 70

Nottaway-Broadback-Rupert project,
320; *see also* Great Whale River

Nottaway River, 34, 56, 70, 102
and James Bay project, 21, 81, 92,
112–13
and James Bay II, 346–55 *passim*

Nouveau Quebec, federal government
and, 113–15

Nunaga, 65

Obalski, Joseph, 74

Obatagamau River, 56

Obidjun rapids, 247; *see also* La Grande
River

Ogoki, Hudson's Bay Company post at,
74

Ojibway Indians, 47

Old Crow, 260

Olga Lake, 56

Opinaca Lake, 56

Opinaca River, 201
and James Bay project, 20–1, 206–7

Oppenheimer, Harry, 64

O'Reilly, James, 23–32 *passim*, 200
  and Court of Appeals, 300, 302, 311,
    315
  and native testimony, 33, 40, 41–2, 44,
    50, 119–21, 126–7, 140, 177–9,
    193–4, 246
  and scientist's testimony, 59, 69, 71,
    72–3, 88, 94–5, 123, 128–32
    passim, 155, 160–3, 169–75
    passim, 223, 249, 258–9
  and settlement, 303–4
Oskelaneo, 77
Otish mountians, 212
Ottereyes, Joseph, 320–1
Ouge-Bougamau, 334
Owen, Mr. Justice George, 311, 316–17

Pachano, George, 41, 116–17
Pachano, Norman, 127
Pachano, Thomas, 179
Pachano family, 125–6, 130–1
Pachanos, Violet, 339–40
Pageau, Therese, 243–4, 313
Paint Hills, 25, 113
  Fort George compared to, 153–6
  testimony from, 44–6
Paneman Lake, 269, 284
Parti Quebecois, 328
Pash, Mrs. Samuel, 185
Pash, Samuel, 110, 187–8
Paul, Mr. Justice Rejean, 324
Penn, Alan, 341, 343, 344, 348
Pentecostal sect, 52, 104, 263–4
Pepabano, Joseph, 246–7
Petagumskum, John, 358–9
Petawabano, Bella, 336–7
Petawabano, Buckley, 96, 230–2
Petawabano, Joe, 336–7
Petawabano, Philip, 95–7, 230
Petawabano, Smally, 52, 75–8, 86, 91,97,
    102
Plitipi Lake, 212
Pontax River, 34
Power, Geoffrey, 142–4, 210–11
power stations, 21, 81, 233–4
  and La Grande River, 127–8, 143–4,
    160
  see also energy needs; reservoirs
Pryde, Duncan, *Nunaga*, 65
Public Service Board, 356

Puisseaux Lake, 334

Quebec Boundaries Extension Act
    (1912), 26
Quebec Court of Appeals, 300–1, 303,
    310–17, 318
Quebec Department of Fish and Game,
    66
Quebecois, 80; *see also* French-Canadians
Quebec Superior Court, 18–20, 24–32
  French-Canadians and, 21–4, 27
  government and James Bay
    Development Corportation
    witnesses at, 243–59, 272
  Great Whale River project, 350
  judgment of, 296–300; *see also* Malouf,
    Mr. Justice Albert H.; Quebec
    Court of Appeals
  native testimony at
    from Eastmain, 36–9, 42–3
    from Fort George, 41, 45–6,
      118–21, 126–7, 135–6, 140,
      177–9, 193–4
    from Great Whale River, 42
    from Mistassini, 29–30, 40, 46,
      50, 85, 103–4
    from Paint Hills, 44–6
    from Rupert House, 33–5, 40–1
    from Waswanipi, 39–40
  scientists' testimony at, 123–4, 128–30,
    152–6, 173–4, 195–6
  anthropologists, 59, 69, 71, 72–3, 88,
    95–5, 247–50, 269
  biologists, 107–8, 113, 131–2, 143–4,
    169–71, 175, 180–2, 210–11,
    212, 276–8
  engineers, 127–8, 147–8, 167–8, 191–2
Queen's University, 152

racism, 259
Rat, William, 246
Ratté, Roger, 57
rejection concept, 154–5
religion, 269–70; *see also* missionaries
reservoirs, 20–1, 81, 108, 147–8, 160–5,
    207, 307, 315
  and animals, 173–5, 179, 182, 277
  and fish populations, 254, 344–5
  see also environment
Revillion Frères, 124
Robertson, 199–200
Rogers, Edward S., 72–3, 94–5
Royal Ontario Museum, 72
Royal Proclamation (1763), 25
Rupert, Willie, 127

Rupert Bay, 32, 56, 102
  and James Bay project, 112–13,
  and James Bay II, 347
Rupert House, 14, 25, 62, 68–9, 70, 77,
    87, 129, 151, 280, 281
  vs. James Bay project, 101–17 *passim*,
    132, 242–3
  missionaries at, 263, 270
  shaking-tent cermony at, 89–90
  testimony from, 33–5, 40–1
Rupert River, 5, 33–4, 40, 68–9, 70,
    76–7, 101, 102, 198, 202–3, 281
  and James Bay project, 21, 81, 104,
    112–13
  and James Bay II, 346–55 *passim*

St. Felicien Airways, 122

St. Pierre, Guy,

Sakami Lake, 21, 323

Sakami River, 225
  and James Bay project, 206–7

Salisbury, Dr. Richard, 245

Sam, Josie, 115, 166, 299
  and Superior Court hearing, 135–6,
    177–9, 193–4

Sam, Norman, 177, 178, 339, 345, 347

Sam, Rose, 128, 130, 195, 197

Sandy, David, 247

Savage, Joseph, 244–5

Schecapio, Chief Isaac, 76, 321

Schecapio, Philip 98, 202

Schefferville, 180, 282, 323
  railway, 212

seal, freshwater, 347, 358–9

Senneterre, 39, 55

separatism issue, 21–4

Sept Isles, 109

Sex-Fox, 16

sex life, 283–4

Shanush, Chief Matthew, 36–9, 305

Shanush, Noah, 220

Shashaweskum, Clifford, 44–5

Shem, George, 196–7

Shively, Tom, 133, 134

Sierra Club, 354–5

Sierra Club Foundation, 354

Sierra Leone, Timney people of, 152

Sindell, Peter, 57

Skinnarland, Einar, 127–8, 160–2, 257,
    259

Soscumica Lake, 56

and James Bay project, 21

Speck, Frank G., 220, 221

Speers, Glen, 62, 65–6, 68, 72, 92, 224,
    238, 244, 295
  and Fort George-Kanaaupscow, 140,
    165, 166, 179
  and Lac Trefart, 262, 267, 270,
    279–82, 284

Spence, John, 122, 242, 255
  and negotiations,
  and Superior Court hearing, 131–2,
    169

sports hunting, 306
  and settlement, 325, 329

spring floods, 250–2, 315

Stockholm environment conference, 320

stock market crash of 1929, 111

sturgeon project of Indian Affairs
    Department, 198–200

Supreme Court of Canada, 301, 319
  and Eskimo, 114
  and Indian rights, 25

Tanner, Adrian
  and Court of Appeals, 313–14
  on Cree education, 235–6
  and land ownership concept, 219–23
    passim
  and Superior Court hearing, 71, 88,
    223, 269–70

Tapiatic, Stephen, 45–6, 70, 126–7, 140,
    161

Taylor, Colin, 162–3, 191–2

Thibodeau, Roger, 27, 311

Thunder Bay, 48

Timney people, 152

Tremblay, Mr. Justice Lucien, 300

Trent University, 103

Tsimshian Indians, 152–5 *passim*

Turgeon, Mr. Justice Jean, 300, 310–16
    passim

unemployment, 22; *see also* welfare
    payments

Ungava Bay, 20, 25, 251, 253

Ungava Peninsula, 18, 20, 45, 109
  caribou and moose of, 180–2, 212
  and Nouveau Quebec, 114

United Nations Social and Economic
    Council, 336

United States
  energy exported to, 348, 351–6 *passim*

Federal Power Commission, 256
  and Indian rights, 24, 25
University of Toronto, 71

Val d'Or, 104
Vincelotte Lake, 157
Voyageur, Abraham, 204, 208, 209,
  217–41 *passim*, 266–7, 290, 333
Voyageur, Louise, 208–9, 267, 333
Voyageur, Reuben, 221, 225, 275
Voyageur, Shirley, 208–9, 211, 267, 275,
  276, 290, 333
Voyageur, Solomon, 69
Voyageur family, 208, 288

Waasheskum Airline, 331
Waconichi Hotel, 16, 49, 52
Wapachee, Bertie, 112
Wapachee, Daniel, 109
Washburn, Eric, 355
Waskaganish, 347; *see* Rupert House

Waswanipi, 25, 55–60, 69, 70, 74, 111,
  249, 330–1, 338, 348
  vs. James Bay project, 85, 104
  testimony from, 39–40
Waswanipi Lake, 55, 56, 329
Waswanipi River, 56, 329, 348
Watt, James, 108–10
Watt, Maud, 108–9
Weapiniccapo, Abraham, 335, 345
Weapiniccapo, John, 42–3
Webb, Tom, 115
welfare payments, 51, 59, 78
  Nemaska people and, 112
Wemindji, 323
Whiskychan, Johnny, 104
Wintrob, Ronald, 57
Wolfe, General James, 25
World Health Organization, 344
World War I, 68
World War II, 75, 86, 105

A NOTE ABOUT THE AUTHOR

Born in New Zealand in 1928, Boyce Richardson has worked as a journalist and editor in New Zealand, Australia, Britain and Canada. He first became interested in the Cree Indians while he was on the staff of the Montreal *Star*, and subsequently produced three documentary films about them: *Job's Garden* (1972), *Cree Hunters of Mistassini* (1973), and *Our Land Is Our Life* (1974). He is the author of several other books and has contributed articles to many magazines in the United States, Canada and Britain. Richardson now lives in Ottawa.

Strangers Devour the Land was originally published by Alfred A. Knopf in 1976. It was designed by Earl Tidwell and typeset in Baskerville by Compucomp Corporation and Dartmouth Printing Company. It was printed on an acid-free Glatfelter paper by Princeton University Press.